AIR POLLUTION

ENVIRONMENTAL SCIENCE AND TECHNOLOGY

A Wiley-Interscience Series of Texts and Monographs

Edited by **ROBERT L. METCALF**, *University of Illinois*
JAMES N. PITTS, Jr., *University of California*

PRINCIPLES AND PRACTICES OF INCINERATION
Richard C. Corey

AN INTRODUCTION TO EXPERIMENTAL AEROBIOLOGY
Robert L. Dimmick

AIR POLLUTION CONTROL, Parts I and II
Werner Strauss

APPLIED STREAM SANITATION
Clarence J. Velz

PHYSICOCHEMICAL PROCESSES FOR WATER QUALITY CONTROL
Walter J. Weber, Jr.

ENVIRONMENTAL ENGINEERING AND SANITATION
Joseph A. Salvato, Jr.

AIR POLLUTION, Second Edition
W. L. Faith and Arthur A. Atkisson

AIR POLLUTION

Second Edition

W. L. FAITH

**Consulting Chemical Engineer
San Marino, California**

ARTHUR A. ATKISSON, JR.

**Professor of Urban Health and Administrative Sciences,
University of Texas
Houston, Texas**

WILEY-INTERSCIENCE

a division of John Wiley and Sons, Inc.

New York   London   Sydney   Toronto

Library of Congress Cataloging in Publication Data

Faith, William Lawrence, 1907–
   Air pollution.

   (Environmental science and technology)
   1959 ed. published under title: Air pollution control.
   Includes bibliographies.
   1. Air—Pollution. I. Atkisson, Arthur A., joint author. II. Title.
TD883.F23 1972   628.5'3   72–4225
ISBN 0–471–25320–0

Printed in the United States of America

10 9 8 7 6 5 4 3 2 1

SERIES PREFACE
Environmental Science and Technology

The Environmental Science and Technology Series of Monographs, Textbooks, and Advances is devoted to the study of the quality of the environment and to the technology of its conservation. Environmental science therefore relates to the chemical, physical, and biological changes in the environment through contamination or modification, to the physical nature and biological behavior of air, water, soil, food, and waste as they are affected by man's agricultural, industrial, and social activities, and to the application of science and technology to the control and improvement of environmental quality.

The deterioration of environmental quality, which began when man first collected into villages and utilized fire, has existed as a serious problem under the ever-increasing impacts of exponentially increasing population and of industrializing society. Environmental contamination of air, water, soil, and food has become a threat to the continued existence of many plant and animal communities of the ecosystem and many ultimately threaten the very survival of the human race.

It seems clear that if we are to preserve for future generations some semblance of the biological order of the world of the past and hope to improve on the deteriorating standards of urban public health, environmental science and technology must quickly come to play a dominant role in designing our social and industrial structure for tomorrow. Scientifically rigorous criteria of environmental quality must be developed. Based in part on these criteria, realistic standards must be established and our technological progress must be tailored to meet them. It is obvious that civilization will continue to require increasing amounts of fuel, transportation, industrial chemicals, fertilizers, pesticides, and countless other products and that it will continue to produce waste products of all descriptions. What is urgently needed is a total systems approach to modern civilization through which the pooled talents of scientists and engineers, in cooperation with social scientists and the medical profession, can be focused on the development of order and equilibrium to the presently disparate segments of the human environment. Most of the skills and tools that are needed are already in existence. Surely a technology that has created such manifold environmental problems is also capable of solving them. It is our hope that this Series in Environmental Sciences and Technology will not only serve to make this challenge more explicit to the established professional, but that it also will help to stimulate the student toward the career opportunities in this vital area.

ROBERT L. METCALF
JAMES N. PITTS, JR.

PREFACE

Since the publication of the first edition of this book in 1959, the pollution of man's environment has escalated in public importance and become an issue at the highest levels of government. New programs for control of air pollution sources have been organized, and countless citizens, legislators, engineers, scientists, planners, administrators, technicians, and law enforcement officials have become involved in the multidisciplinary field of air quality management.

This second edition has been prepared to meet the needs of these groups and also to serve the function of the first edition as a textbook for college air pollution courses. Like the first edition, it aims to give the reader an understanding of the types, origin, sources, atmospheric movement, and effects of air pollutants, and of the basic concepts and methods of air pollution control. Discussion of the socio-economic and legal constraints on community control programs has been expanded and new material has been included on the nature and components of effective air quality management programs.

The basic organization of the second edition is essentially the same as the first edition. Some chapters have been expanded, some rearranged for clarity, others redirected in their goals. All have been rewritten.

Chapter 2, "Meteorology" now places major emphasis on design and evaluation of tall stacks for air quality control. The chapters "Dusts, Fumes, and Mists," and "Gases" have been augmented by consideration of current air quality standards and suggested emission regulations. The chapter on "Automobile Exhaust" in the first edition has been updated and expanded into a chapter on "Transportation Sources." A parallel chapter on "Stationary Sources" has been added. This chapter describes the problems of specific industries and the state of the art with regard to means of control.

The treatment of photochemical smog has been removed from the transportation chapter and presented in a new chapter on "Photochemical Air Pollution." The treatment of radioactive pollutants has also been expanded.

New chapters have been added to deal with "The Social Origins of Air

Pollution" and the "Organization and Operations of Air Pollution Agencies." The first edition chapter on "Legal Aspects of Air Pollution Control" has been expanded and is presented here under the title, "Air Quality Management." The reader will find, however, that the legal aspects of air pollution control are discussed also in other sections of the book when appropriate to understanding of the focal subject.

Finally, pertinent tabular material and lists of literature citations have been revised and expanded so that the second edition may serve as a desktop reference for professionals in the field of air pollution control.

Because of its expanded coverage of the field and its broader focus on the many community factors that lead to the development of air pollution problems, we think this second edition will be of interest to students and professionals in the field of urban planning and design and to those with an interest in the management of urban affairs.

W. L. FAITH
A. A. ATKISSON

CONTENTS

1

THE AIR POLLUTION PROBLEM

The first incidence of air pollution is lost in unrecorded history, but it certainly goes back to the discovery of fire. Undoubtedly, restricted meteorological conditions have occurred from time to time somewhere on the face of the earth since the beginning. Sooner or later, a brush or forest fire must have occurred in an area of restricted ventilation, with subsequent contamination of the air by a thick pall of smoke. Similarly, heavy fogs and sandstorms preceded the dawn of history. The worldwide polluted atmosphere that resulted from Krakatao's "blowing its top" in 1883 was the best known of similar explosions. Sulfide-laden gases emitted from hot springs or fissures in the earth have made life unbearable in restricted localities at one time or another. Although all these examples are of natural origin, they still produced polluted air.

When we think of the *air pollution problem,* however, we associate its source with some activity of man, whether it be farming, manufacturing, or just moving about in this world of ours. Practically all air is contaminated to some extent or other, so some reasonable definition of the term "air pollution" is a prerequisite to an orderly discussion.

Basically, air pollution is the presence of foreign substances in the air. An air pollution problem arises when the concentration of these substances interferes with the well-being of people. A more specific definition of air pollution has been developed by the Engineers Joint Council [9]:

1

Air pollution means the presence in the outdoor atmosphere of one or more contaminants, such as dust, fumes, gas, mist, odor, smoke, or vapor, in quantities, of characteristics, and of duration such as to be injurious to human, plant or animal life or to property, or which unreasonably interfere with the comfortable enjoyment of life and property.

These interferences have been classified and are discussed later in this chapter.

The public has chosen the word *smog* to define objectionable air pollution. Originally the word was a contraction of *smoke* and *fog*, but recently it has become descriptive of any air pollution event accompanied by a decrease in visibility. In some cases it has been used to describe malodorous or vegetation-damaging conditions where visibility was no problem. To all intents and purposes, then, the words *smog* and *air pollution* may be considered synonymous.

NOTABLE AIR POLLUTION EPISODES

The London Smog

Historically, the longest record of intermittent air pollution problems belongs to the city of London, England. The notorious pea-soup fogs become particularly offensive when mixed with coal smoke. The word *smog* (smoke and fog) was coined to describe this foul condition.

Sir Hugh E. C. Beaver, Chairman of the Government Committee of Enquiry into the Nature, Causes, and Effects of Air Pollution, says in his review of the growth of public opinion [32]:

It strikes a sympathetic chord, I think, to learn that 700 years ago almost to a year the then Queen of England moved out of the city to Nottingham where she was residing because of the insufferable smoke; and that some 300 years later the brewers of Westminster offered to use wood instead of coal because of Queen Elizabeth's allergy to coal smoke. But it was only about the end of her reign that feeling began to lead to action; and then there was a prohibition—probably ineffective—of the use of coal in London while Parliament was sitting.

In 1661, John Evelyn published his well-known pamphlet, *"Fumifugium: or The Inconvenience of the Aer and Smoake of London Dissipated."* His major recommendation was the removal of all smoke-producing plants from London. But London did little about it until the famous London smog of December 1952, truly a major air pollution disaster. The smog lasted 5 days (December 5 through 9) and caused 4000 deaths (principally among the old, the infirm, and those with respiratory diseases). The onset of the fog was followed by acute respiratory symptoms in a number of cattle at the Smithfield Club's livestock show; about 60 required major veterinary treatment, 12 of the more serious cases were slaughtered, and one died. Just what the lethal agent was is still a matter of conjecture [8].

Almost exactly ten years later, December 3 to 7, 1962, London experienced another black fog, with 340 excess deaths. The improvement over the 1952 episode was laid to smoke reduction brought about by the Clean Air Act and public awareness of the harmful effects of smog which restrained many respiratory cripples from going outdoors [33].

The Donora Smog

Donora, Pennsylvania (1950 population—12,186), is an industrial town on the banks of the Monongahela River about 30 miles south of the heart of Pittsburgh. The major industrial installations were a steel and wire mill, a zinc smelter, and a sulfuric acid plant. During a particularly calm and meteorologically stable period from October 27 to 31, 1948, air pollutants accumulated, and as a result many persons were hospitalized and 20 died. Illnesses of several thousand persons were blamed on the episode, and over 130 separate lawsuits were filed.

As in the London smog of 1952, the causative agent of the deaths and illnesses was never determined incontrovertibly, but in both instances sulfur compounds (SO_2, SO_3, H_2SO_4, inorganic sulfates) were present in the air in abnormally high quantities [7, 44].

Meuse Valley, Belgium

A strong atmospheric inversion settled over the Meuse Valley on December 1, 1930, and remained until December 5. Effluents from the several factories in the Valley, chiefly oxides of sulfur, various inorganic acids, metallic oxides, and soot, were then trapped in the stable atmosphere. Sixty-three persons (generally the old and infirm) died, and several hundred others became ill. Although sulfur oxides and hydrofluoric acid are suspected by many, the actual lethal substance was never proved [30].

Ducktown, Tennessee

In the early 1900's, gases from short stacks at two copper smelters near the Georgian border of Tennessee caused widespread damage to vegetation in the surrounding countryside. When taller stacks were built, damage extended 30 miles into the forests of Georgia. An interstate suit resulted, which was finally carried to the United States Supreme Court. The problem was eventually solved by means of a by-product sulfur dioxide recovery plant [46].

Trail, British Columbia

Two decades later, a similar case involved the lead and zinc smelter of the Consolidated Mining and Smelting Company of Canada at Trail, B.C. The smelter was located on the west bank of the Columbia River, 11 miles north of the international boundary between Canada and the United States. When extensive damage to vegetation occurred on the United States side of the border, a damage suit, finally settled by an international tribunal, was instigated. In this case, after damages were assessed, the problem was solved partly by sulfur recovery and partly by operating the smelter according to a plan based on meteorological considerations [18].

Pittsburgh (Allegheny County), Pennsylvania

Prior to 1948 the nickname, "Smoky City," was appropriate for Pittsburgh. A black pall of smoke and soot often turned day into night, blackened the brightest buildings in a few months, and made washday a nightmare. Finally, the activity of the civic-minded Allegheny Conference on Community Development brought about a smoke-control ordinance that dramatically changed the condition of the atmosphere. The change resulted largely from regulations prohibiting the sale, transportation, and use of high-volatile solid fuels except where adequate mechanical stoking equipment is available. The dieselization of locomotives and the extinguishing of burning gobpiles also helped. Smoke reduction between 1945 and 1953 was estimated at 70% by the Department of Public Health [31].

St. Louis, Missouri

Actually, the first effective control of solid-fuel quality for the prevention of air pollution was initiated in St. Louis. Prior to 1940, when the law went into effect, the need for street lamps and automobile headlights at midday in the winter was not at all unusual. Strict enforcement of the law resulted in a 75% reduction in smoke and a consequent economic boon in reduced dry cleaning, lighting bills, building maintenance, and vegetation damage [31]. More recent regulations in both Pittsburgh and St. Louis have further improved air quality.

Los Angeles, California

Probably the most publicized smog problem in the United States is that of Los Angeles. Meteorological conditions in the 1600-square-mile Los Angeles Basin are conducive to stable atmospheric conditions. In 1542 Juan Rodriguez Cabrillo, after observing the smoke made by Indians burning brush, called the San Pedro Bay the "Bay of Smokes." The tremendous increase in population of the Los Angeles Basin (from less than 1 million in 1920 to 2.86 million in 1940 and more than 6 million in January 1958) and concurrent increases in industrial and human activity brought about an intolerable atmospheric condition. In 1947 the Los Angeles County Air Pollution Control District was organized and appropriate regulations restricting emissions of smoke and sulfur dioxide were passed. Only slight relief resulted. Eye irritation, damage to vegetation, restricted visibility, and the peculiarly high oxidant content of the air continued to increase. In the early 1950's it was shown that these conditions resulted largely from a reaction between organic compounds and nitrogen dioxide activated by sunlight. Both reactants are emitted in large quantities in the exhaust gases from internal-combustion engines. As a consequence, the state of California passed its first motor vehicle pollution control law in 1959 and has strengthened it several times. Smog in Los Angeles has improved somewhat since then but would have been intolerable had controls on new motor vehicles not been required.

Other Cities

The cities previously mentioned are unique only in that their air pollution problems are notorious. One could add other cities, whose names have been attached to specific diseases peculiar to the locality. Thus "Tokyo-Yokohama asthma" affected many U.S. servicemen stationed there [42]. Periodic outbreaks of "New Orleans asthma" [48] puzzled medical authorities for years, but it is now laid to dust from grain elevators which was held near the earth by stagnant air. Some cities have recognized the seriousness of air pollution episodes only well after their occurrence when statistical studies brought excess mortality and morbidity to light. Thus, Dr. Leonard Greenburg, then New York City's Commissioner of Air Pollution only in 1967 attributed 800 excess deaths to smog during a 1963 episode [24].

Emphasis on urban air pollution problems may be misleading, inasmuch as air is not confined by political boundaries. In recognition of this fact, the U.S. Air Quality Act of 1967 directed the Secretary of Health, Education, and Welfare to designate air quality control regions in which the potential air pollution problem was common to several municipalities, even interstate. Currently designated air quality control regions are listed in Table 1.1.

EFFECTS OF AIR POLLUTION

One of the difficulties in coping with air pollution lies in the variety of its effects on people. A farmer is most interested in its effect on his crops; a housewife will complain that dirt and soot soil clothing and furniture; a

TABLE 1.1
Summary of Air Quality Control Regions

	Interstate AQCRs	Intrastate AQCRs	Remaining area AQCRs	Total AQCRs affecting each state
Alabama	3	4	0	7
Alaska	0	4	0	4
Arizona	3	1	0	4
Arkansas	4	3	0	7
California	0	11	0	11
Colorado	1	7	0	8
Connecticut	2	2	0	4
Delaware	1	1	0	2
District of Columbia	1	0	0	1
Florida	2	4	0	6
Georgia	5	4	0	9
Hawaii	0	1	0	1
Idaho	1	2	1	4
Illinois	7	4	0	11
Indiana	5	5	0	10
Iowa	6	6	0	12
Kansas	1	6	0	7
Kentucky	5	4	0	9
Louisiana	3	0	0	3
Maine	1	4	0	5
Maryland	2	4	0	6
Massachusetts	3	3	0	6
Michigan	2	4	0	6
Minnesota	3	4	0	7
Mississippi	2	2	0	4
Missouri	2	3	0	5
Montana	0	5	0	5
Nebraska	2	1	1	4
Nevada	1	1	1	3
New Hampshire	2	0	1	3
New Jersey	3	0	1	4
New Mexico	3	5	0	8
New York	2	6	0	8
North Carolina	1	7	0	8
North Dakota	1	0	1	2
Ohio	6	8	0	14
Oklahoma	2	6	0	8
Oregon	1	4	0	5
Pennsylvania	3	3	0	6
Rhode Island	1	0	0	1
South Carolina	3	7	0	10

TABLE 1.1 (*continued*)

	Interstate AQCRs	Intrastate AQCRs	Remaining area AQCRs	Total AQCRs affecting each state
South Dakota	2	1	1	4
Tennessee	4	2	0	6
Texas	3	9	0	12
Utah	1	1	1	3
Vermont	1	0	1	2
Virginia	2	5	0	7
Washington	2	4	0	6
West Virginia	4	6	0	10
Wisconsin	4	4	0	8
Wyoming	0	2	1	3
American Samoa	0	0	1	1
Guam	0	0	1	1
Puerto Rico	0	1	0	1
U.S. Virgin Islands	0	1	0	1

traveler may be inconvenienced by low atmospheric visibility; a large segment of the general public is concerned with the possible health effects of polluted air.

The five most common effects of air pollution are visibility reduction, economic damage to property, annoyance to human senses, damage to health, and substantive changes in the ecology of the natural environment.

Limited Visibility

Restriction of visibility is the most widely noticed and probably least understood of all effects of air pollution. Smoke and dust clouds that are sufficiently dense to darken the sky will obviously limit visibility, but there are many other times when horizontal visibility is restricted and the sky overhead is bright. The most noteworthy case of this type is the so-called "Los Angeles smog." The sun shines brilliantly with horizontal visibility less than a quarter of a mile on many occasions (Fig. 1.1). The effect is similar to the low horizontal visibility in a ground-hugging fog, except that the relative humidity is very low. Another familiar example of limited visibility with little or no sky-darkening is the white or blue smoke from burning brush or leaves in open fires, or from the wigwam-type incinerators used for burning wood waste.

The difficult problem with respect to visibility restriction is the determination of whether or not it is independent of natural phenomena, i.e., fog and desert or mountain haze. Combination effects like London's smog (smoke and fog) Denver's smaze (smoke and haze), El Paso's smust (smoke and dust) confound the issue.

FIGURE 1.1

Sunshine and limited visibility in Los Angeles; low inversion. (*Courtesy Los Angeles County Air Pollution Control District.*)

Restricted visibility is actually caused by the forward scattering of light by minute solid or liquid particles (aerosols) in the size range of 0.4 μ to 0.9 μ. Smoke, fog, and industrial fumes all contain particles in this range and thus restrict visibility in proportion to the number of particles present in this size range. Much smaller particles, which are emitted from various sources, may also grow sufficiently in the open atmosphere to become important in light-scattering. Thus, minute salt nuclei in an ocean breeze may absorb moisture under proper conditions (usually above 70% relative humidity) to produce fog. Chemical condensation of the reaction products of pollutants in the air may undergo similar growth. In fact, it is believed that the formation of smog particles in Los Angeles is a phenomenon of this type (see Chapter 8). Under suitable conditions sulfur dioxide may be oxidized atmospherically to sulfur trioxide and then condensed with moisture to yield droplets of sulfuric acid (H_2SO_4). The resulting blue haze is a familiar sight in the plumes from many industrial stacks.

Measurement of the particulate matter and aerosols in the atmosphere is a complex problem and is described in Chapter 4. Measurement of the *effect* of these solid or liquid particles, i.e., visibility, is simpler. At United

States Weather Bureau stations, visibility is commonly estimated by an observer by viewing prominent landmarks at known distances from the point of observation. This method is not useful at night or where landmarks are few.

Numerous attempts have been made to relate visibility quantitatively to the concentration of particles in the atmosphere. Obvious drawbacks are the vast range in size of atmospheric particles and measurement of concentration at only one site and often over too long a period of time. Nevertheless, Charlson et al [15] and Noll et al [40] have proposed generalized formulas relating visibility with atmospheric dust concentration measured by high-volume samplers (see Chapter 4) over a 24-hour period. The several formulas are each in the form

$$L_v \times M = K$$

where L_v = visual range in kilometers
M = mass concentration in micrograms per cubic meter
K = a constant

In his most recent publication, Charlson [14] suggests 1800 as the value of the constant, K, when relative humidity is less than 70%. As mentioned previously, the equation has obvious limitations and should be used only as a general guide.

A similar formula, corrected for relative humidity, has been proposed for sulfuric acid mist [39].

Public objection to reduced visibility stems from two factors, transportation hazards and delays, and aesthetic considerations. Aircraft landing hazards at two Kansas City airports led federal authorities, in 1967, to recommend curtailment of open burning and further control of industrial process dust emissions which were said to be responsible for low visibility at the airports. The hazard to motor vehicle traffic caused by a smoke or dust plume dipping across a highway is well known.

Only recently have aesthetic considerations been taken seriously by control agencies. The blotting of the horizon in a smoke-filled valley can well discourage tourist trade and reduce the land value of spectacular-view sites. Obviously, aesthetics have an economic value in this case. A truly aesthetic consideration receiving more attention is public objection to visible plumes. Eventually the problem must be faced on a cost-benefit basis.

ECONOMIC DAMAGE TO PROPERTY

Air pollution damage to property includes damage to materials, vegetation, and animals, as well as interference with production and services.

Materials Damage

Air pollution damages materials chiefly by *corrosion* of metals, presumably from acidic compounds in polluted atmospheres. The most important acid-forming pollutant is sulfur dioxide (SO_2). It is released in greatest quantities by the combustion of sulfur-burning fuels (see Chapter 5). In the presence of oxygen, sulfur dioxide is slowly converted to sulfur trioxide (SO_3), which, in turn, may react with moisture in the air to form sulfuric acid. Deposition of this acid on the metal parts of building roofs, eaves, downspouts, and other metal equipment results in a considerable loss from atmospheric corrosion in most urban communities. Losses through atmospheric corrosion in New York City were estimated in 1955 as 6 million dollars per year. Hydrogen fluoride (H_2F_2) and hydrogen chloride (HCl) will also react with water vapor to form highly corrosive droplets of fog.

Another form of property damage which is becoming increasingly apparent in many communities is *rubber cracking*. Principal areas of economic importance are the sidewalls of tires and various forms of electrical insulation. Damage of this type is caused by ozone [25]. The problem has become so acute in areas like Los Angeles that tire manufacturers add a special anti-ozone compound to all tires sold in the area. More rapid failure of rubber insulation in atmospheres of high ozone content has also been noticed in power-transmission substations and in telephone exchanges.

Painted surfaces are likewise subject to deterioration by a variety of air contaminants. Hydrogen sulfide is a common cause of the darkening of surfaces covered with paint containing white lead. In one southern city periodic release of caustic soda by a chemical manufacturer has several times dissolved the paint on a nearby bridge. Pitting and scaling of paints and enamels, both on buildings and automobiles, have often been traced to accidental release of specific chemicals.

The most common form of permanent damage to *textiles* has been deterioration of nylon hose. This has usually been traced to sulfuric acid mist. Dyes [3, 49], paper [23], and leather are also affected deleteriously by various pollutants.

Temporary property damage results from the *soiling* of surfaces by pollutants. Here the economic impact is the cost of cleaning. Losses most commonly encountered are the additional cost of laundering and dry cleaning clothing and other fabrics and of redecorating buildings, both inside and out. Most of these losses are caused by smoke, soot, and dustfall. Various estimates of soiling costs have been published [34, 49]. *Vegetation damage* as a result of air pollution has already been referred to (see Ducktown, Tennessee and Trail, British Columbia episodes, pp. 3, 4). Such widespread and complete plant damage is seldom encountered anymore, but economic damage to at least a minor degree appears to be omnipresent.

The most frequently encountered air contaminants toxic to vegetation are sulfur dioxide (SO_2), hydrogen fluoride (H_2F_2), chlorine (Cl_2), hydrogen chloride (HCl), nitrogen oxides (NO, NO_2, etc.), hydrogen sulfide (H_2S), ammonia (NH_3), hydrogen cyanide (HCN), mercury vapor (Hg), ethylene (C_2H_4), sprays of weed killers, and constituents of photochemical smog.

The nature of the damage varies with the toxicant, but is usually some form of chlorotic marking (disappearance of green color), banding, or silvering or bronzing of the underside of the leaf. Typically damaged plants are shown in Fig. 1.2. In extreme cases, defoliation and death of the plant may result.

The extent of the damage to an individual plant also depends on many factors, e.g., the pollutant, type of soil, relative humidity, amount and type of plant food available, stage of growth, viability of the plant, concentration of the pollutant, time of exposure, and amount of light. In general, the phytotoxicant is absorbed through the stomata of the leaf, so any factor that tends to hold the stomata open increases the susceptibility of the plant. Typical SO_2 damage to two plant species is shown in Fig. 1.3.

A specific type of plant damage of recent origin is caused by the drifting spray of *weed-killing solution.* Cases have been reported where these sprays have drifted several miles and then damaged vegetation severely. In one report, dust from a rice field sprayed with a 2,4-D formulation (an ester of 2,4-dichlorophenoxyacetic acid) was subsequently blown into a field 15 to 20 miles away and caused considerable injury to a cotton crop.

FIGURE 1.2a.
Effect of fluoride accumulation on grape leaves near Fontana, Calif. (*Courtesy University of California, Riverside.*)

FIGURE 1.2b.
Effect of naturally occurring air pollutants (possibly ethylene) on roses collected in San Francisco, Calif. (*Courtesy University of California, Riverside.*)

In addition to visible injury to plants there has been much speculation about *invisible injury* or growth retardation of various plants as a result of air pollution. So many factors affect growth that it is difficult to determine the effect of air pollutants. Nevertheless, Hull and Went [27] have reported that sublethal fumigations with Los Angeles smog have retarded the growth of alfalfa, sugar beet, endive, oats, spinach, and tomato plants. How widespread growth retardation from air pollution occurs, is not known [11].

FIGURE 1.3a.
SO_2 **damage to tomato plants.** (*Courtesy Moyer Thomas.*)

FIGURE 1.3b.
Typical SO$_2$ damage to oats. (*Courtesy Moyer Thomas.*)

A variety of estimates have been made of the cost of vegetation damage from air pollutants. In 1961, Middleton at the University of California, Riverside, estimated the annual loss in fruit and vegetable crops in California at eight million dollars, and on the eastern seaboard of the United States at eighteen million dollars. Brandt and Heck [12] refer to a more recent estimate of California losses as 132 million dollars. This last figure includes an estimate of loss of citrus crops by growth reduction.

A true assessment of vegetation damage resulting from air pollution is complicated by many other environmental factors. Climate, soil, insects, diseases, genetic history, and lack of care affect crop yields and in some cases cause leaf markings almost identical to those caused by an air pollutant. Often one can be sure of air pollution damage only if a specific pollutant is present or is known to have been present in sufficient dosages to have caused the damage.

Several excellent reviews of the effects of air pollution on vegetation have been published [6, 13, 28, 47]. For effects of specific pollutants on vegetation, see discussions in Chapter 5, Gases.

The economic effects of pollution on *animals* is normally restricted to effects on domestic animals raised for profit.

The most widely publicized animal problem is damage from grazing in areas where grasses are contaminated by fluoride dusts or have absorbed fluoride compounds from the atmosphere. The toxicity of the fluoride particulates depends, of course, on their solubility, sodium fluoride being much more toxic than calcium fluoride or rock phosphate. The chief effect of ingested fluorides on animals is fluorosis, an accumulation of fluoride in the bone structure of the animal, which in time may lead to weight loss and lameness. Safe levels of dietary fluoride for various types of livestock are shown in Table 1.2.

TABLE 1.2

Safe Levels of Fluoride in Daily Total Ration of Livestock*

Species	Soluble fluoride	Rock phosphate or phosphatic limestone
Dairy cow	30–50 ppm F	60–100 ppm F
Beef cow	40–50	65–100
Sheep	70–100	100–200
Swine	70–100	100–200
Chicken	150–300	300–400
Turkey	300–400	–

* Reference: National Academy of Sciences, National Research Council, Publ. No. 381 (1955).

Interference with production and services includes a variety of secondary effects occasioned primarily by other air pollution effects. Included would be automobile and air-traffic delays caused by poor visibility, and a general lethargy in human activities because of the depressing nature of some effects. Many high school principals in southern California have reported a general student lethargy during periods of heavy smog. Controlled tests in telephone exchanges have also shown a relationship between eye irritation and the alertness of operators. These are not unexpected results, because nearly everyone has suffered a loss in mental or physical activity under similar uncomfortable conditions of high temperature and high relative humidity.

Related interferences to services of more direct economic consequence are loss of retail trade, loss of tourist trade, and reduction in land and improvement values. No estimates of economic loss are available.

Another economic item that must be charged to air pollution is the cost of control equipment. The number and cost of control installations are growing every year and obviously are a factor in the cost of production of goods and services. Annual control costs for stationary sources only are estimated to be $1.88 billion by 1975 [22].

Annoyance to the Senses of People

This category of air pollution effects includes a multitude of reactions that can be generally divided into two classes: (1) eye, nose, and throat irritation, and (2) odors. Just where annoyance stops and danger to health begins is controversial, but in this discussion annoyance will be limited to the two classes of effects mentioned. Headache, allergies, nausea, and similar effects will be classified as health effects.

Eye irritation is probably the most exasperating of all effects of air pollution. Unfortunately, its extent has never been surveyed widely; only isolated newspaper reports are available. Even in those areas where eye irritation is a commonplace phenomenon, e.g., Los Angeles, its frequency and severity are known only qualitatively. This lack of knowledge concerning actual episodes of atmospheric eye irritation is related to the nature of the occurrences. It is seldom widespread or of long duration. A slight breeze will dissipate the eye irritant or move the air parcel containing it to another location. But if the air is calm or the irritant continually added to a moving stream of air, it is a real nuisance.

Two forms of atmospheric eye irritation are recognized: (1) the emission of an irritating substance, such as tear gas, into the atmosphere, and (2) the formation of an eye irritant in the atmosphere by reaction of otherwise nonirritating pollutants. Cause and effect are easy to relate when known irritants escape. It sometimes happens that mixtures of unknown composition and of unsuspected irritability are released into the atmosphere. A source of this type is difficult to trace, particularly when air parcels containing high concentrations are carried several miles and then descend to ground level. During World War II an incident of this type supposedly created considerable public indignation in Los Angeles. The irritant gases were emitted from an open cooling tower treating quench water from a surface condenser. The plant was manufacturing butadiene by an oil-cracking process. Supposedly, a mixture of volatile by-products of low solubility was released from the cooling water and created a widespread nuisance. Interestingly enough, even after the plant was shut down, episodes of eye irritation were blamed on "that synthetic rubber plant" for at least ten more years. Sometimes the public has a long memory, but it's often a false one. Nowhere is this more true than in the field of air pollution.

The second type of eye irritants, those caused by atmospheric reactions, are becoming a major problem in urban communities. The photochemical (light-induced) reaction between certain organic materials and nitrogen dioxide (NO_2), chiefly from automobile exhaust, has been shown to be responsible for the high incidence of eye irritation that occurs in Los Angeles County, and probably for the less frequent episodes in other large cities, particularly New York, Philadelphia, and Detroit. The exact nature of the eye irritant is not known. Some believe it to be a gaseous material; others, an aerosol.

Formaldehyde, acrolein, and peroxyacetylnitrate (PAN) are products of the atmospheric photochemical smog reaction (see Chapter 8), but other constituents undoubtedly also contribute to eye irritation. Some authorities believe aerosols potentiate the effect of the eye-irritating gases.

A case where an atmospherically produced eye irritant was known was reported in 1952 [1]. Here, bromine gas from a chemical plant moved over a sewer outfall containing a small amount of styrene dissolved in water. Reaction of the bromine and vaporized styrene, presumably under the influence of sunlight, to produce a brominated styrene to the extent of 25 parts per hundred million caused noticeable eye irritation.

A third type of eye irritation brought about by pollen in the air is considered a health effect and not discussed here.

One of the problems in assessing eye irritation is the subjectivity of the measurement. Neither instrumental nor objective methods are available. To have any validity at all, panels of at least five subjects (see Fig. 1.4) must

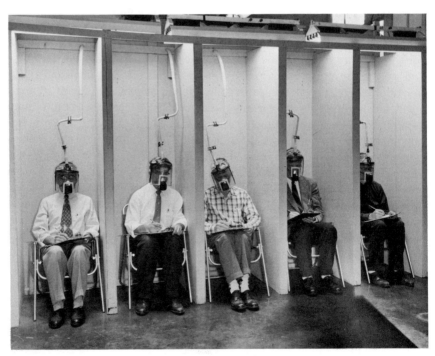

FIGURE 1.4.
Eye irritation panel, Stanford Research Institute.

be used. Each member of the panel should have been pretested to determine if he can distinguish different levels of eye irritation in a reasonably reproducible manner. Relative eye irritation is reported either on an arbitrary point scale [43] or time in seconds to first notice eye irritation [19].

Nose and throat irritation have often been reported as effects of air pollution. The effect is similar to eye irritation and just as difficult to measure.

Odor is also a subjective response of people, but it is even more difficult to define than eye irritation. This is particularly true because an odor objectionable to one person may be pleasing to another. The odor of night-blooming jasmine is a case in point; milady's perfume is another.

Other problems are the extreme sensitivity of the sense of smell, tremendous variations among individuals, and desensitization of the olfactory nerve by some substances. Despite the variation among people, there is fairly good agreement as to which odors are most objectionable. No one likes the odor of a slaughterhouse, a pigsty, an excited skunk, or a fish market. Similarly, industrial odors, such as hydrogen sulfide, mercaptans, phenolic compounds, and compounds related to butyric acid, are objectionable to almost everyone.

Another factor that must be considered is the persistence of an odor. Even a pleasant odor becomes tiresome when it continues for long periods of time.

Many attempts have been made to classify odors, but most of these have resulted in very complicated systems. These systems may be related to the chemical structure of the odorous material, to one constituent, to the general nature of odor, to its effect, or even to its source. Common descriptions of odors include sulfurous (hydrogen sulfide and mercaptans), nitrogenous (decaying plant and animal life), oxidizing (ozone and chlorine), nauseating (skatole), aldehydic, sweet, and aromatic (coffee roasters).

Measurement of the degree of odor may be made by a panel using an arbitrary scale of the type used for eye irritation. Another method widely used in the laboratory involves progressive dilution of the odor until it is no longer distinguishable. Various aspects of the odor problem are treated in detail in Chapter 6.

DAMAGE TO HEALTH

Of all air pollution effects, damage to health is undoubtedly foremost in most peoples' minds. Air is necessary for the survival of man. Five minutes without *life-supporting* air is certain death. Potentially, air may also have *life-damaging* properties if the balance between its important constituents, oxygen, nitrogen, and carbon dioxide, is sufficiently disturbed or if sufficient amounts of contaminants are present. There is no such thing as absolutely "pure" or "clean" air, nor is there any such thing as an absolutely safe contaminant. At some concentration in air any gas or aerosol will damage health, if through no other means than dilution of oxygen. The specific concentration at which a contaminant will damage health depends on how the word "health" is defined, the nature of the contaminant (pollutant), the length of time the air containing the specific pollutant (at a given concentration) is breathed or in contact with the receptor, and the state of health of the receptor.

Herein lie the problems of air pollution control. To some, health is the absence of disease. To others any deviation from an enjoyable state of being is a health effect. In a recent hearing before the California State Board of Health, a physician testified that if the sight of a smoke plume a mile away annoyed her, her health was affected. In this vein the World Health Organization has defined health as "A state of complete physical, mental, and social well-being and not merely the absence of disease or infirmity." A further complication is the subjective nature of health, which is probably the reason for much of the emotional approach to matters concerning air pollution.

The state of health of a receptor is obviously a highly important variable. The so-called "normal" man is exceedingly resistant to environmental insult and readily adapts to marked changes in atmospheric conditions. But the "sensitive" individual or one weakened by disease, exposure, or a variety of stresses may be affected markedly by concentrations of atmospheric pollutants that would be unnoticed by the normal healthy individual.

The foregoing are some of the problems that must be faced by public health officers, medical scientists, and others (even legislators) in specifying the nature of air pollution standards and control regulations.

Keeping in mind the variations among population groups, there are certain facts that appear to be incontestable.

1. People have died as a result of polluted air as evidenced by the several air pollution episodes in London, Donora, and other communities (see pp. 2, 3). Many of these may not have died as a direct result of air pollution, but certainly air pollution was a contributing factor.

2. There is a concentration or dosage of specific air pollutants which evokes various physiological responses from receptors.

3. Certain concentrations of atmospheric pollutants aggravate the symptoms of chronic pulmonary disease and possibly other diseases.

4. Some pollutants irritate sensitive membranes, those of the eyes, nose, throat, and lungs particularly, but the irritation varies greatly from one person to another.

5. Health effects that may be caused or aggravated by air pollution

may also be caused or aggravated by overcrowding, occupation, working environment, nutrition, habits (such as smoking), climate, and other variables.

6. It is probable but not proved conclusively that mixtures of pollutants will affect some people more than would be predicted by the additive effects of individual pollutants.

The problem facing public health authorities and medical research specialists is to determine what concentrations and dosages of specific pollutants are responsible for specific effects on different segments of the population. These effects theoretically can be balanced against other effects to the community caused by reducing pollution to various levels.

Determination of Health Effects of Pollutants.   Three common means are available for determining the effect of various concentrations and dosages of pollutants on people. They are (1) experimental exposures of men and animals, (2) clinical studies, and (3) epidemiology.

The experimental exposure of different types of animals under controlled conditions to various concentrations and dosages of atmospheric pollutants can yield valuable clues as to the mode of action of a pollutant, the dependence of effects on other variables, and cumulative effects over several generations. Like all biological experiments, results must be subjected to rigorous statistical analysis. The difficult problem that remains is the extrapolation of results to human populations.

Experimental exposures of men must of necessity be limited to concentrations and dosages of pollutants that will not result in serious illness. Studies of this kind have usually shown that there is a wide variation in measurable effects among subjects. Types of effects noticed and measured include detection of odor; eye, nose, and throat irritation; variations in lung capacity, breathing frequency, pulse rate; reduction in physical activity, dark adaptation, and many other physiological responses.

Clinical studies predominately involve observations made on subjects who are or were exposed to atmospheric pollutants under uncontrolled conditions. Thus, patients visiting physicians to complain of symptoms supposedly caused by living in polluted areas may upon observation yield clinical data of importance. Studies of downtown traffic police, truck drivers, and other occupational groups may yield valuable information. Other important information has been gathered by follow-up of the acute air pollution episodes mentioned previously. These data indicate beyond doubt a relationship between air pollution and disease, particularly pulmonary disease, but no conclusive information concerning relationships to specific pollutants.

In *epidemiological studies* one looks for relationships between the distribution of specific diseases in a human population and the factors that determine the distribution. Thus one may compare mortality (death) records, hospital admissions, morbidity (illness) records, absenteeism, and other health-related data from several geographical areas with levels of atmospheric pollutants in the same areas to determine if significant correlations may be discovered. To avoid misinterpretation, populations under study must be carefully screened for smoking habits, occupational exposures, periods of residence in the area, population density, and any other factor that might prejudice the results of the study. Similarly, data on atmospheric pollution should be sufficiently complete that the likelihood of spurious results is minimal. In actual fact, the weak points in many epidemiologic air pollution studies are neglect of some meteorological variable closely related to pollutant concentration or insufficient atmospheric pollutant data.

Specific Health Effects. Although there may not be adequate data to relate specific diseases to specific concentrations of pollutants, certain general relationships appear sound. Principal relationships are:

1. Chronic pulmonary disease (bronchitis, asthma, emphysema) is aggravated by sufficiently high concentrations of SO_2, NO_2, particulate matter, and photochemical smog.

2. It has been difficult to relate atmospheric pollution to lung cancer; at least data are not so convincing as the relationship between heavy cigarette smoking and lung cancer.

3. Cardiovascular diseases are related to atmospheric pollution in that any pollutant placing sufficient stress on the pulmonary function may affect the heart.

4. Carbon monoxide ties up the hemoglobin in the blood to a sufficient amount to put added stress on those suffering from cardiovascular and pulmonary disease.

5. Photochemical smog irritates the eyes but apparently does not damage them. Persons with chronic pulmonary disease may suffer aggravation of symptoms.

6. A variety of particulates, particularly pollens, initiate asthmatic attacks.

7. Various gases, e.g., hydrogen sulfide, mercaptans, and ammonia, have noticeable odors at low concentrations.

8. Certain heavy metals, e.g., lead, may enter the body through the lungs and accumulate in the bones and various tissues. In the case of lead, however, the amounts which enter by way of food and drink are more important [29].

Air Quality Criteria Based on Health Effects

It is generally agreed that the basic criteria for acceptable air quality should be based on health effects and that compromise in this regard is unacceptable. Since passage of the Air Quality Act of 1967, the Secretary of Health, Education, and Welfare (later, the Administrator of the Environmental Protection Agency) is charged with publication of air quality criteria for important atmospheric pollutants. As of February 1, 1971, criteria had been published for suspended particulate matter, sulfur oxides, carbon monoxide, photochemical oxidants, hydrocarbons, and oxides of nitrogen [20, 35–39]. These publications list and discuss various published documents relating atmospheric concentrations and dosages of the specified pollutants to their effects on health as well as other factors. Similar reviews have been published by other organizations [2, 41]. There is no *general* agreement as to specific numerical values defining acceptable air quality from a health standpoint. However, the air quality standards published by the Environmental Protection Agency in April, 1971 have the force of law (see Table 1.3). Considerable margins of safety are included (see Table 1.4).

The state of the art with respect to air pollution medical studies has been the subject of an extensive review by Heimann [26].

Ecosystem Changes

Not all of the effects of air pollution emissions are as dramatic and instantly discernible as the episodes which have occurred in Donora, London, and the Meuse Valley. Some effects may occur far from the site at which the emission occurred and result from a series of changes induced in the environment as a result of the initial pollution emission.

For example, storms, cloud formations, and increased rainfall in areas downwind from large pollution sources now seem to be well documented. In the State of Washington, a 30% increase in average long-term precipitation has been ascribed to particulate emissions from pulp and paper mills [21]. The Council on Environmental Quality has reported that similar emissions "from the stack of a single wood pulp mill in Pennsylvania causes fog formations which sometimes fill a valley several miles wide and twenty miles long and spills out into adjacent valleys" [21].

Other recent studies suggest that pollution emissions from a variety of community sources are resulting in increased rainfall and storm activity in areas downwind from large metropolitan areas. In the Netherlands and in Sweden, measurements have shown an increase in the acidity of rain droplets, and this increase has, in turn, been linked to an increase in the acidity of small lakes and rivers which now threatens the stability of their

TABLE 1.3

Federal Air Quality Standards
(as published by Environmental Protection Agency, April 30, 1971)

	SO_2	Particulate matter	CO	Oxidant	NO_x	HC
	(micrograms per cubic meter*)					
PRIMARY STANDARDS						
Annual						
(arith. mean)	80				100	
(geom. mean)		75				
24-hr. max	365	260				
8-hr. max			10,000			
3-hr. max						160
1-hr. max			40,000	160		
SECONDARY STANDARDS (where different from primary)						
Annual						
(arith. mean)	60					
(geom. mean)		60				
24-hr. max	260	150				
3-hr. max	1300					

* Conversion factors:
SO_2: $\mu g/m^3 \times 3.82 \times 10^{-4} = ppm$
CO: $\mu g/m^3 \times 8.73 \times 10^{-4} = ppm$
Oxidant: $\mu g/m^3 \times 5.094 \times 10^{-4} = ppm$
(As O_3)
NO_x: $\mu g/m^3 \times 5.319 \times 10^{-4} = ppm$
(As NO_2)
HC: $\mu g/m^3 \times 20.33 \times 10^{-4} = ppm$
(As carbon atoms)

ecosystems and the survival capacity of certain aquatic animals [45]. The increased acidity apparently is a result of the emission of sulfur products from industrial sources in western Europe [45].

In Denver, the fraction of the sky which is covered by high cloud formations has doubled in less than ten years, apparently as a result of water vapor and particulate emissions from high-flying jet aircraft [21].

Beyond these kinds of changes in the quality of regional environmental systems, other evidence suggests that man-caused pollution discharges may now be leading to ultimate changes in global weather patterns and to other alterations in the planetary ecologic system. One ecologist has observed that the "earth's atmosphere, hydrosphere, biosphere, and the superficial layers

TABLE 1.4
Hazardous Air Pollution Levels*

Pollutant	Concentration	Time period (average)
Sulfur dioxide	2,620 $\mu g/m^3$	24-hr.
Particulate matter	1,000 $\mu g/m^3$	24-hr.
	8COHs	24-hr.
Sulfur dioxide plus particulates	Product of SO_2 conc. in $\mu g/m^3$ times particulate conc., each in $\mu g/24$-hr. average $= 490 \times 10^3$ or SO_2 conc. in $\mu g/m^3$ times COHs $= 1.5$	
Carbon monoxide	57.5 mg/m^3	8-hr.
	86.3 mg/m^3	2-hr.
	1,400 mg/m^3	1-hr.
Photochemical oxidants	800 $\mu g/m^3$	4-hr.
	1,200 $\mu g/m^3$	2-hr.
	1,400 $\mu g/m^3$	1-hr.
Nitrogen dioxide	3,750 $\mu g/m^3$	1-hr.
	938 $\mu g/m^3$	24-hr.

* Defined by Environmental Protection Agency as levels that might cause "significant harm" to human health during episodes of high air pollution.

of its lithosphere all together constitute a vast ecosphere within which a change in any one component evokes changes in all the others" [17]. Exchanges of energy, minerals, carbon dioxide, oxygen, water vapor, and heat between organisms, rocks, the seas, and the atmosphere constitute a seamless web of interactions within an "indivisible" planetary ecologic system (ecosystem).

It now appears that some types of pollution emissions may trigger important changes in the operations of that system. That such changes can be triggered is neither surprising nor necessarily threatening. Within most complex ecosystems substantial changes can take place without threatening the survival capacity of the focal organisms. So it is with man and his planetary environment. Ecologic changes have taken place throughout the course of his upward evolution. Regional climatologic patterns have been altered; whole land masses have disappeared; and a great many of man's biotic partners have expired.

It also appears that man's potential for producing other harmful modifications in the workings of his ecosystem are far larger than earlier theorists believed, if somewhat less certain than contemporary alarmists are suggesting. Principal among such possible modifications are changes in the planet's albedo, or average radiation reflecting capacity, and increases in the atmospheric concentrations of carbon dioxide.

Under normal circumstances, approximately 30% of all solar radiation which reaches the earth is either scattered back into outer space by the constituents of the atmosphere or directly reflected there by clouds at the earth's surface [45]. Processes leading to increased cloud formations or to escalations in the atmospheric burdens of suspended particulate matter therefore might increase the planet's albedo, thereby lower the amount of radiation received from the sun, and thus lower the mean temperatures of the globe. If severe enough, such a process could lead to another ice age. At present, about 31% of the earth's surface is covered by low clouds, and one study indicates that an increase in coverage to 32% would lower global temperatures by 1.4°F [21]. Speculation that such a process may now be operative has been prompted by the drop in global temperatures which has been measured in recent decades and by an apparent southward movement of the northern frostline [45].

Increases in atmospheric burdens of carbon dioxide also could lead to global climatologic changes—but in the direction of higher, rather than lower, mean temperatures. The carbon dioxide molecule is able to absorb thermal energy that otherwise would radiate outward from the earth and into outer space. Increases in atmospheric burdens of this substance therefore could lead to a reduction in the amount of heat energy lost by the earth to outer space and to a consequent increase in global temperatures. However, as global temperatures increased through the workings of this process, increased evaporation of water also could take place and this, in turn, could lead to increased cloud formations that might alter the earth's albedo and thereby counteract the warming that otherwise would occur.

Substantial increases in global levels of carbon dioxide have been measured since 1870 and presumably are a result of the dramatic increases in fossil fuel consumption. Several studies have been directed toward quantifying the degree of increase in atmospheric carbon dioxide and projecting it into the future. A study by a Presidential Science Advisory Committee in 1965 pointed out that the carbon dioxide content of the atmosphere was increasing at a rate of 0.23% annually, and suggested that a 25% increase may come about by the year 2000. Bolin [10] predicts that atmospheric levels of carbon dioxide will reach 375 to 400 ppm by the year 2000; Singer suggests 500 to 540 parts per million by the year 2040 [45]. These estimates assume that in the future, as in the past, only one-third of the increased carbon dioxide emissions will be accounted for by rising atmospheric burdens, and that two-thirds will be taken into the oceans or consumed by growing vegetation.

That carbon dioxide levels have increased, that emissions of particulate materials has risen, that increased cloud coverage has been measured over some sections of the globe, and that mean temperatures of the globe have dropped in recent years, are facts. That these apparent effects of pollution will lead to substantial alterations in the planetary ecosystem, is speculation based on our still imperfect understanding of the workings of that system. After reviewing the evidence related to pollution emissions and the processes which might lead to alterations in the planetary ecosystem, Cloud and Gibor recently stated that "He who is willing to say what the final effects of such processes will be is wiser or braver than we are. Perhaps the effects will be self-limiting and self-correcting, although experience should warn us not to gamble on that" [16].

The fragmentary knowledge of the fate of these and other contaminants in the atmosphere led a committee of the Amercan Chemical Society to recommend that "Systematic measurement should be undertaken for a number of relatively long-lived substances in the atmosphere, including carbon monoxide, nitrous oxide, methane, carbon dioxide, and sulfur hexafluoride. The general turbidity of the atmosphere should be measured systematically on as wide a basis as possible, and more effort should be devoted to determining the nature of the aerosols that cause such turbidity" [4].

Sound, adequate data can lead to sound ecologic action.

Economics and Politics

Much of what has been said thus far concerning air pollution emissions and effects has been cast in general terms. Control activities, however, must be most specific. Legal, political, and engineering considerations require that emission standards be expressed in precise and clearly understandable language. Moreover, the public interest demands that the pollution problems of a community be defined in terms specific enough that policy-makers and the general public can identify the culpable parties and determine the adequacy of proposed control measures.

To guide community efforts, minimum national air quality standards now have been published for a few contaminants. *Primary standards* are those intended to protect the public from adverse health effects, while *secondary standards* are those intended to provide protection from other, less serious consequences. In both cases, the standards represent the lowest level at which available evidence suggests that an effect will be produced, *plus a substantial safety margin.*

The standards are listed in Table 1.3.

As will be seen later, however, air quality standards are not the complete answer to the air pollution problem. They only define the goals of a community. To attain these goals, source emissions must be controlled. With sufficient knowledge of the amount and nature of current and projected pollutant emissions, one can calculate the amount of pollutants that may be emitted and still meet the air quality standard. Up to this point, the scientific approach has been used, but when it comes to establishing rules and regulations to control emissions, economics and politics step in.

The initial economic impact of an air pollution rule or regulation is on the person responsible for the emission. Pollution control equipment costs money (sometimes more than the basic process equipment emitting the pollutant) and seldom offers an immediate return to the owner of the pollution source. So he complains bitterly. He can't afford it. His competitor will gain an unfair advantage. He will have to shut his plant down and lay off his workers. Sometimes his cry is true, sometimes not. But in the end the community faces an economic penalty in higher prices, less tax receipts, an increase in unemployment, or other economic effects. What it is may be difficult to assess. An even more difficult assessment to make is the value of the atmospheric improvement brought about by the reduction in emissions. Nevertheless, some public body has to make a decision, must balance the equities. Theoretically, increased *costs* can be balanced against increased *benefits* to attain an optimum *cost-benefit ratio*. But air pollution control costs vary with the nature of the pollutant source, its size, location, and many other factors. Quantitative data on benefits are almost nonexistent. So when one comes to evaluating increased mortality or the tenuous "enjoyment of life," objectivity disappears. Only the subjective "value judgment" remains. The judgment is made by a political body responsible to the public. *Which* political body should make the judgment is a matter of controversy. See Chapter 12.

REFERENCES

1. Adams, E. M., and E. J. Schneider, "Eye Irritants Formed by the Interaction of Styrene and Halogens in the Atmosphere," *Proc. Air Pollution and Smoke Prevention Assoc. Am.* **45,** 61–63 (1952).

2. Air Pollution Control Association, "Toxicologic and Epidemiologic Bases for Air Quality Criteria," *J. Air Pollution Control Assoc.* **19,** 629–732 (1969).

3. Ajax, R. L., C. J. Conlee, and J. B. Upham, "The Effects of Air Pollution on the Fading of Dyes," *J. Air Pollution Control Assoc.* **17,** 220–224 (1967).

4. American Chemical Society, "Cleaning our environment—The Chemical Basis for Action," American Chemical Society, Washington, D.C., (1969).

5. American Industrial Hygiene Assoc., "Air Pollution Manual," 49–61 (1961).

6. Ibid., pp. 63–71.

7. Anon., "Air Pollution in Donora, Pa.," *Public Health Bull.* 306 (1949).

8. Anon., "Mortality and Morbidity During the London Fog of December, 1952," Ministry of Health Rept. 95, H.M. Stationary Office, London (1954).

9. Bishop, C. A., "EJC Policy Statement on Air Pollution and Its Control," *Chem. Eng. Progr.* **53**, No. 11, 146–152 (1957).

10. Bolin, Bert, "The Carbon Cycle," *Scientific American,* **223**, No. 3, 125–131 (Sept. 1970).

11. Brandt, C. S., and W. W. Heck, "Effects of Air Pollutants on Vegetation," Chapter 12, Volume I, 407–408 of *Air Pollution,* A. C. Stern, ed., Academic Press, New York (1968).

12. Ibid., p. 402.

13. Ibid., pp. 401–443.

14. Charlson, R. J., "Atmospheric Visibility Related to Aerosol Mass Concentration," *J. Envir. Sci. and Technology,* **3**, 913–917 (1969).

15. Charlson, R. J., N. C. Ahlquist, and H. Horvath, "On the Generality of Correlation of Atmospheric Aerosol Mass Concentration and Light Scatter," *Atmospheric Environment,* **2**, 455–464 (1968).

16. Cloud, Preston, and Pharon Gibor, "The Oxygen Cycle," *Scientific American,* **223**, No. 3, 111–123 (Sept. 1970).

17. Cole, Lamont C., "Man's Ecosystem," *Bioscience,* **16**, 243–248 (1966).

18. Dean, R. S., and R. E. Swain, "Report Submitted to the Trail Smelter Arbitral Tribunal," *U.S. Bur. Mines Bull.* 453 (1944).

19. Doyle, G. J., "Design of a Facility (Smog Chamber) for Studying Photochemical Reactions under Simulated Tropospheric Conditions," *Environ. Sci. and Tech.,* **4**, 907–916 (1970).

20. Environmental Protection Agency, "Air Quality Criteria for Nitrogen Oxides," Publ. No. FS 2.300 AP-84, Supt. of Documents, U.S. Government Printing Office, Washington D.C. (1971).

21. Executive Office of the President. Environmental Quality: The First Annual Report on Environmental Quality. U.S. Government Printing Office, Washington, D.C. (August 1970).

22. Fogel, M. E. et al, "Comprehensive Economic Cost Study of Air Pollution Control Costs for Selected Industries and Selected Regions," Research Triangle Institute, Durham, North Carolina (Feb. 1970). Reproduced by Federal Clearinghouse, Springfield, Va. as PB 191054.

23. Gibson, Weldon B., "The Economics of Air Pollution," *Proc. Natl. Air Pollution Symposium. 1st Symposium, Pasadena, Calif., 1949,* pp. 109–114.

24. Greenburg, Leonard, F. Field, C. L. Erhardt, M. Glasser, and J. I. Reed, "Air Pollution, Influenza, and Mortality in New York City," *Arch. Environ. Health*, **15**, 430–438 (1967).

25. Haagen-Smit, A. J., C. E. Bradley, and M. M. Fox, "Ozone Formation in Photochemical Oxidation of Organic Substances," *Ind. Eng. Chem.* **45**, 2086 (1953).

26. Heimann, H., "Status of Air Pollution Health Research," *Arch. Environ. Health,* **14,** 488–503 (1967).

27. Hull, Herbert M., and Frits W. Went, "Life Processes of Plants as Affected by Air Pollution," *Proc. Natl. Air Pollution Symposium. 2nd Symposium, Pasadena, Calif., 1952,* pp. 122–128.

28. Jacobson, J. S., and A. C. Hill, editors, "Recognition of Air Pollution Injury to Vegetation: A Pictorial Atlas," Air Pollution Control Assoc., Pittsburgh, Pa. (1970).

29. Kehoe, R. A., "The Metabolism of Lead in Man in Health and Disease." The Harben Lectures, reprinted from *J. Royal Inst. Public Health and Hygiene* (1961).

30. Mage, J., and G. Batta, "Results of the Investigation into the Cause of the Deaths Which Occurred in the Meuse Valley during the Fogs of December, 1930," *Chim. & Ind. (Paris)*, **27**, 961–975 (1932).

31. Mallette, Frederick S., "A New Frontier: Air-Pollution Control," *Proc. Inst. Mech. Engrs. (London)*, **169**, No. 22, 598–599 (1954).

32. Mallette, Frederick S., editor, *Problems and Control of Air Pollution,* Reinhold, New York, 2 (1955).

33. Meethum, A. R., *Atmospheric Pollution,* Pergamon Press, Ltd., 3rd. Ed., 234 (1964).

34. Michelson, I., and B. Tourin, Public Health Report (U.S.), **81,** No. 6, 505 (1966).

35. National Air Pollution Control Administration, "Air Quality Criteria for Carbon Monoxide," Publ. No. FS 2.300 AP-62, Supt. of Documents, U.S. Government Printing Office, Washington, D.C. (1970).

36. National Air Pollution Control Administration, "Air Quality Criteria for Hydrocarbons," Publ. No. FS 2.300 AP-64, Supt. of Documents, U.S. Government Printing Office, Washington, D.C. (1970).

37. National Air Pollution Control Administration, "Air Quality Criteria for Photochemical Oxidants," Publ. No. FS 2.300 AP-63, Supt. of Documents, U.S. Government Printing Office, Washington D.C. (1970).

38. National Air Pollution Control Administration, "Air Quality Criteria for Particulate Matter," Publ. No. FS 2.300 AP-49, Supt. of Documents, U.S. Government Printing Office, Washington, D.C. (1969).

39. National Air Pollution Control Administration, "Air Quality Criteria for Sulfur Oxides," Publ. No. FS 2.300 AP-50, Supt. of Documents, U.S. Government Printing Office, Washington, D.C. (1969).

40. Noll, K. E., P. K. Mueller, and M. Imada, "Visibility and Aerosol Concentration in Urban Air," *Atmospheric Environment, 2,* 465–476 (1968).

41. Occupational Health Institute, "Proceedings of the Symposium on Air Quality Criteria," *J. of Occupational Medicine, 10,* No. 9, 1–135 (1968).

42. Phelps, H. W., "Air Pollution Asthma Among Military Personnel in Japan," *J. Amer. Med. Assoc., 175,* 146 (1961).

43. Schuck, E. A., "Eye Irritation from Irradiated Auto Exhaust," Air Pollution Foundation (San Marino, Calif.), Rept. 18 (1957).

44. Shilen, Joseph, et al, "The Donora Smog Disaster," Bureau of Industrial Hygiene, Pennsylvania Department of Health, Harrisburg.

45. Singer, S. Fred, "Human Energy Production as a Process in the Biosphere," *Scientific American, 223,* No. 3, 175–190 (Sept. 1970).

46. Swain, R. E., "Smoke and Fume Investigation," *Ind. Eng. Chem.* **41,** 2384–2388 (1949).

47. Thomas, M. D., "Effects of Air Pollution on Plants," *in* "Air Pollution," 233–278, World Health Organization, Columbia University Press, New York (1961).

48. Weill, Hans, M. M. Siskind, R. C. Dickerson, and V. J. Derbes, "Allergenic Air Pollutants in New Orleans," *J. Air Pollution Control Association,* **15,** No. 10, 467–471 (1965).

49. Yocom, John E., and R. O. McCaldin, "Effects of Air Pollution on Materials and the Economy," Chapter 15, Volume I, *Air Pollution,* A. C. Stern, ed., Academic Press, New York (1968).

2

METEOROLOGY

Every air pollution problem has three requisites:

1. There must be an emission of the pollutant or its precursor into the free atmosphere.
2. After emission, it must be confined to a restricted volume of air.
3. The polluted air must interfere with the physical, mental, or social well-being of people.

The third requirement was discussed in Chapter 1. The second requirement will be considered here.

Emissions of pollutants are usually the same from day to day; the weather is the variable that triggers the air pollution episode. We cannot do much about the weather, but we can understand how it affects the air pollution problem if we have a knowledge of meteorology, which is that branch of physics that treats of the atmosphere and its phenomena. Then we can apply this knowledge to solve the problem.

The chief areas of application of meteorology to air pollution control are:

1. Determination of allowable emission rates.
2. Planning and interpreting air pollution surveys.
3. Stack design.
4. Plant-site selection.
5. Prediction of the air pollution potential of an area.

The *atmosphere* is the layer of air that surrounds the earth. It is probably 100 miles thick, but more than one-half of its total weight is in the four miles just above the surface of the earth. The upper portion of the atmosphere is called the stratosphere; the lower portion, the troposphere. The stratosphere is an isothermal region ($-60°F$ in polar regions and $-90°F$ in the tropic region) where clouds of vapor do not form and where little convection (air movement caused by temperature differences) takes place. Conversely the troposphere varies in temperature from the surface of the earth to the stratosphere; it contains clouds of water vapor; considerable air movement, caused by temperature differences, takes place. Generally, the troposphere is 10 to 11 miles deep at the equator and only 4½ miles deep at the north and south poles.

The chemical composition of dry air is shown in Table 2.1 [8]. Thus dry

TABLE 2.1*
Nonvariable Components of Atmospheric Air

Constituent	Content (%)	Content (ppm)
N_2	78.084 ± 0.004	
O_2	20.946 ± 0.002	
CO_2†	0.033 ± 0.001	
A	0.934 ± 0.001	
Ne		18.18 ± 0.04
He		5.24 ± 0.004
Kr		1.14 ± 0.01
Xe		0.087 ± 0.001
H_2		0.05
CH_4		2
N_2O		0.5 ± 0.1

* Ref. 8.   Publication of Compendium of Meteorology was made possible through contract support of the Geophysics Research Directorate, Air Force Cambridge Research Center, Air Research and Development Command.
† Extrapolated to 1950.

air is approximately 78% nitrogen; 21% oxygen; and 0.9% minor elements, principally argon. In addition to these and water vapor, minute amounts of ozone (O_3), sulfur dioxide (SO_2), nitrous oxide (N_2O), nitrogen dioxide (NO_2), nitric oxide (NO), formaldehyde (HCHO), iodine (I_2), sodium chloride (NaCl) or salt, ammonia (NH_3), carbon monoxide (CO), methane (CH_4), and some dust and pollen are present. The origin of these materials and their variations are discussed in Chapters 4 and 5.

The chief physical variations of air are its density and its humidity. Density is expressed in weight per unit volume, e.g., pounds per cubic foot, and varies both with temperature and pressure. At sea level, the pressure exerted by the weight of the column of air above that level (standard barometric pressure) is 29.92 in. Hg (inches of mercury), or 14.7 lb per in.² Meteorologists commonly express this pressure as 1013.3 mb (millibars) where 1 mb = 1000 dynes per cm². Obviously, as one ascends from the earth the concentration of gas molecules becomes less dense, and, because of this thinner atmosphere, the pressure exerted per unit area by the air becomes less. With this relationship in mind, we can understand one reason why a column of smoke, for instance, tends to rise. Any gas, free to expand, will tend to move from a region of high pressure to one of lower pressure. As it does so, it expands and if there is no interchange of heat with the surroundings, the gas will become cooler and denser. At a point where the density of the air parcel is the same as the density of the surrounding air, it is in equilibrium with the surrounding air and has no tendency to rise or sink. In the process of rising as described above, the particular parcel of air is said to have undergone adiabatic cooling.

A complication enters this adiabatic relationship because of the pressure of water vapor in any actual air sample. The partial pressure exerted by water vapor is dependent only on the temperature of the air and is independent of the presence of the other gases. However, unless the temperature is decreased until liquid water droplets begin to form (the dew point), the adiabatic relation for dry air may be applied to moist air.*

Adiabatic Lapse Rate

The change in air temperature with pressure, and hence with altitude (adiabatic lapse rate), is an important consideration in the incidence of air pollution. In well-mixed air the dry adiabatic lapse rate is 5.4°F per 1000 ft (0.98°C per 100 m). In other words, air temperature decreases 5.4°F for each 1000 ft above the earth's surface. When this condition is extant, a smoke plume will rise directly into the atmosphere (by virtue of its low density, in turn occasioned by high temperature) until it reaches air of similar density. In many situations, however, because of external heating or cooling effects, the lapse rate may be greater or less than the adiabatic. Some of these conditions are shown in Fig. 2.1.

The two most important conditions from an air pollution standpoint are the superadiabatic lapse rate and the negative lapse rate (inversion). On a clear summer day, rapid heating of the earth by the sun warms the air near the surface to the point where the lapse rate is superadiabatic. The

---

* The amount of water vapor in air is usually expressed as per cent relative humidity (R.H.). At any given temperature the vapor pressure of water is a constant. For example, at 20°C, the vapor pressure of water is 23.4 mb. The partial pressure of water vapor in *saturated* air at 20°C is therefore 23.4 mb. If the partial pressure of water vapor in a sample of moist air at 20°C were only 12.3 mb, the air would be said to have a relative humidity of 52.6% (12.3/23.4 × 100). If this sample of air were cooled adiabatically to 15°C, its R.H. would be 12.3/17.0 × 100 = 72.4%, because the vapor pressure of water at 15°C is 17.0 mb. If adiabatic cooling were continued to 10°C, where the vapor pressure of water is 12.3 mb, the relative humidity would then be 100% (12.3/12.3 × 100). The air would be saturated, and if it were cooled below 10°C, droplets of liquid water would begin to form. Thus, 10°C is the dew point of air at 20°C and 52.6% relative humidity.

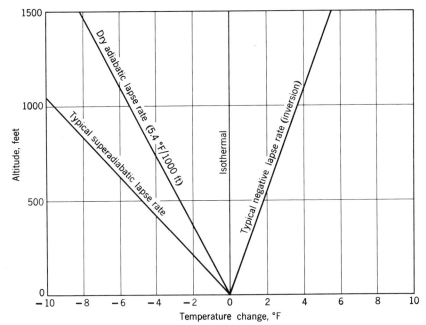

FIGURE 2.1.

Typical atmospheric temperature lapse rates.

decrease in air temperature with height is greater than the normal adiabatic lapse rate. Under this condition the atmosphere is said to be in unstable equilibrium, and marked vertical mixing of the air results. This is a condition where pollutants are dispersed rapidly.

The opposite condition is a negative lapse rate or temperature inversion, in which case the air temperature *increases* with height. It may be caused by cooling of the air near the ground because of the earth's losing heat by long-wave radiation at night. Under conditions of an atmospheric temperature inversion, the atmosphere is said to be stable and very little mixing or turbulence takes place, because the denser air is near the ground. Under such conditions, pollutants in the air do not disperse.

In many actual cases, different types of lapse rates may exist at the same time. Some of these are shown in Fig. 2.2. Common conditions contributing to intense air pollution episodes are shown in curves A and B. In both cases, vertical mixing of polluted air near the ground with the atmosphere above is hindered by the stable inversion layer. Only when the slope of the lapse rate is reversed will the air become turbulent and vertical mixing be resumed.

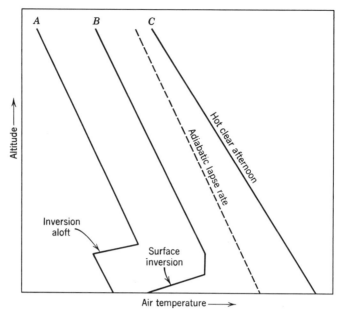

FIGURE 2.2.
Typical atmospheric soundings.

Horizontal Motion of the Atmosphere

So far, only vertical motion of air, induced by temperature differences, has been discussed. In most cases the effect of horizontal motion, or wind, is even more important. Wind movements, like other weather events, may be divided into two classes, depending on their magnitude. Both the large-scale (macrometeorology) and small-scale (micrometeorology) events are important in air pollution studies.

Macrometeorological events encompass the movement of large masses of air brought about by differences in temperature and pressure and the rotation of the earth. Basic movements of air are the heating of air masses at the equator and cooling at the poles, with a consequent north-south circulation. The rotation of the earth from west to east tends to drag the atmosphere along with the surface so that the winds in the Northern Hemisphere tend to curve toward the right and those in the Southern Hemisphere toward the left. As a consequence, winds in the United States are generally from west to east.

The condition of the atmosphere at any one time is important in weather forecasting and is usually shown on a weather map (Fig. 2.3). These maps are prepared in the United States by the National Weather Service, formerly the U.S. Weather Bureau, and are based on measurements and observations

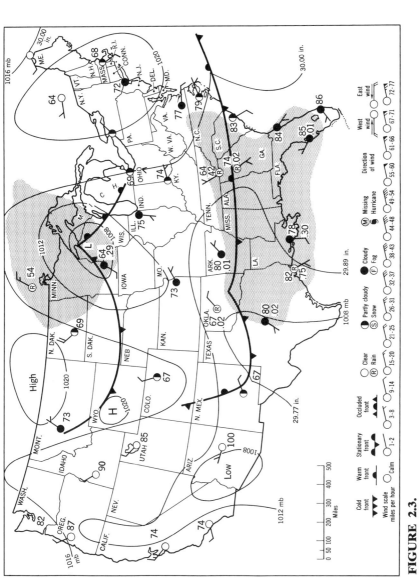

FIGURE 2.3.
Typical United States weather map.

at many stations throughout the country. These measurements include atmospheric pressure; temperature; humidity; wind speed and direction; amount, type, and height of clouds; precipitation; visibility; and other special conditions in the atmosphere. After these data are plotted they are analyzed by drawing isobars, i.e., lines of constant pressure, at 3-millibar intervals. The boundary lines (fronts) between different air masses are drawn and the type of front indicated. Shaded areas show incidence and type of precipitation; arrows indicate direction of the wind. Isotherms are usually drawn for 0°F and 32°F.

Even a cursory examination of these maps will indicate whether atmospheric conditions in a given area are likely to be favorable or unfavorable for dispersion of pollution. For instance, when a cold front moves into a warmer area the air next to the ground will become heated and result in an unstable atmospheric condition (pollutants will disperse readily). But an air mass approaching a colder area will be cooled near the ground, and a stable atmosphere will develop in the surface layers.

The wind velocities shown on the map are those of the gradient wind, which is due to the rotation of the earth. These winds are generally along the isobars and at a height (2000 ft and above) sufficiently great to be unaffected by surface friction. Flow is clockwise around high-pressure areas and counterclockwise around low-pressure areas. Near the ground, friction affects the speed and direction of the gradient wind so that surface winds generally blow between 20° and 30° across the isobars toward the low-pressure center and at half the speed of the gradient wind.

Further details on the preparation and interpretation of weather maps may be found in any good textbook on meteorology [5, 34]. In addition to daily weather maps, monthly maps of average weather conditions are also available. These are particularly useful in determining areas and periods of stable high-pressure atmospheric conditions conducive to serious air pollution problems. The high-pressure area (anticyclone) is the condition always responsible for extreme cases of local air pollution. When a high-pressure area becomes stagnant in one locality for several days, conditions for pollution buildup are extremely favorable. In the central portion of the "high," winds are very light and the atmosphere quite stable. This was the situation at Donora and the disaster potential could have been foreseen, at least at the end of the second day of these conditions. The Environmental Science Services Administration (ESSA) makes regional forecasts of periods of high air pollution potential. These forecasts are based on the likelihood of occurrence of stagnating anticyclones over an area favorable for the accumulation of air pollutants for at least 36 hours. The forecasts are available to interested parties upon request.

In addition to large-scale effects of air movement, those of a *micro-meteorological* nature must not be overlooked. Local circulation and temperature variations in valleys and on the slopes of hills and mountains are extremely important from an air pollution viewpoint; similarly, the presence of buildings (with resultant eddies) and type of ground cover affect air movement.

Movement of Pollutants

Any factor that restricts the movement of air will of course prevent the movement and dispersion of pollutants entering the atmosphere. As mentioned previously, one very important factor in producing a stable atmosphere is a negative temperature lapse rate, or atmospheric inversion. Two types of inversions are important. The *radiation inversion* occurs, usually at night, when the earth's surface loses heat by radiation, thus cooling the layer of air next to the ground. The inversion may be localized or spread over a wide area; in both cases it is more common in winter than in the summer because of longer nights and less solar heating during the day. In a valley the condition may be further accentuated by restricted air movement. Height of the inversion may be several hundred feet. This is quite noticeable when the air is humid and its temperature drops below the dew point, thus forming fog. In some areas (southeastern United States) these inversions occur more than two-thirds of the nights in the year.

The *subsidence inversion* is commonplace on the west coast of continents in the temperate latitudes. A typical subsidence inversion is that over the west coast of the United States. Air circulating around the stationary high-pressure area over the eastern north Pacific Ocean descends as it flows south along the California coast. As it subsides it becomes compressed and gains heat, much as the air from a bicycle pump becomes heated. In southern California the inversion is always present during the warm months of the year; its height above the ground may vary from the surface to over 4800 ft. The temperature gradient in the inversion layer may be as much as 10° to 12°C. A typical sounding in Los Angeles is shown in Fig. 2.4. When the inversion drops below 500 ft, extreme smog conditions may be expected.

The subsidence inversion is not restricted to the west coasts of continents. Sutton [37] has pointed out that the typical winter "gloom" of Great Britain is a result of the presence of an anticyclone, in which there is a general slow descent of air over a wide area. As the air settles, it becomes compressed and heated and forms an effective lid, which prevents the vertical movement of pollutants. In the London smog of December 1952, both subsidence and radiation inversions were present, and the two were probably very close together over London.

The importance of *wind direction* and *speed* on the dispersal of pollutants is obvious, but the variations of these two parameters with time of day and season of the year is an even greater factor. These variations are compara-

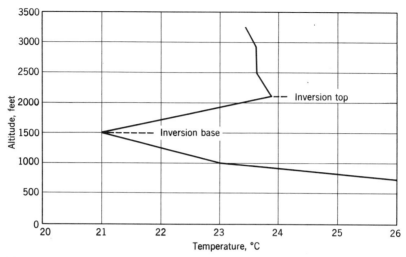

FIGURE 2.4.
Altitude-temperature sounding over Los Angeles (Sept. 23, 1954).

tively constant at a given location and are important considerations in plant location and in the degree of control necessary for emissions of air pollutants. The variations in speed and direction may be plotted in the form of a wind rose for a day, a month, or any other period of time. Figure 2.5 shows the hourly resultant winds in wind-rose form for monthly periods at the Los Angeles Federal Building. These show clearly the diurnal pattern characterized there by the sea breeze during the day and the land breeze at night. This diurnal pattern is the result of the relative constant temperature of air over the ocean and alternate heating and cooling of air over the land. During the daylight hours, land air becomes heated and rises with consequent replacement with cooler air from the ocean (sea breeze). At night the condition reverses, and the land breeze (toward the ocean) sets in. The latter is most well developed during the winter months and lighter during the rest of the year.

Fig. 2.6 shows a seasonal (summer) wind rose for a site in a broad river valley in the Mojave Desert. Wind-channeling because of topography is apparent, as is the diminution of wind speed at night.

In many valley locations, the diurnal pattern is often an updraft during the day, from heating the slopes, and a "drainage" wind at night. As an example, under inversion conditions the surface layer of air on a slope is colder and heavier than air on the same level nearer the center of the valley. Gravity flow results in what is called mountain winds, canyon winds, or *katabatic* winds, depending on the topography. Light gravity winds may

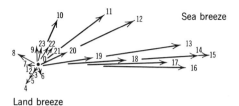

September 1947

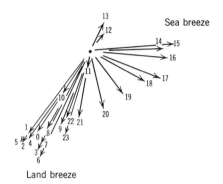

December 1947

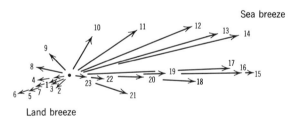

April 1948

Los Angeles Federal Bldg.

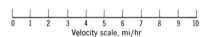

FIGURE 2.5.

Hourly winds in wind-rose form for monthly periods at Los Angeles Federal Building.

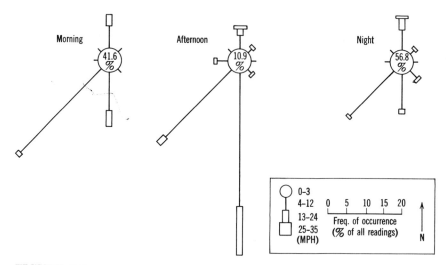

FIGURE 2.6.
Valley site summer wind roses.

even occur on the shaded slope of a hill or mountain during the afternoon. This filling of level valleys with cold air during the night is an important factor in many air pollution problems.

Wind variability or *gustiness* is another important meteorological concept in air pollution studies, since it measures the degree of turbulence in the air. Each location has a pattern of its own [9]. Another measure of atmospheric stability is the *vertical wind-velocity profile* [35]. In stable air the increase in wind velocity with height is much less than it is in a turbulent atmosphere.

Measurement of Meteorological Variables

For most air pollution problems only a few types of meteorological measurements need be made. Generally, these are pressure, temperature, and humidity of the atmosphere; speed and direction of the wind; and, in some cases, the amount of insolation (sun intensity).

Pressure.   The usual instrument for pressure measurement is the barometer. Two types are commonly used, the mercury barometer and the elastic or aneroid barometer. The *mercury* barometer is essentially a U-tube with one end hermetically closed and the other open to the atmosphere. The tube, from which the air has been exhausted, is filled with mercury and atmospheric pressure acts on the mercury in the open end of the tube, causing it to rise in the closed end until the column of mercury equals the weight of the column of air. Thus atmospheric pressure is stated in terms of the height of the mercury column in inches or millimeters. At sea level standard pressure is 760 mm or 29.92 in. Hg (when the height of the column is corrected to 0°C or 32°F). Meteorologists usually express atmospheric pressure in millibars (1000 dynes per cm²). Standard pressure (760 mm or 29.92 in. Hg) is 1013.2 mb. Thus 3 in. Hg is about 100 mb.

The *aneroid* barometer operates on the principle that an elastic membrane, held at the edges, will be deformed if the pressure is greater on one side than the other. In the aneroid barometer two elastic membranes form a chamber. The change in thickness of the chamber with change in atmospheric pressure is magnified and transferred to a scale by means of a pointer. This unit is not so accurate as a mercury barometer, but it is portable and more convenient.

Temperature. The common instrument for measurement of temperature is, of course, the thermometer. Mercury and spirit thermometers depend on thermal expansion; bimetallic thermometers are based on the differential expansion of two metals; the electric resistance thermometer is based on the variation in electrical resistance of a metallic wire with temperature change; and the thermoelectric thermometer or thermocouple is based on the electric current which flows (varies with temperature) when two dissimilar metal wires are joined. All four types are used.

To obtain accurate readings, the thermometer must be shielded from radiant energy. This is usually done by means of a louvered wooden box or polished metal screens. Further, the air surrounding the thermometer should be kept moving.

Humidity. A variety of means are available for determining the humidity of the atmosphere [21]. One of the simplest and most reliable instruments is the *sling psychrometer.* Two thermometers, one with a wetted cloth surrounding the bulb, are whirled in the air. From the temperature difference between the dry bulb and wet bulb thermometers, the humidity of the air can be calculated. The wet-and-dry-bulb principle has been adapted to a recording instrument. A useful recording instrument is the hair hygrometer. This instrument is based on the principle of the change in length of a human hair with relative humidity. Temperature and humidity are often measured and recorded simultaneously, as in the hygrothermograph (Fig. 2.7).

Wind Speed. Instruments for measuring wind speed are called anemometers; if they are recording instruments they are known as anemographs. The most common type is the cup anemometer (Fig. 2.8). The rate of rotation of the shaft to which the cups are attached is indicated or recorded electrically. In gusty winds the instrument may give a high mean reading, because it accelerates more rapidly than it slows down. Wind vanes are commonly combined with an anemometer so that both speed and direction can be recorded on the same chart.

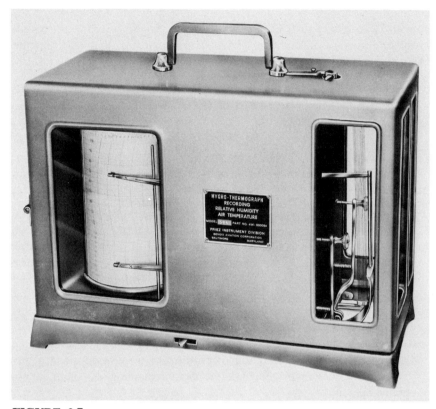

FIGURE 2.7.

Hygro-Thermograph recording relative humidity and air temperature. (*Courtesy Friez Instrument Division, Bendix Aviation Corp.*)

Care must be exercised in installing anemometers so that exposure is free from obstructions, such as nearby trees, buildings, hills, and the like. Installations on the roofs of buildings usually give high readings because of crowded streamlines as the wind passes the building. Further, unless the building is symmetrical, gustiness will vary considerably with wind direction.

Although the cup anemometer is widely used, other types are available. Middleton and Spilhaus [21] have described four other types. Where upper wind measurements are desired, pilot balloons (*pibals*) are released and followed by theodolites or transits. In cloudy weather, radiosonde and radar methods may be used.

Upper air measurements are not restricted to wind speed and direction; any measurement required at the surface may also be important in the upper air. Instruments especially designed for light weight are used and are

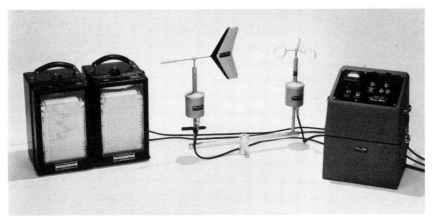

FIGURE 2.8.
Cup anemometer and wind speed and direction recorder. (*Courtesy Beckman and Whitley, Inc.*)

carried aloft by appropriate means: captive balloons, free balloons, kites, aircraft, even rockets. Pertinent data may either be recorded aloft or transmitted back to earth by radio.

Solar Radiation. In some cases, e.g., where photochemical smog is important (see Chapter 8), a knowledge of sunlight intensity is important. Several instruments are available; a typical instrument is the Eppley pyrheliometer (Fig. 2.9). This instrument measures the temperature difference (by thermocouples) between two concentric silver rings—one with a blackened surface, the other with a whitened surface. Both 10- and 50-junction-type pyrheliometers are available.

APPLICATION OF METEOROLOGY
IN AIR POLLUTION PROBLEMS

The utilization of meteorological data in the air resource management program poses formidable problems. We know that meteorological conditions vary tremendously from one day to another, from one season to another, and from one location to another. There are times and places when almost any amount of pollution can be dispersed readily, while at other times and in other places, normally minor emissions can be a problem. We have yet to learn how to modify the weather so that it will be most favorable to good ventilation and adequate dispersion of air pollutants. We must take it as it comes and adapt man's activities to its variations. Theoretically all major pollution sources not subject to control at the source could be located in those areas where atmospheric dispersion processes were most

FIGURE 2.9.
Pyrheliometer. (*Courtesy Eppley Laboratory, Inc.*)

favorable, or could be so operated that emissions were nil or at a minimum during times of adverse conditions. These alternatives are commonly known as land-use control and meteorological control, respectively. Land-use control entails consideration of air pollution factors in zoning decisions.

Meteorological control of pollutant emissions is simple in concept: maximum emissions are allowed when good atmospheric dispersion may be expected; they are minimized when the air is stagnant. Controls of this nature have often been used where emissions could be interrupted or postponed without unbearable economic penalty. For instance, it is common practice in many areas to restrict open burning of trash and other waste material to those days or hours when atmospheric conditions are favorable to good dispersion. Some localities never allow open burning at night because of the great likelihood of stagnant air. Other localities prohibit plowing and harrowing when the wind velocity exceeds a certain value. Similar restrictions may be applied to land clearing and building demolition.

Where processes are continuous in nature as at chemical plants, pulp mills, etc., meteorological controls are seldom applicable. In fact, shut-down procedures often produce more pollution than operating does; nearly always the economic penalty is great. There have been a few cases where courts have ordered cessation or curtailment of certain emissions under specified wind and weather conditions. The most widely known case is that of a smelter at Trail, British Columbia where an International Tribunal directed the plant to vary operating rate according to meteorological conditions (see Ref. 18, Chapter 1).

Prediction of Atmospheric Diffusion

The fundamentals of meteorology are particularly useful in predicting the degree of air pollution that will result from specific sources. Knowledge of the type of smoke plume that will result under a given condition and the degree of diffusion that may be expected enables one to design stacks of proper height and to select satisfactory sites for industrial plants.

Plume types are classified by appearance, as shown in Fig. 2.10. A *looping* smoke plume occurs in a highly unstable atmosphere because of rapid mixing. If, however, the top of the stack is not sufficiently high, considerable pollution may reach the ground level in the vicinity near the base of the stack. *Coning* takes place in a neutral atmosphere when wind velocity is greater than 20 mph. Rapid diffusion of the smoke results. *Fanning* is characteristic of a plume emitted under extreme inversion conditions where little vertical mixing takes place. In flat country such a plume may be seen for miles downwind from the source.

Diurnal changes in atmospheric turbulence will, of course, change the nature of a plume markedly. At night under a radiation inversion, a fanning plume will probably be emitted, but shortly after dawn when the sun breaks the inversion, the plume will begin to loop energetically. When this atmospheric change is taking place, it is interesting to watch the plumes from low-level stacks loop (the air near the ground is heated first) while taller stacks are still sending out a fanning plume. When this transition takes place and the fanning plume breaks up, momentary high concentrations of smoke may be carried to the ground level even well downwind from the stack. This phenomenon is called "fumigation" [11].

Several formulas have been developed to predict atmospheric diffusion. The most commonly used formulas are those developed by Sutton [36] and

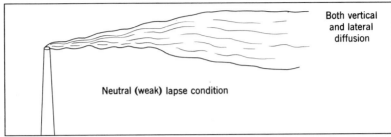

Both vertical
and lateral
diffusion

Neutral (weak) lapse condition

Coning

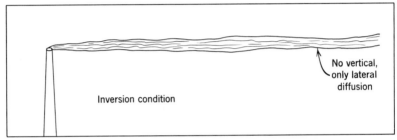

No vertical,
only lateral
diffusion

Inversion condition

Fanning

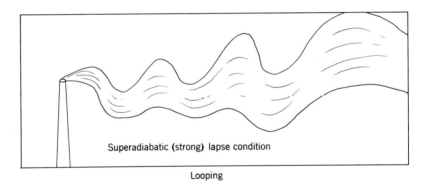

Superadiabatic (strong) lapse condition

Looping

FIGURE 2.10.

Smoke plume types.

by Bosanquet and Pearson [2]. The method proposed in 1961 by Pasquill
is also widely used [27].

Each of these equations is based on the assumption that the spread of a
plume from a point source takes the form of a normal or gaussian distribution and thus may be expressed as:

$$\chi_{(x,y,o)} = \frac{Q}{\pi\sigma_y\sigma_z\bar{u}} \exp - \{(h^2/2\sigma_z^2) + (y^2/2\sigma_y^2)\} \qquad (2\text{-}1)$$

where χ = concentration (units/m³) (for example gm/m³)
   Q = pollutant release rate (units/sec) (for example, gm/sec)
   σ_y, σ_z = crosswind and vertical plume standard deviations (m)
   $\bar{u}$ = mean wind speed (m/sec) at h_s, the height of the stack
   h = effective stack height (m)
   x, y = downwind and crosswind distances (m).

Detailed solutions of this equation are explained in several excellent manuals [32, 40] and will not be explained here, but it is well to examine the equation and understand the importance of the various parameters. For instance, the equation shows that χ, the ground-level ($z = 0$) concentration of a pollutant at a point (x, y) downwind from a source is proportional to the weight-rate of the emission (Q). It is *inversely* proportional to the wind speed ($\bar{u}$), and to the parameters, σ_y and σ_z. It may also be shown that the *maximum* ground-level concentration (glc) of a gaseous pollutant is inversely proportional to the square of the effective stack height.

Dispersion Calculations. To solve the equation, one must obtain suitable values of σ_y and σ_z and calculate the effective stack height. All three values are dependent on the stability of the atmosphere. For best degree of accuracy, σ_y and σ_z should be estimated from 3-dimensional wind measurements at the site or at least from wind-vane measurements. Often these are not available so values are estimated from generalized charts (see Ref. 32, pp. 45, 46; Ref. 40, pp. 8, 9; Ref. 31, pp. 102, 103) in which the parameters are shown in terms of atmospheric stability and distance downwind from the stack. Again, when actual atmospheric stability data are not available, one may resort to generalized stability classes such as those prepared by Pasquill (see Table 2.2). In using charts for the estimation of σ_y

TABLE 2.2
Pasquill Stability Types*

Surface wind speed, m/sec	Daytime insolation			Nighttime conditions	
	Strong	Moderate	Slight	Thin overcast or $\geq \frac{4}{8}$ cloudiness	$\leq \frac{3}{8}$ cloudiness
<2	A	A–B	B		
2	A–B	B	C	E	F
4	B	B–C	C	D	E
6	C	C–D	D	D	D
>6	C	D	D	D	D

* From Ref. 31.
   A—Extremely unstable conditions   D—Neutral conditions
   B—Moderately unstable conditions   E—Slightly stable conditions
   C—Slightly unstable conditions   F—Moderately stable conditions

and σ_z, it is imperative that one note the time period involved, i.e., hourly-mean values as in Ref. 32, or 10-minute values as in Ref. 40.

Estimation of Plume Rise.   In order to calculate downwind concentrations the probable plume rise must be estimated so that effective stack height (actual stack height + plume rise) under various conditions may be calculated.

A variety of plume rise formulas have been developed. Some give good results from small sources with comparatively cold plumes; others are better adapted to large sources emitting gases with considerable heat content. In the absence of a valid reason for using one formula or another, the empirical formula of Moses and Carson [24] is often used. The equation is:

$$\Delta h = C_1 \frac{V_s d}{\bar{u}} + C_2 \frac{Q_h^{1/2}}{\bar{u}} \tag{2-2}$$

where Δh = plume rise (meters)
V_s = stack exit velocity (m/sec)
$\bar{u}$ = wind speed (m/sec)
d = stack diameter (meters)
Q_h = heat emission rate (kilocalories/sec)
C_1, C_2 = plume rise regression coefficients, dependent on atmospheric stability.

Inspection of the formula reveals that the first term evaluates the vertical momentum of the gases leaving the stack, and the second term rates the buoyancy force of the plume which is a function of heat content of the plume. Thus, the magnitude of the plume rise is inversely proportional to wind speed, directly proportional to the sum of the mass ejection rate, and the square root of the heat content of the gases. Values for the regression coefficient C_1 and C_2 are listed in the reference cited. The nature of the values are such that much greater plume rises may be expected in unstable atmospheres than in stable ones. Two other plume rise formulas with much backing for specific cases are those of Lucas, Moore, and Spurr [17], and the so-called CONCAWE equation [29]. A formula proposed by Briggs [4] appears to be highly suitable to large power plants.

Having calculated plume rise, one can determine the effective stack height and calculate downwind ground-level concentrations for the time period of interest. Very often, however, if one calculates a 10-minute maximum, he may also want to estimate hourly maxima or daily maximum values. There is considerable variation from one area to another but as a general case for large power plants, one may assume hourly maxima are one-half 10-minute maxima, and 3 times daily maxima. To calculate *average* concentrations in the vicinity of an elevated source, the frequency of various wind speeds and stabilities must be considered.

Utility of Dispersion Equations.   The foregoing discussion of dispersion calculations is, to some extent oversimplified, but the calculations give excellent clues to what might be expected by use of high stacks to disperse otherwise intolerable emissions. Accordingly, the equations listed are used along with experience and common sense in the design of stacks for new installations and for estimating the ground-level concentrations of pollutants that would be expected from existing stacks operated under various conditions.

Occasionally, all that one may desire is an estimate of a series of maximum concentrations. For this purpose the equation

$$X_{max} = \frac{Q}{e\pi\bar{u}h^2}\frac{\sigma_z}{\sigma_y} \qquad (2\text{-}3)$$

may be used. The symbol e represents the base of natural logarithms, 2.718. The maximum concentration occurs at a distance downwind where $\sigma_z = h/\sqrt{2}$. Different values of σ_z/σ_y and a new effective stack height must be used for each change in meteorological conditions and wind speed. A nomograph for relating some of these parameters has been published by Turner [40, p. 29]. The atmospheric dispersion manuals previously cited should be consulted for discussions of various factors which may affect the utility of the equations. The more important of those factors are considered only cursorily here.

It is important that *mean wind speed* at the centerline of the plume be used in dispersion equations. Where data, at least at stack height, are not available, upper wind speeds may be estimated by Equations 2-4 or 2-5,

$$\bar{u}_h = \bar{u}_1 \left(\frac{h}{z_1}\right)^{.50} \text{ stable} \qquad (2\text{-}4)$$

$$\bar{u}_h = \bar{u}_1 \left(\frac{h}{z_1}\right)^{.25} \text{ unstable} \qquad (2\text{-}5)$$

where subscripts represent two different heights.

Nearby buildings may affect plume behavior by introducing mechanical turbulence which may bring portions of the plume to ground level near the stack. The effect is most noticeable when the stack is *downwind* of the building and wind speeds are high. The effect may be avoided by use of the highly conservative rule of thumb stating that stacks 2 to 2½ times the height of nearby buildings will clear the building wake.

When the stack is *downwind* of a large building "downwash" (part of the plume creeping down the outside of the stack) may result from wake gusts unless the stack is sufficiently high. Any downwash caused by the stack itself may be avoided by ensuring that the exit velocity from the stack exceeds the horizontal wind speed.

Irregular terrain may also affect ground-level concentrations. The dispersion formulas commonly used are designed for flat, level country. Various methods of estimating concentrations on hillsides, bluffs, etc., are used, but very little experimental data are available. Any calculational method is misleading if common sense is neglected in interpreting results. Generally, rough terrain promotes dispersion.

In *urban areas,* there are both "heat island" and mechanical turbulence effects which affect the dispersion of plumes. The effect usually enhances vertical dispersion.

The presence of an atmospheric *inversion* imposes certain restraints on the dispersion of plumes and thus on the methods of estimating dispersion. For plumes emitted into an inversion layer, dispersion calculations for stable atmospheres are used. Special consideration must be given to cases where an inversion layer is located well above ground level, as is often the case with a subsidence inversion. The problem is complicated by the extent to which a plume will penetrate the inversion. Obviously, the portion that penetrates the inversion will level off in or above the inversion layer and no longer be of interest at the ground level. That portion of the plume which does not penetrate the inversion will disperse in the layer beneath the inversion base in a manner dependent on wind speed and stability below the inversion base. The manuals cited show methods for calculating the dispersion. Formulas for determining if a heated plume will penetrate an inversion have been suggested by Scorer [30] and by Briggs [4].

The process called *fumigation* must also be considered at times. When a low-level inversion breaks up, say by virtue of the ground being heated by solar radiation, momentary high concentrations of pollution may reach the ground as any pollution stratified in the inversion layer is rapidly mixed with the unstable air below the rising inversion base. See Ref. 32 and 40 for suitable equations. Formulas must be used with caution, particularly for very high stacks and deep inversions. Investigations by Sporn and Frankenberg [33] indicate that in such cases fumigation concentrations are not experienced at ground level, probably because the inversion layer dissipates from the top rather than the bottom.

Another common practical problem is the estimation of the combined contribution of *multiple stacks.* Turner [40] suggests a method for making such estimations. The relationship between the maximum concentration from a single stack and multiple stacks of the same strength is given in Equation 2-6, suggested by Smith [32],

$$X_{max(n)} = X_{max(1)} N^{0.8} \qquad (2-6)$$

where $X_{max(1)}$ is the maximum concentration for a single stack from equation 2-3, and N is the number of stacks. Experiments of Thomas [38] at

TVA indicate lower concentrations from multiple stacks than those predicted by Equation 2.6.

At times it may be required to estimate the downwind concentration from a *line source* rather than from an isolated point source. An equation describing this situation has been suggested by Turner [40].

Consideration to this point has been limited to elevated sources emitting continuously. In some cases, such as short-term releases or explosions, the *instantaneous* or *"puff" release* is important. A case of particular importance is calculation of radioactivity doses which would result from instantaneous release of nuclear fission products. Both the characteristics of the instantaneous release and calculation of the radioactive cloud dose are discussed in detail in "Meteorology and Atomic Energy" [31].

METEOROLOGICAL MODELS

An important tool of air quality management is the meteorological model of a city or air pollution control region. Basically the model is a mathematical description of the meteorological transport and dispersion processes of an area on which are superimposed rates of emissions of pollutants from various sources. If the model is sound and input data are accurate, then one could obtain from the model the concentration of any pollutant at any point in the area under consideration at any time.

A suitable model for a locality would allow the air pollution control agency to make a great many decisions on a valid basis, e.g., how much an increase or decrease in emissions from a given location would affect downwind atmospheric concentrations; where to locate samplers to measure the greatest effect of a given source; how to relate emission regulations to air quality standards; where to locate new pollution sources; how much a new zoning ordinance will affect air quality; and many others.

The most important parameter in model development is the description of the dispersion itself as affected by stability, wind direction and speed, source height, buildings, topography, etc. Many models make use of a gaussian distribution which varies between elevated and ground-level sources, and choose suitable plume deviation constants for various stability types. Of course, no model can yield good results with poor input data, and it may well be that present difficulties with urban modeling are related to poor emission data, particularly variations with time.

Several models for specific areas have been proposed [6, 15, 16, 18, 22, 28, 39]. Most of these are for either sulfur dioxide or nitrogen oxides. In both cases allowances must be made for atmospheric decay of the pollutant, even though our knowledge in this field is minimal.

Models may be prepared for a variety of effects of air pollution. Thus

Friedlander and Ravimohan [7] have proposed a model to relate pollution to mortality, as experienced in the December, 1952 London smog. Several simplified correlations between atmospheric suspended patriculate matter and visibility (see Refs. 14 and 40, Chapter 1) have been proposed. These could serve as the bases for a pollutant-visibility model.

As models become more complex, much more time is required for tedious, repetitive calculations. Accordingly, the use of computers is growing in the field. Even for stack height determination studies, many consulting and design organizations have developed codes for sorting out meteorological data and calculating concentrations for a variety of assumed conditions.

The importance of atmospheric modeling is only now on the verge of wide use for air management decisions. In fact, regional air pollution control agencies (see next section) were requested by NAPCA (and its successor, EPA) to use simulation models to develop air pollution control implementation plans under the Air Quality Act of 1967.

In many cases, the nature of available meteorological data or the complexity of atmospheric dispersion may be such that more knowledge of atmospheric movement is necessary. Here *tracer studies* are invaluable. Solid particles or gases of unique composition are released at a stack or other source at a known rate and for a known period of time. The air at downwind points is then passed through a filter or other suitable equipment to collect the unique material. Analysis of the material collected at a sufficient number of stations allows one to determine diffusion patterns of the stack effluent under existing weather conditions. Finely divided fluorescent particles, such as zinc and cadmium sulfides, have been been found to be suitable for this purpose. In one test of this type [3], concentrations of more than 200 particles per cubic meter were found 107 miles from the dispersal point. The method was also used successfully in tracing the wind paths in Los Angeles County [26].

Hemeon has developed a similar method [10] in which antimony sulfide powder is used as the tracer. The material is collected downwind on filter paper, irradiated in an atomic pile, and the particles counted with a Geiger tube.

When plume movement may be affected by nearby structures or topography *wind tunnel studies* often yield information of value. The chief use of wind tunnels to date has been in determining the effect of the shape of powerhouse and adjacent buildings on the behavior of stack effluents. A photograph of the wind tunnel at New York University is shown in Fig. 2.11.

Determination of the height and location of new stacks on industrial buildings can best be determined by simulated smoke studies. A source of smoke, such as an aerosol bomb containing titanium tetrachloride, may be

FIGURE 2.11.
Wind tunnel at New York University. (*Courtesy Industrial Wastes, November–December, 1955.*)

attached to a collapsible mast and the smoke liberated under various weather conditions. The course of the plume is then followed visually and recorded by motion pictures. Kite balloons may be used in place of masts for the same purpose [25].

AIR POLLUTION CONTROL REGIONS

The importance of meteorology in air pollution control programs was accented by the Air Quality Act of 1967. The act directed the U.S. Department of Health, Education, and Welfare to designate air quality control regions on the basis of meteorological, topographical, and other factors. Subsequently, through its National Air Pollution Control Administration, HEW divided the United States into eight atmospheric areas or meteorological regions (Fig. 2.12). Each area is a segment of the country in which climate, meteorology, and topography are generally homogeneous. Five areas outside the contiguous United States were also designated.

After designation of atmospheric areas, NAPCA embarked on a program of designating air pollution control regions on the basis not only of meteoro-

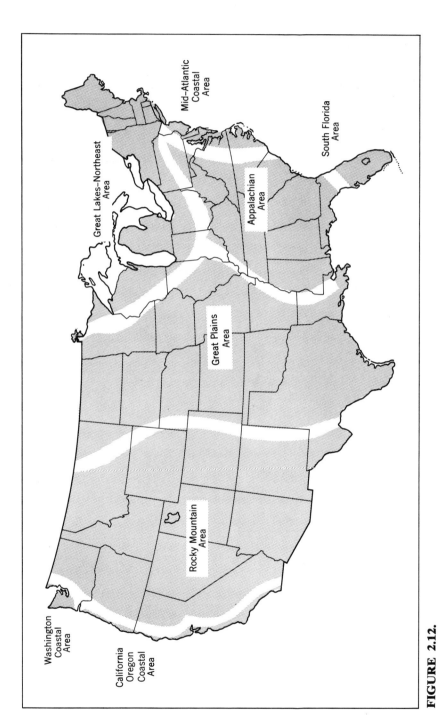

FIGURE 2.12.

Atmospheric areas of the United States. (*As designated by the National Air Pollution Administration.*)

logical and topographical factors but also the pattern of urbanization and industrialization, with some consideration for jurisdictional factors, as required by the Air Quality Act. The basic regional boundaries were selected partially by simulation modeling procedures. Two hundred forty-seven control regions have been identified (see Table 1.1).

Several states have also divided their areas into air quality control basins. Those adopted by the state of California are shown in Figure 2.13.

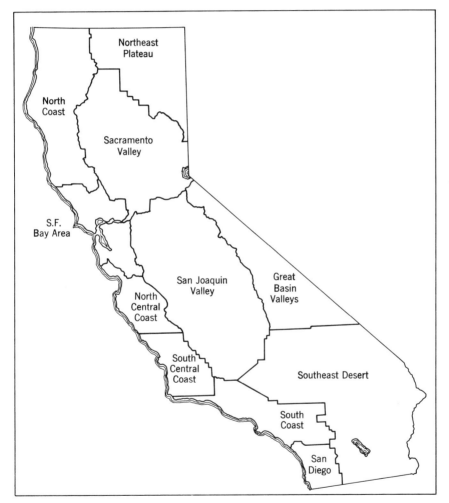

FIGURE 2.13.
State of California air basins. (*California Air Resources Board, adopted November 1968.*)

DATA SOURCES

The utility of meteorological calculations in defining the nature of an air pollution problem is largely dependent on the quality of available atmospheric measurements. The most dependable measurements, of course, are those which have been made in the specific area of interest. To be most useful, the measurements should have been made often and over a sufficient period of time to represent normally expected variations. As a minimum, one should have available data taken over a two-year period.

In the United States, data on wind speed and direction are available for some 300 weather stations operated by the Environmental Science Service Administration (ESSA). Supplementary wind speed and direction data are taken at most airports and at many industrial installations. Thus, where on-site data are not available, wind roses prepared from data collected nearby may be used with caution. Most of the wind data are obtained near the surface, although more meteorological towers taking upper wind measurements and temperature measurements (for determination of atmospheric stability) are being erected. More extensive data on atmospheric stability measured by radiosonde techniques are available for only a few stations. Most of the data mentioned above, plus climatological data (temperature, humidity, visibility, rainfall, etc.) are available from the National Weather Records Center, Asheville, North Carolina. Various analyses from original records will also be made at cost. Similar data are available outside the United States from the National Meteorological Offices in many countries.

Special summaries of nationwide data on inversions, atmospheric stagnation, visibility, and solar radiation have been published. A generalized summary of these diffusion climate data has been tabulated by McCormick [19].

REFERENCES

1. Anon., "Atmospheric Pollution in Leicester—A Scientific Survey," Department of Scientific Industrial Research (British Government), Atmospheric Pollution Research Tech. Paper 1 (1945).
2. Bosanquet, C. H., "The Flow of Chimney Gases," in *Air Pollution,* M. W. Thring, ed., Butterworths Scientific Publications, London (1957).
   ——, and J. L. Pearson, "The Spread of Smoke and Gases from Chimneys," *Trans. Faraday Soc.* 32, 1249 (1936).
3. Braham, Roscoe R., Ben K. Seeley, and W. D. Crozier, "A Technique for Tagging and Tracing Air Parcels," *Trans. Am. Geophys. Union,* 33, 825–833 (1952).
4. Briggs, G. A., "Plume Rise," US Atomic Energy Commission (1969).

5. Byers, H. R., *General Meteorology,* 3rd Edition, McGraw-Hill, New York, (1959).

6. Clarke, J. F., "A Simplified Diffusion Model for Calculating Point Concentrations from Multiple Sources," *J. Air Poll. Control Assoc.,* **14,** 347–352 (1964).

7. Friedlander, S. K., and A. L. Ravimohan, "A Theoretical Model for the Effect of An Acute Air Pollution Episode on a Human Population," *J. Environmental Science and Technology,* 2, 1101–1108 (1968).

8. Glueckauf, E., "The Composition of Atmospheric Air," in *Compendium of Meterology,* Thomas F. Malone, ed., American Meteorological Society, Boston (1951).

9. Gosline, C. A., L. L. Falk, and E. N. Helmers, "Evaluation of Weather Effects," in *Air Pollution Handbook,* Paul L. Magill, Francis R. Holden, and Charles Ackley, eds., McGraw-Hill, New York, 5–15 to 5–17 (1956).

10. Haines, George F., Jr., Herman Cember, and W. C. L. Hemeon, "Stack Gas Tracer Technique Employing Neutron Activation Analysis," Final Rept. to American Petroleum Institute Smoke and Fumes Comm. (June 1956).

11. Hewson, E. Wendell, "The Meteorological Control of Atmospheric Pollution by Heavy Industry," *Quart. J. Roy. Meteorol. Soc.,* **71,** 266 (1945).

12. Holzworth, G. C., in "Air Over Cities," Rept. SEC TR62-5. Robert A. Taft Sanitary Engineering Center, Cincinnati, Ohio (1962).

13. Holzworth, G. C., "Estimates of Mean Maximum Mixing Depth in the Contiguous United States," *Monthly Weather Rev.,* **92,** 235–242 (1964).

14. Hosler, C. R., "Low-Level Frequency in the Contiguous United States," *Monthly Weather Rev.,* **89,** 319–339 (1961).

15. Koogler, J. B., R. S. Sholtes, A. L. Davis, and C. I. Harding, "A Multivariable Model for Atmospheric Dispersion Predictions," *J. Air Poll. Control Assoc.,* **17,** 211–214 (1967).

16. Leavitt, J. M., S. B. Carpenter, J. P. Blackwell and T. L. Montgomery, "Meteorological Program for Limiting Power Plant Stack Emissions," *J. Air Poll. Control Assoc.,* **21,** 400–405 (1971).

17. Lucas, D. H., Moore, D. J., and Spurr, G., "The Rise of Hot Plumes from Chimneys," *Intern. J. Air and Water Poll.,* **7,** 473–500 (1963).

18. Martin, D., and J. Tikvart, "General Atmospheric Diffusion Model for Estimating the Effects on Air Quality of One or More Sources," Paper No. 68-148, 61st Annual Meeting, Air Pollution Control Association, St. Paul, Minn. (June 1968).

19. McCormick, R. A., "Air Pollution Climatology," Chapter 9, Volume I, *Air Pollution,* A. C. Stern, ed., 305 (1968).

20. McDonald, T. H., "Estimated Mean Daily Ultraviolet Radiation for Wavelengths Equal to and Less than .3192 Microns," Office Meteorol. Res., U.S. Dept. of Commerce, Washington, D.C. (1959).

21. Middleton, W. E. Knowles, and Athelstan F. Spilhaus, *Meteorological Instruments,* 3rd Ed., Rev., University of Toronto Press, 96–117 (1953).

22. Miller, M. E., and G. C. Holzworth, "An Atmospheric Diffusion Model for Metropolitan Areas," *J. Air Poll. Control Assoc.,* **17,** 46–50 (1967).

23. Miller, M. E., and L. E. Niemeyer, "Air Pollution Potential Forecasts—A Year's Experience," *J. Air Poll. Control Assoc.,* **13,** 205–210 (1963).

24. Moses, H., and Carson, J. E., "Stack Design Parameters Influencing Plume Rise," *J. Air Poll. Control Assoc.,* **18,** 454–457 (1968).

25. Munger, H. P., "Techniques for the Study of Air Pollution at Low Altitudes," *Chem. Eng. Progr.,* **47,** 436–439 (1951).

26. Neiburger, M., "Tracer Tests of Trajectories Computed from Observed Winds," Air Pollution Foundation (San Marino, Calif.) Rept. 7 (1955).

27. Pasquill, F., "Atmospheric Diffusion," D. Van Nostrand Co., Ltd., London (1962).

28. Pooler, F., Jr., "A Tracer Study of Dispersion Over a City," *J. Air Poll. Control Assoc.,* **16,** 677–81 (1966).

29. Round Table on Plume Rise and Dispersion, *Atmospheric Environment,* **2,** 193–196 (1968).

30. Scorer, R., "Air Pollution," 58–63, Pergamon Press, London (1968).

31. Slade, D. H., ed., "Meteorology and Atomic Energy," U.S. Atomic Energy Commission (1968).

32. Smith, M. E., ed., "Recommended Guide for the Prediction of the Dispersion of Airborne Effluents," American Society of Mechanical Engineers (1968).

33. Sporn, P., and T. T. Frankenberg, "Pioneering Experience with High Stacks on the OVEC and American Electric Power Systems," Proceedings of the International Clean Air Congress, London, Oct. 4–7, 1966, pp. 102–105.

34. Sutton, O. G., "The Challenge of the Atmosphere," Harper and Brothers, New York (1961).

35. Sutton, O. G., *Micrometeorology,* McGraw-Hill, New York (1953).

36. Sutton, O. G., "The Theoretical Distribution of Airborne Pollution from Factory Chimneys," *Quart. J. Roy. Meteorol. Soc.,* **73,** 426–436 (1947).

37. Sutton, O. G., "Meteorology and Air Pollution," Proc. Hastings Conf. 1957, National Society for Clean Air (London), pp. 78–87.

38. Thomas, F. W., S. B. Carpenter, and F. E. Gartrell, *J. Air Poll. Control Assoc.,* **13,** 198–204 (1963).

39. Turner, D. B., "A Diffusion Model for An Urban Area," *J. Appl. Meteorol.,* **3,** 83–91 (1964).

40. Turner, D. B., "Workbook of Atmospheric Dispersion Estimates," Public Health Service Publication No. 999-AP-26, National Air Pollution Control Administration, Cincinnati, Ohio (1969).

3

SMOKE

Public Enemy No. 1 in the air pollution field is smoke, particularly black smoke produced by the incomplete combustion of carbonaceous fuels. Physically, it consists of fine particles (0.01 to 1.0 μ) of soot borne in a stream of air and flue gases. Under the microscope the carbonaceous particles appear as agglomerates of porous balls (Fig. 3.1). In many cases the particles are not pure carbon, but contain highly condensed aromatic hydrocarbons of very low hydrogen content, such as anthracene and pyrene and their derivatives [31].

As soot or carbon in smoke is nongaseous material, it eventually shows up in the atmosphere as particulate matter, which term also embraces dusts, fumes, and mists. Although the particulate matter in smoke is largely particles of carbon, other particulate matter is also usually present. Both coal and oil smoke contain fly ash from the noncombustible mineral in the fuel; coal smoke often contains liquid aerosols from the distillation of volatile matter. The latter material is particularly visible on coal fires and is yellowish-brown in color. The blue color of smoke from burning wood and trash reflects very finely divided aerosols of pyroligneous acid and wood tar; free carbon is not present in large amounts. The term "white smoke" is a misnomer. Such plumes are usually water vapor or finely divided organic aerosols, such as atomized fuel oil.

Historically, however, because of its source (fuel burning), its smudging properties, and its widespread occurrence, black smoke is considered a separate air pollutant. In fact, the first air pollution control laws were called smoke abatement ordinances. Similarly, the present-day Air Pollution Control Association is an outgrowth of the older Smoke Prevention Association of America.

59

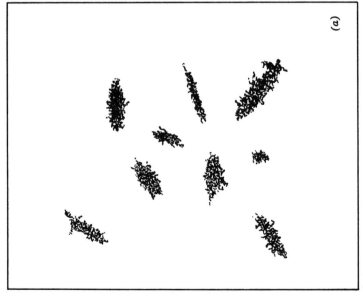

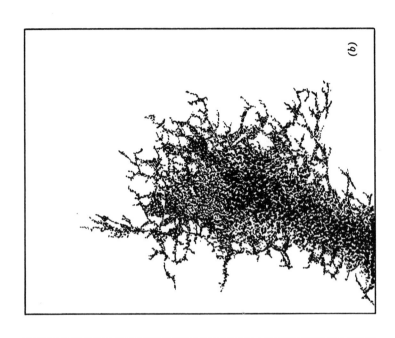

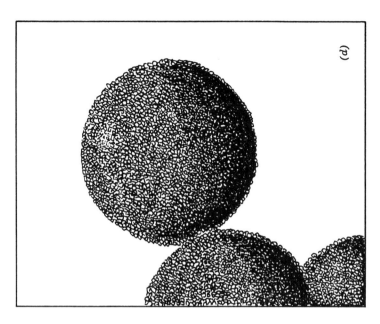

(d)

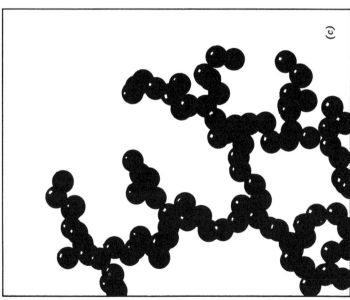

(c)

FIGURE 3.1.

Particles of soot (0.01 μ to 1.0 μ) under the microscope (Ref. 27). (a) Particles of soot. $\times$ 86. (b) Edge of soot particle. $\times$ 860. (c) Edge of soot particle showing chainlike structure of carbon particles. $\times$ 86,000. (d) Nature of packed crystallite structure making up carbon particle. $\times$ 860,000.

SMOKE FORMATION

Smoke is formed when the combination of fuel used and method of burning is such that incomplete combustion results. All fuels can be burned smokelessly; some with more difficulty than others. A few so-called "smokeless fuels," such as coke and charcoal, normally never smoke. Others, such as natural gas, can be made to smoke, but are usually burned smokelessly with ease.

A fuel produces smoke when it has been heated to a sufficiently high temperature to decompose a portion of the fuel to finely divided carbon and other products, but not high enough or in the presence of sufficient oxygen to burn the carbon. For instance, when high-volatile coal is thrown on a hot bed of coals, volatile combustible gases are evolved. Sufficient air, at a high enough temperature for combustion, must reach these gases rapidly or they will decompose to form the free carbon of black smoke. Gaseous and atomized liquid fuels smoke when insufficient air is present, when fuel and air are poorly mixed, or when the flame is chilled before combustion is complete. In each of these cases, the unburned fuel is cracked to carbon and hydrogen, with the carbon forming the basic smoke constituent.

Thus, the fundamental requirements for smokeless combustion are sufficient temperature, sufficient time, adequate turbulence for good mixing, and a sufficient supply of oxygen. These are commonly referred to as "T, T, T, and O." To obtain the proper conditions of time, temperature, turbulence, and oxygen, one must have equipment designed and operated for the specific fuel under consideration.

Proper combustion is important, not only from an air pollution standpoint, but also as an economic matter. Smoke represents a loss of heat; hence, a waste of energy. So a dual motive exists for its abatement.

NATURE OF FUELS

Since the composition and physical state of a fuel are variables met in combustion processes, a knowledge of the characteristics of different fuels is a prerequisite to good combustion practice. Fuels are generally classified according to their physical state, i.e., solid, liquid, and gas. Solid fuels are represented by coal, coke, wood, charcoal, and various organic wastes. The common commercial liquid fuels are derived from petroleum and include products with a wide range of vapor pressure from gasoline to heavy residual fuel oils. Gaseous fuels are represented by natural gas, liquefied petroleum gases (LPG), and various manufactured and by-product gases.

Solid Fuels

Coal is by far the most important commercial fuel. It is a geologically aged plant material that has progressively changed in form from plant material (wood) to peat, lignite, subbituminous, bituminous, semianthracite, and anthracite coal, in sequence. The final stage would be graphite or carbon. Coal is classified by *rank* and by *grade*. Rank classification, standardized by the American Society of Testing and Materials, is shown in Table 3.1. This

TABLE 3.1[a]

ASTM Classification of Coals by Rank[b]

Rank	Fixed Carbon, %[c]	Volatile Matter, %[c]	Btu/lb[d]
Meta-anthracite	98–100	0–2	
Anthracite	92–98	2–8	
Semianthracite[e]	86–92	8–14	
Low-volatile bituminous	78–86	14–22	
Medium-volatile bituminous	69–78	22–31	
High-volatile bituminous A	31–69	31+	14,000+
High-volatile bituminous B			13,000–14,000
High-volatile bituminous C[f]			11,000–13,000
Sub-bituminous A[g]			11,000–13,000
Sub-bituminous B			9,500–11,000
Sub-bituminous C			8,300–9,500
Lignite[h]			8,300−
Brown coal[i]			8,300−

[a] Ref. 35.
[b] Conforms with ASTM D388–38.
[c] Dry, mineral-matter-free.
[d] Moist, mineral-matter-free.
[e] Nonagglomerating.
[f] Agglomerating or nonweathering.
[g] Both nonagglomerating and weathering.
[h] Consolidated.
[i] Unconsolidated.

classification is based on a dry, mineral-matter-free or unit-coal basis [4].

In addition to rank classification, several subclasses have been set up for bituminous and subbituminous coals. The ASTM defines four special types [35]:

> *Common banded coal* is the commonest variety of bituminous and subbituminous coal. It consists of a sequence of irregularly alternating layers of black vitreous material, black or grayish striated material with a satiny luster, and a soft powdery mineral charcoal. *Splint coal* is grayish black in color, has a dull luster, and has a hard, tough, compact

structure. It usually breaks with a splintery fracture (thus "splint" coal). *Cannel coal* is usually dark gray to black in color, greasy appearing, has a fine-grained compact structure, and breaks with a conchoidal fracture. *Boghead coal* is similar to cannel coal in appearance and combustion behavior, but it gives a greater quantity of tar and oils on destructive distillation.

Classification by grade includes standardized size range, calorific value, ash content, ash-softening temperature, and sulfur content [5]. Common commercial sizes of bituminous and anthracite coals are shown in Table 3.2.

TABLE 3.2*
Commercial Sizes of Coal

Bituminous		Anthracite	
Designation	Size	Designation	Size
Run of mine	Fines to large lumps	Broken	$3\frac{1}{4}$ to $4\frac{3}{8}$ in.
Lump	Fines to 5 in.	Egg	$2\frac{7}{16}$ to $3\frac{1}{4}$ in.
Egg	2 to 5 in.	Stove	$1\frac{5}{8}$ to $2\frac{7}{16}$ in.
Nut	$1\frac{1}{4}$ to 2 in.	Chestnut	$1\frac{3}{16}$ to $1\frac{5}{8}$ in.
Stoker	$\frac{3}{4}$ to $1\frac{1}{4}$ in.	Pea	$\frac{9}{16}$ to $1\frac{3}{16}$ in.
Slack	0 to $\frac{3}{4}$ in.	Buckwheat No. 1	$\frac{5}{16}$ to $\frac{9}{16}$ in.
		Buckwheat No. 2 (rice)	$\frac{3}{16}$ to $\frac{5}{16}$ in.
		Buckwheat No. 3 (barley)	$\frac{3}{32}$ to $\frac{3}{16}$ in.
		Buckwheat No. 4	$\frac{3}{64}$ to $\frac{3}{32}$ in.

* From Ref. 35.

Ash content is usually in the range of 4 to 20%; ash-softening temperatures, 2000° to 2800°F; and sulfur, 0.5 to 5.0%. Some typical analyses are shown in Table 3.3.

Solid fuels are normally burned on grates through which primary air flows to reach the burning fuel. Everything else being equal, the tendency of a solid fuel to smoke is a function of the amount and character of the volatile matter in the fuel.

One important exception to burning on grates is the use of *pulverized coal*. Here, the powdered coal is finely ground and burned in a specially designed burner similar to those used for fuel oil. Smoke is no problem, but considerably more fly ash is formed and must be removed from the flue gases. Some of the newer type burners, e.g., the cyclone burner, remove a larger portion of the fly ash as molten slag directly from the burner body.

Coke (specifically, *gas coke*) is manufactured from bituminous coal by heating the latter material to high temperatures in the absence of air. All volatile material is driven off and a hard, gray, porous product (almost en-

TABLE 3.3*

Typical Analysis of Coals

	Proximate analysis (dry)				Ultimate analysis (dry)						Gross heating value, Btu/lb (dry)
	Moisture as received, %	Vol. matter, %	Fixed C, %	Ash, %	C, %	H₂, %	O₂, %	N₂, %	S, %		
Lignite	33.7	44.1	44.9	11.0	64.1	4.6	18.3	1.2	0.8		11,084
Subbituminous C	16.3	42.8	39.0	18.2	60.7	4.6	14.5	0.9	1.1		10,582
Subbituminous B	17.0	42.8	54.4	2.8	75.0	4.9	15.5	1.3	0.5		13,248
Subbituminous A	10.6	38.9	56.4	4.7	75.1	5.0	12.8	1.4	1.0		13,595
High-volatile bituminous, C	7.8	36.4	54.5	9.1	73.1	4.8	8.9	1.5	2.6		13,469
High-volatile bituminous, B	3.6	39.2	55.4	5.4	78.3	5.2	8.2	1.5	1.4		14,108
High-volatile bituminous, A	1.4	34.3	59.2	6.5	79.5	5.2	6.1	1.4	1.3		14,396
Bituminous, med. vol.	3.4	22.2	74.9	2.9	86.4	4.9	3.6	1.6	0.6		15,178
Bituminous, low vol.	3.6	16.0	79.1	4.9	85.4	4.8	2.6	1.5	0.8		15,000
Semianthracite	2.4	13.0	74.6	12.4	78.3	3.6	2.3	1.4	2.0		13,580
Anthracite	3.3	3.4	87.2	9.4	84.2	2.8	2.2	0.8	0.6		13,810
Meta-anthracite	2.8	1.2	90.7	8.1	86.8	1.6	2.0	0.6	0.9		13,682

* From Ref. 35.

tirely fixed carbon and ash) remains. Because of the absence of volatile materials, coke is a smokeless fuel. The other commercial coke is *petroleum coke*, a residue of certain petroleum-cracking processes. Because of the nature of the raw material, petroleum coke has a low ash content. A similar material made by the destructive distillation of wood or wood wastes is charcoal. The characteristics of cokes and charcoal are shown in Table 3.4.

TABLE 3.4*

Characteristics of Cokes and Charcoals

	Petroleum Coke	Gas Coke†	Charcoal
Proximate analysis, %			
Fixed carbon	79–95	85.3–88	80–90
Volatile matter	1–18	1.4–2.5	5–15
Ash	0–2	9.8–12.0	1–2
Moisture	0–3.3	0.7–1.3	3–12
Ultimate analysis, %			
Carbon	. . .	84.6–86.8	85–90
Hydrogen	. . .	0.6–1.1	1.5–3.5
Oxygen	. . .	0.9–1.5	0.1–0.4
Nitrogen	. . .	1.1–1.4	4–10
Sulfur	0.2–4.2	0.7–0.8	0–0.1
Ash	. . .	0.7–1.3	3–12
Heating value, Btu/lb			
(as received)	14,500–16,400	12,500–12,800	12,000–13,500

* From Ref. 35.
† Ref. 15.

Wood and wood wastes are also occasionally used as fuels. See Table 3.5 for characteristics typical of woody fuels.

A special form of woody fuel is *combustible trash*. This is usually a mixture of lumber scrap, wastepaper and cardboad, tree limbs, leaves, grass trimmings, flower clippings, and even garbage. This is not used as fuel, but is often burned in order to dispose of it. In many cases, combustion is poor and a smoke problem of considerable magnitude results.

The average composition, analysis, and calorific value of municipal refuse is shown in Table 3.6.

Liquid Fuels

All common liquid fuels are produced by refining petroleum crude oil. These fuels are classified as motor fuels and burner fuels, depending upon their method of utilization. Liquid motor fuels are called gasolines. Burner

TABLE 3.5*

Typical Characteristics of Woody Fuels

	Wood, air-dried	Woody wastes	Tanbark
Proximate analysis, %			
Moisture	15–25	36–60	60–75
Volatile matter	60–70	30–55	20–25
Fixed carbon	9–12	8.5–11	4–6
Ash	0.5–1.5	0.5–1.0	1–2
Ultimate analysis, %			
Hydrogen	6–8	8–9	8–10
Carbon	35–50	21–35	14–16
Nitrogen	0.1–0.3	0–0.1	0
Oxygen	50–70	58–68	72–75
Sulfur	0	0	0
Ash	0.5–1.0	0.5–1.0	1–2
Heating value, Btu/lb			
Green	2370–4000	2300–4000	
Air-dried	5200–6000	3600–5800	2500–3000

* Refs. 3, 21, and 35.

fuels include kerosene, distillate fuel (includes diesel fuel), and residual fuels. These are mixtures of various hydrocarbons of different specific gravities, volatility, viscosity, sulfur content, carbon-hydrogen ratio, etc.

Gasoline is used almost entirely as a fuel for internal-combustion engines. Consequently, its specifications are related to the power requirements of the engine. Smoke in engine exhaust is a specialized problem and is discussed in Chapter 9.

Specifications for *diesel fuel* may include ash content, carbon residue, cetane number (a measure of ignition quality), distillation range, flash point, pour point, sulfur, viscosity, and water and sediment. Distillation range (particularly the end point) and carbon residue determine the smoking tendency of the fuel, although poor maintenance or operating practices may cause any fuel to smoke. With suitable fuel and proper maintenance and operating methods, no diesel engine need smoke.

The specifications for various common grades of *fuel oils* are shown in Table 3.7. Proper burning of fuel oil requires that the oil be completely vaporized and well mixed with sufficient air for complete combustion. This may be accomplished by vaporization prior to mixing with air or by atomizing the oil into fine droplets which vaporize during the mixing operation. Under proper conditions, not difficult to attain, any fuel oil may be burned smokelessly. Again, the tendency for fuel oils to smoke is indicated by the carbon residue test.

TABLE 3.6

Composition and Analysis of an Average Municipal Refuse from Studies Made by Purdue University

Component	Percent of all refuse by weight	Moisture (percent by weight)	Volatile matter	Carbon	Hydro-gen	Oxygen	Nitro-gen	Sulfur	Noncom-bustibles*	Calorific value (Btu/lb)
					Analysis (percent dry weight)					
			Rubbish, 64%							
Paper	42.0	10.2	84.6	43.4	5.8	44.3	0.3	0.20	6.0	7572
Wood	2.4	20.0	84.9	50.5	6.0	42.4	0.2	0.05	1.0	8613
Grass	4.0	65.0	—	43.3	6.0	41.7	2.2	0.05	6.8	7693
Brush	1.5	40.0	—	42.5	5.9	41.2	2.0	0.05	8.3	7900
Greens	1.5	62.0	70.3	40.3	5.6	39.0	2.0	0.05	13.0	7077
Leaves	5.0	50.0	—	40.5	6.0	45.1	0.2	0.05	8.2	7096
Leather	0.3	10.0	76.2	60.0	8.0	11.5	10.0	0.40	10.1	8850
Rubber	0.6	1.2	85.0	77.7	10.4	—	—	2.0	10.0	11330
Plastics	0.7	2.0	—	60.0	7.2	22.6	—	—	10.2	14368
Oils, paints	0.8	0.0	—	66.9	9.7	5.2	2.0	—	16.3	13400

Linoleum	0.1	2.1	65.8	48.1	5.3	18.7	0.1	0.40	27.4	8310
Rags	0.6	10.0	93.6	55.0	6.6	31.2	4.6	0.13	2.5	7652
Street sweepings	3.0	20.0	67.4	34.7	4.8	35.2	0.1	0.20	25.0	6000
Dirt	1.0	3.2	21.2	20.6	2.6	4.0	0.5	0.01	72.3	3790
Unclassified	0.5	4.0	—	16.6	2.5	18.4	0.05	0.05	62.5	3000
Food wastes, 12%										
Garbage	10.0	72.0	53.3	45.0	6.4	28.8	3.3	0.52	16.0	8484
Fats	2.0	0.0	—	76.7	12.1	11.2	0	0	0	16700
Noncombustibles, 24%										
Metals	8.0	3.0	0.5	0.8	0.04	0.2	—	—	99.0	124
Glass and ceramics	6.0	2.0	0.4	0.6	0.03	0.1	—	—	99.3	65
Ashes	10.0	10.0	3.0	28.0	0.5	0.8	—	0.5	70.2	4172
Composite refuse, as received										
All refuse	100	20.7	—	28.0	3.5	22.4	0.33	0.16	24.9	6203

* Ash, metal, glass, and ceramics.

TABLE 3.7*

Fuel Oil Specifications

	Grade				
Characteristics	No. 1 volatile	No. 2 medium volatile	No. 3 low viscosity	No. 5 medium viscosity	No. 6 high viscosity
Flash point, °F					
Minimum	100 or legal	110 or legal	110 or legal	130 or legal	150
Maximum	165	190	230	. . .	. . .
Pour point, °F, max.	0	10	20	. . .	. . .
Water and sediment, % by vol., max.	trace	0.05	0.10	1.00	2.00
Carbon residue, % max.	†	‡	0.15	. . .	. . .
Ash, % max.	. . .	. . .	. . .	0.10	. . .
Distillation temperature					
10%, max., °F	410	440	. . .	. . .	. . .
90%, max., °F	. . .	600	675	. . .	. . .
90%, min., °F	. . .	. . .	600	. . .	. . .
End point, max., °F	560	. . .	. . .	. . .	. . .
Viscosity, Saybolt Universal					
At 100° F, max., sec	. . .	. . .	45	. . .	. . .
At 100° F, min., sec	. . .	. . .	. . .	50	. . .
Viscosity, Saybolt Furol					
At 122° F, max., sec	. . .	. . .	. . .	40	300
At 122° F, min., sec	. . .	. . .	. . .	. . .	45
Sulfur, max., %	0.5	0.5	0.75	. . .	. . .

* Refs. 5 and 35.
† 0.05% on 10% residuum.
‡ 0.25% on 10% residuum.

Gaseous Fuels

Commercial gas fuels are classified as natural gas, manufactured gas, by-product gas, and liquefied petroleum gas (LPG). *Natural gas* is essentially methane, CH_4, plus small varying amounts of carbon dioxide, and the lower gaseous hydrocarbons. The common *manufactured gases* are coal gas, from the destructive distillation of coal; oil gas, from the destructive distillation of petroleum oils; water gas, made by reacting steam with hot coke to produce a mixture of carbon monoxide and hydrogen; carbureted water gas, a mixture of water gas and oil gas; and acetylene gas, made by severe cracking of gaseous hydrocarbons or by treating calcium carbide with water. *By-product gases* include various petroleum refining gases that are produced in processing petroleum; coke-oven gas, a by-product of the manufacture of coke and similar to coal gas; and blast-furnace gas, principally carbon monoxide and nitrogen from iron-blast furnaces. *Liquefied petroleum gases* are a mixture of propane and butane recovered from natural gasoline or from refinery operations. The composition of these various gases is shown in Table 3.8.

The fuel gases may be burned smokelessly with comparative ease, since no difficulty is encountered in mixing them with adequate air for combustion. Since the combustibles in most fuel gases are hydrocarbons, improper mixing or insufficient air will lead to cracking of the hydrocarbon molecule and formation of carbon. The smoking tendency of hydrocarbons increases with increasing stability of the carbon skeleton of the molecule. A report released by the National Advisory Committee for Aeronautics [29] showed that the rate at which hydrocarbon types could be burned without smoking decreased in the approximate order: n-paraffins $>$ isoparaffins $>$ cycloparaffins $>$ olefins $>$ cycloolefins $\gtrsim$ diolefins $\sim$ alkynes $>$ n-alkylbenzenes.

SMOKE CONTROL

If any one activity is indicative of modern civilization, it is the tremendous use of heat and power derived from the combustion of fuel. Yet myriad columns of black smoke from chimneys and open burning are silent monuments to our inefficiency. So the need for smoke control is not only to relieve an air pollution problem, but also to harbor our natural resources. The emission of smoke is a lost opportunity to convert our fuel resources to useful energy.

The more important sources of smoke are domestic heating plants, industrial power plants, refuse incinerators, open fires, railroad locomotives, ships, diesel engines, and automobiles. The amount of smoke evolved from each of these sources varies considerably. It varies with type of fuel, type of combustion equipment, and equipment operating procedures. The types of fuels used in the United States by general type of consumers are shown in Table 3.9 [25].

Principles of Combustion

In order to eliminate smoke, it is necessary to apply the well-developed principles of combustion to the design and operation of combustion equipment. Basically, combustion is an oxidation reaction in which light and large quantities of heat are liberated. For all practical purposes, carbon (C) and hydrogen (H) are the elements that are oxidized, although sulfur (S) and others also meet the requirements stated. Air is the oxidizing agent. The principal reactions taking place in practical combustion processes are listed in Table 3.10.

TABLE 3.8*
Typical Composition Ranges of Fuel Gases

	Volume %								Specific gravity Air = 1.0	Gross heating value, Btu/cu ft
	Hydrogen, H_2	Methane, CH_4	Ethane, C_2H_6	Illuminants	Carbon monoxide, CO	Carbon dioxide, CO_2	Oxygen, O_2	Nitrogen, N_2		
Natural gas	0	66–98	0–32	0	0	0–2.7	0–0.4	0.1–7.5	0.59–0.67	1047–1210
Mixed refinery gas†	2.0–20.9	21.6–53.3	13.4–19.3	10–16.2	0	1.5–8.0	1.5–8.0	1.5–8.0	0.83–1.15	1380–1828
Oil gas	24–55	26–60	0	3–15	0–10	1–4	0.3–0.5	2.5–4.0	0.37–0.50	540–700
Coal gas	37–50	27–40	0–3	3–4	6–8.5	1.6–2.4	0.4–1.0	3.2–11.0	0.45–0.55	540–700
Coke oven gas	48–53	30–35	0–1	3–4	4–6	1–3	0–1.1	4–6	0.4–0.5	550–650
Producer gas	10–19	0.3–6.3	0	0–0.4	17–30	2.5–7.3	0.3–0.7	49–58	0.82–0.87	135–170
Water gas	45–70	0–4.5	0	0–1.1	32–41	3.0–5.0	0.2–0.7	4–10	0.5–0.6	290–320
Carbureted water gas	32–37	11–14	0	8–10	26–30	4–6	0.4–0.9	4.5–12.0	0.55–0.6	528–545
Blast furnace gas	1–4	0	0	0	26–27.5	11.5–13.0	0	57.6–60	0.95–1.0	90–93
Acetylene (commercial)	0	0	0	95–96	0	0	0–0.8	3–4	0.91–0.96	1300–1400

* From Refs. 33 and 35.
† Contains 1.3 to 4.2% propane and 0.2 to 3.5% butane.

TABLE 3.9

Estimated 1966 United States Energy Consumption
by Selected Consumer (10^{12} Btu) (Ref. 25)

	Consumer			
Energy source	Household and commercial	Industrial	Power generation	Total
Anthracite coal	143	41*	56	240
Bituminous and lignite coal	575	2,206*	6,341	9,122
Natural gas	5,945	5,674*	2,692	14,311
Petroleum†	2,247*	2,512*	905	5,664
Hydroelectric	0	0	2,060	2,060
Nuclear	0	0	58	58
TOTAL	8,910	10,433	12,112	31,455

* Excludes noncombustion consumption.
† Excludes naphtha, kerosene, and liquefied petroleum gases.

TABLE 3.10*

Chemical Reactions

Combustible substance	Reaction	Btu/lb-mole
Carbon to carbon monoxide	$C + \frac{1}{2}O_2 = CO$	+54,000
Carbon to carbon dioxide	$C + O_2 = CO_2$	+174,000
Carbon + carbon dioxide	$C + CO_2 = 2CO$	−70,200
Carbon + water vapor	$C + H_2O = CO + H_2$	−71,000
Carbon + water	$C + 2H_2O = CO_2 + 2H_2$	−72,000
Carbon monoxide	$CO + \frac{1}{2}O_2 = CO_2$	+122,400
Carbon monoxide + water	$CO + H_2O = CO_2 + H_2$	−700
Hydrogen	$H_2 + \frac{1}{2}O_2 = H_2O$	+123,100
Sulfur to sulfur dioxide	$S + O_2 = SO_2$	+126,000
Sulfur to sulfur trioxide	$S + \frac{3}{2}O_2 = SO_3$	+20,500
Methane	$CH_4 + 2O_2 = CO_2 + 2H_2O$	+382,000
Ethane	$C_2H_6 + \frac{7}{2}O_2 = 2CO_2 + 3H_2O$	+667,000
Propane	$C_3H_8 + 5O_2 = 3CO_2 + 4H_2O$	+950,000
Butane	$C_4H_{10} + \frac{13}{2}O_2 = 4CO_2 + 5H_2O$	+1,232,000
Acetylene	$C_2H_2 + \frac{5}{2}O_2 = 2CO_2 + H_2O$	+550,000
Ethylene	$C_2H_4 + 3O_2 = 2CO_2 + 2H_2O$	+600,000
Benzene	$C_6H_6 + \frac{15}{2}O_2 = 6CO_2 + 3H_2O$	+1,360,000

* From *Gaseous Fuels*, Louis Shnidman, Ed., American Gas Association, New York, 1954, p. 116.

Solid Fuels

In the burning of solid fuels on grates, the basic chemical reactions taking place are represented by the first, third, and sixth equations shown in the table. Two types of operations are generally used, depending on the relative direction of flow of air and fuel. In the *overfeed fuel bed,* air and fuel flow countercurrently, i.e., the fuel is fed on top of the bed of coals and air is fed through the grates in the opposite direction. This method is shown diagrammatically in Fig. 3.2 [20]. It is exemplified in practice by hand-fired furnaces. In the *underfeed fuel bed,* air and fuel flow concurrently, i.e., in the same direction. This method is used in the first stage of burning in traveling-grate stokers [8]. There is also a *cross-feed bed,* which is actually a combination of the two previously mentioned methods.

The combustion reactions that take place in the overfeed fuel bed may

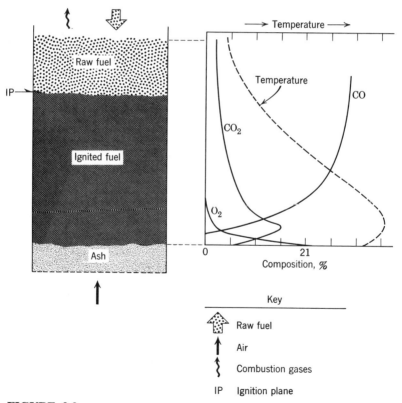

FIGURE 3.2.

Idealized overfeed fuel bed and relative distribution of temperature and products of combustion. (*Courtesy U.S. Bureau of Mines.*)

be visualized by reference to the diagram in Fig. 3.2 and the temperature-composition profile alongside the diagram. Air passes through the grate and the bed of ash above it and comes in contact with ignited fuel (glowing coals). Half of the oxygen necessary to burn the carbon completely reacts here according to the equation: $C + \frac{1}{2}O_2 \rightarrow CO$; $\Delta H = -54,000$ Btu per lb-mole of carbon consumed. Then the carbon monoxide formed reacts in the void spaces between the fuel particles with the remainder of the required oxygen: $CO + \frac{1}{2}O_2 \rightarrow CO_2$; $\Delta H = -122,400$ Btu per lb-mole CO consumed. Next, the carbon dioxide formed then reacts with more hot carbon as follows: $CO_2 + C \rightarrow 2CO$; $\Delta H = 70,200$ Btu per lb-mole of carbon consumed. Note that ΔH is positive. This means that heat is absorbed by the reaction.

The change in composition of the gas phase as it progresses through the fuel bed is shown in an idealized way in the graph in Fig. 3.2. Oxygen disappears a short distance above the base of the fuel bed. Thereafter, the percentage of carbon dioxide and the temprature decrease, whereas the percentage of carbon monoxide increases. In the zone where oxygen is still present, heat evolution is 174,000 Btu per lb-mole of carbon consumed and the temperature of the bed reaches a maximum. Thereafter, as carbon dioxide reacts with hot carbon, the bed temperature decreases.

Further cooling occurs as the hot gases pass through a layer of raw fuel. It is here that the most common source of smoke arises in firing domestic furnaces. If a high-volatile fuel is used, volatile hydrocarbons and tars are distilled from the raw fuel and pass into the space above the fuel bed as white, yellowish, or dense black smoke, depending upon the degree of cracking. Unless adequate heat and air are made available to burn these gases and tars, smoke will issue from the stack.

The characteristics of the underfeed fuel bed can be understood by examination of Fig. 3.3. In this case, the hot combustion products pass through ignited fuel and ash rather than through raw fuel. Consequently, any volatile matter in the fuel has an opportunity to burn in the hot fuel bed. In both cases, however, some carbon monoxide and unburned volatile matter escape from the fuel bed. Secondary air must therefore be supplied to complete the combustion.

The cross-feed fuel bed was developed primarily to minimize smoke emissions. Primary air moves across the fuel bed at right angles to the downward flow of fuel. When strong coking coals are used, air also is passed over the top of the fuel bed to prevent "hanging up" of the fuel in the fuel chamber. Combustion products that still contain combustible matter pass under an arch to a separate combustion chamber where secondary air is added. A detailed description of this method of burning has been published by Landry and Sherman [22].

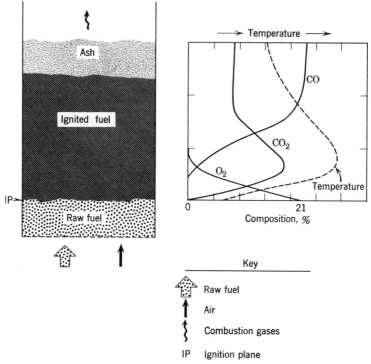

FIGURE 3.3.

Idealized underfeed fuel bed and relative distribution of temperature and products of combustion. (*Courtesy U.S. Bureau of Mines.*)

Open Fires

No discussion of methods of burning solid fuels would be complete without some consideration of open fires. In this case, smoking is particularly bad during ignition and refueling. The obviously poor mixing of fuel and air results in considerable evolution of volatile matter. Rapid quenching of the combustion gases by large unrestricted volumes of cold air is responsible for the excessive smoke evolved. Consequently, only nonvolatile fuels, such as coke and charcoal, can be burned smokelessly in the open. The problem has been discussed in detail by Fox [18].

A few estimates have been made of the amount of smoke and other pollutants produced by open burning [7, 14].

Liquid Fuels

Compared with solid fuels, the burning of liquid and gaseous fuels is simple. Fuel oils are generally preheated to decrease viscosity and then atomized mechanically or with steam.

When mixed with 15 to 20% excess air in a properly designed firebox, complete combustion takes place according to the equation:

$$C_xH_y + \left(x + \frac{y}{2}\right) O_2 \rightarrow xCO_2 + \frac{y}{2} H_2O$$

Heat evolution is 54,000 Btu per lb atom of carbon and 61,550 Btu per lb atom of hydrogen.

Gaseous Fuels

The combustion of gases requires only premixing with adequate air and no interference with the burning flame. The different burning characteristics (e.g., rate of flame propagation) of the various fuels must also be taken into consideration. As many housewives know, adjustment must be made in atmospheric burners if one changes from manufactured gas to natural gas or a mixture of the two. With proper burner design, however, gas burns cleanly and without smoke. This is apparent to anyone who has lived or visited in both gas-burning and coal-burning communities.

Typical combustion reactions of fuel gases are:

(Natural gas) $CH_4 + 2O_2 \rightarrow CO_2 + 2H_2O$; $\Delta H = -382,000$ Btu/lb-mole

(Propane) $C_3H_8 + 5O_2 \rightarrow 3CO_2 + 4H_2O$; $\Delta H = -950,000$ Btu/lb-mole

Combustion Equipment

It is beyond the scope of this book to describe all the design and operating parameters of combustion equipment. Some of the fundamental considerations of furnace and burner design are discussed, as well as practical operating considerations.

The requirements for smokeless combustion of any fuel are:

1. Proper air-fuel ratio.
2. Sufficient mixing of air and fuel.
3. Sufficient ignition temperature.
4. Sufficient space to permit time for proper burning.
5. Proper distance from the grate.

The requirements listed are, of course, just another way of stating the T, T, T, and O (sufficient time, temperature, turbulence, and oxygen) mentioned previously. Unfortunately, in practice these considerations are not always met. In the combustion of solid fuels, particularly coal on grates, three types of equipment are recognized: open fires (fireplaces), domestic and small industrial furnaces and boilers, and large industrial boiler plants.

Although not widely used in the United States, open fireplaces (estimated at 12 million) are quite common in Great Britain and are the chief cause of the smoke problem there. It is well-nigh impossible to burn high-volatile coal smokelessly in such equipment because of the lack of sufficient temperature above the firebed to burn volatilized gases and tars. The only recourse is to burn smokeless fuels, e.g., coke, anthracite, low-volatile coal, and the various processed briquetted fuels (basically, devolatilized coal).

In furnaces, both domestic and industrial, there is no acceptable reason for smoke with any fuel, because proper design will allow complete combustion of volatile matter. Variations in furnace design begin with the method of feeding coal to the fuel bed. The principal feeding methods are hand-fired, overfeed stoker, chain-grate stoker, and underfeed stoker.

Hand-firing methods are rapidly becoming obsolete in the United States, even in the older house-heating plants. Although good technique can reduce smoke somewhat, the only real cure is use of smokeless fuel. With any other solid fuel, it is impossible to prevent smoking when fuel is fed to the fire.

With the *underfeed stoker,* coal is pushed up through the fuel bed and the cinder falls to the side. Air is admitted over the clinker plate to burn the cinder. High-volatile fuels and coking coals may be burned. The ash must have a high fusion point. Only forced draft is used. The use of underfeed stokers is generally restricted to installations burning less than 3500 pounds of coal per hour.

With the *overfeed stoker,* the inclined grate extends from the hopper into the furnace at a rather steep angle. Coal flow is regulated by a controlled rocking or plunging movement imparted to the grate bars. Almost any type of coal may be used successfully. Natural or forced draft may be used.

Chain-grate (see Fig. 3.4) and *travelling-grate* stokers are essentially moving-grate sections which move from the front to the rear of the furnace,

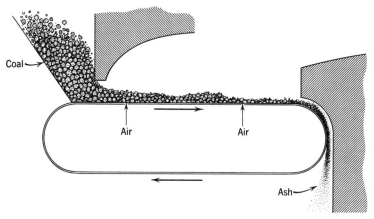

FIGURE 3.4.
Chain-grate stoker.

where the ash is discharged continuously. The fuel bed burns progressively to the rear. Any noncoking, clinkering coal may be used. Although both natural and forced draft are suitable to the unit, modern units all use zone-controlled force draft. Units range in size from 20 million to 300 million Btu per hour input.

Large industrial furnaces commonly use the *spreader stoker* in which a rotating flipper mechanism throws the fuel into the furnace and onto the grate. Coals of wide ranges in grade and type may be used. The coal is burned partly in suspension and partly on the grate. Various types of grates may be used. Response to fluctuations in load is rapid, but high-burning rates increase fly-ash carryover. Smoke may increase at low-burning rates. In the larger units, cinders are often returned to the grate from the fly-ash collector to reduce losses of unburned carbon. The size of spreader stokers ranges from 6 million to 500 million Btu per hour input.

No matter how fuel is fed to a furnace, some volatile matter will escape the fuel bed unburned. Hence, it is necessary to furnish a means for gaseous combustion. Air must be mixed completely and rapidly into the volatile gases; a source of ignition must be provided; and a hot combustion chamber of sufficient size must be available. If insufficient air, poor mixing, or even delayed mixing ensues, the volatile matter will "crack" to lighter hydrocarbons and soot, and black smoke will pour from the chimney. If the combustion chamber is too small or too cool, the flame will reach a cold surface before combustion is complete and smoke will result. It is appalling to consider the number of furnaces that have been sold all over the world without adequate provision for secondary air or with poor combustion-chamber design.

Even with the best design in the world, however, a smoke nuisance can be created by overloading the furnace or boiler and by poor maintenance practices. It is quite likely that more smoke is put into the air by an operator trying to "cut corners" than by any other means. Many boilers are rated too high by the manufacturer in order to get sales in a highly competitive market. In such cases it is impossible to operate the unit at rated capacity without producing smoke. Then, when the operator resorts to low-grade fuels to save on operating costs he only compounds the felony.

William G. Christy, former director of the Bureau of Smoke Control of the City of New York, has listed some of the conditions he found in heating plants and in small commercial and industrial boiler plants [10]. Every one of the items on the following list contributes to fuel waste, smoke, and low efficiency:

Air, oil, steam, or water leaks.
Broken, burned, or inoperative grates.

Warped, cracked, or missing doors.
Cracked or badly spalled refractory.
Burners or stokers out of adjustment.
Carbon formation in combustion chambers or in oil preheaters.
Dirty oil strainers and nozzles.
Defective or nonoperating air, fuel, and damper controls.
Poor selection of fuel.
Use of fuel not suited for equipment.
Broken, burned-out, clogged, or fused stoker tuyères.
Defective or nonoperating instruments, valves and fittings.
Dirty heating surfaces.
Corroded breechings (smoke pipes).
Broken, defective, or nonoperating dampers.
Accumulation of ash in ash pit.
Cracks and leaks in brick work of brick-set boilers or furnaces.
Hand firing coal on stokers.
Cracked or loose brick or broken tile in chimney.
Corroded steel stack.
Top of chimney lower than nearby buildings.
Leaks at breeching connection to chimney.
Canopy, cap or hood on top of chimney.
Fuel burning equipment of insufficient capacity.

One method of reducing smoke used in central power plants and large boiler-plant stations is burning *pulverized coal.* A fuel-rich mixture of powdered coal and primary air is produced by feeding the mixture at the center of a burner. Secondary air is admitted through an outer annulus so as to create the turbulence and mixing necessary for complete combustion.

The chief problem with pulverized fuel is control of fly ash. Eighty-five per cent of the ash is entrained in the flue gases, so a very efficient collector is required to hold fly-ash emissions to an acceptable level. As shown in Chapter 4, it may be accomplished with suitable equipment.

Wood Burning

The smokeless combustion of wood involves the same general principles as methods required for coal, with proper allowance for the different burning characteristics of wood [27]. Because of the higher moisture content of wood, a primary drying zone must be incorporated in the furnace. Generally this results in a two-zone design. Further, the ratio of overfire air to primary air through the grate is greater than in coal furnaces. The nature of the evolution of volatile matter from wood also makes it necessary to provide more secondary combustion space.

Refuse Burning

A special case of solid fuels burning, which has extremely great importance in air pollution control, is the burning of refuse from households, commercial enterprises, and industry. The process is usually carried out in units called incinerators or destructors. Unlike fuel burning, the objective is not heat and power but reduction in volume of waste material. Combustion efficiency has not been an objective and, as a consequence, smoke, fly ash, and odor emissions have usually resulted. Although smoke is the subject of this chapter, it is difficult to separate the three emissions mentioned, so control of all three will be considered here. In fact, most air pollution control officers agree that the fly ash and odor problems of incineration are greater than the smoke problem.

Probably the biggest problem confronting incinerator designers is the wide variation in the moisture content, density, physical form, and calorific value of common waste materials. These have been classified tentatively by the Air Pollution Control Association as follows:

Type 1—Rubbish. Consists of combustible waste, such as paper, cartons, rags, grass cuttings, leaves, wood scraps, sawdust, and floor sweepings from domestic, commercial, and industrial activities. It may contain up to 10% noncombustible solids and up to 25% moisture and have a heating value between about 5,500 and 8,500 Btu/lb as fired.

Type 2—Refuse. Consists of approximately equal weights of rubbish and garbage and is common to domestic occupancy.

Type 3—Garbage. Consists of animal and vegetable wastes from restaurants, cafeterias, hotels, hospitals, markets, and similar installations. It may contain up to 85% moisture and up to 5% noncombustible solids and have a heating value as low as 1,000 Btu/lb as fired.

Type 4—Human and Animal Remains. Consists of carcasses, organs, and solid organic wastes from hospitals, laboratories, abattoirs, animal pounds, and similar sources.

Type 5—Industrial By-Products. Consists of gaseous, liquid, and semiliquid wastes from industrial operations and includes noxious or toxic materials, such as tar, paint, solvents, and fumes from cooking processes. The heating value will depend upon the particular materials to be handled.

Type 6—Industrial By-Products. Consists of solid wastes from industrial operations, such as rubber, plastics, wood wastes, and noxious or toxic materials. The heating value will depend on the materials to be disposed of.

The tremendous variation is apparent by examination of the above list. Finding a suitable method for disposal of rubbish is an important problem facing most urban communities. In far too many locations rubbish is reduced by burning in open dumps. Fortunately, the practice is on the wane owing to public indignation. In some areas, burial of rubbish by sanitary landfill methods has been adopted. The method is restricted, of course, to communities where suitable sites are available. In the long run, reduction by burning in well-designed incinerators, private and municipal, will probably be the most satisfactory solution in a majority of cases.

It is estimated that Americans generate 4.5 lb of refuse per person per day, of the type normally collected for disposal [6]. Some communities have estimated per capita waste production as high as 8 lb per day. To this should be added 4.5 lb per capita per day of waste generated by industry and disposed of directly.

Waste incinerators are usually classified as (1) domestic, (2) flue-fed, (3) commercial and industrial, (4) municipal, and (5) special purpose incinerators. Obviously this is a very general scheme of classification with a range of types in each class.

Domestic incinerators are used to dispose of refuse and garbage from single-family residences. The most common type consists of an enclosed grate upon which rubbish is piled. Air enters both under the grate and above the charge by natural draft. Studies of a "backyard" burner of this type by Battelle Memorial Institute, Columbus, Ohio, demonstrated clearly the unsuitability of these units from an air pollution standpoint. Many cities now ban the use of this type of "incinerator," and more will probably follow.

One method of increasing the efficiency of domestic incinerators is the use of an auxiliary fuel, preferably natural or manufactured gas, to increase the temperature in the unit and to furnish heat for dehydrating the charge.

The American Gas Association has sponsored research work on the development of gas-fired domestic incinerators and several approved units are on the market [34]. One type is shown diagrammatically in Fig. 3.5.

Tests made at the University of Detroit on fourteen representative gas-fired domestic incinerators indicated, however, that none of the units was satisfactory [32]. The author of the report stated:

> Even if the incinerator is designed and constructed for complete odorless combustion, the operator would be the housewife, who, in most cases, would not be skilled in the firing technique required and she certainly would not invest the time and trouble necessary to follow firing methods required for complete combustion.

It was much this same philosophy which led the Los Angeles County Air Pollution Control District to ban all "single-chamber" domestic incinerators, whether gas-fired or not.

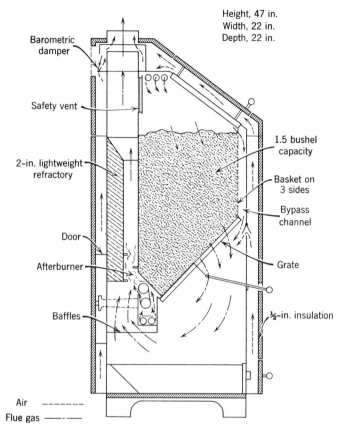

Height, 47 in.
Width, 22 in.
Depth, 22 in.

Barometric damper

Safety vent

2-in. lightweight refractory

Door

Afterburner

Baffles

1.5 bushel capacity

Basket on 3 sides

Bypass channel

Grate

½-in. insulation

Air ------

Flue gas —·—

FIGURE 3.5.

Combination gas-fired incinerator burner and after-burner.

One of the most objectionable types of incinerators from an air pollution standpoint is the *flue-fed incinerator* used in large apartment houses. Three basic types are shown in Fig. 3.6. The major operating problem with these units is caused by the dumping of wet garbage on the grate. Disturbance of the ash and burning rubbish causes fly ash and partially burned material to be expelled from the stack. Further, the wet garbage is partially distilled before complete burning, with consequent odor and smoke emissions. The only satisfactory solution to the problem is the installation of scrubbers and dust-catching devices at the top of the stack (see Chapter 4) and the burning of odoriferous material in catalytic fume burners (see Chapter 6).

Commercial and industrial incinerators include a wide variety of designs from small general-purpose incinerators to large units for burning specific

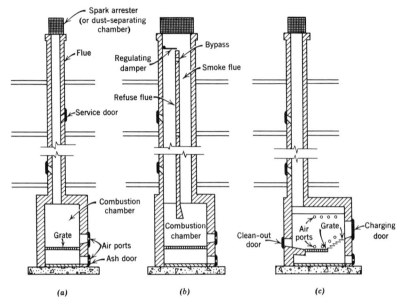

FIGURE 3.6.

Basic types of flue-fed incinerators. (*Courtesy U.S. Bureau of Mines.*)

materials, such as waste lumber, sawdust, sewage solids, etc. Three basic types are shown schematically in Fig. 3.7. In each type the charge is fed onto a grate in a primary combustion chamber; a secondary combustion chamber is provided for burning combustible gases, sometimes with the addition of auxiliary fuel. Even with a secondary combustion chamber, a considerable fly-ash problem ensues unless flue gas velocity is low enough, or sufficient baffles are supplied, to allow settling of particulate matter. This problem is discussed further in the next chapter.

One of the most perplexing problems of urban life is the disposal of garbage and refuse, both combustible and noncombustible. Various methods are available, e.g., sanitary landfill for refuse, and grinding followed by sewage disposal for garbage. But where land area is at a premium, as in most cities, *municipal incineration* has proved the most effective way of reducing the volume of waste.

Many different designs of municipal incinerators are available; in fact, identical units are difficult to find. Basically, all of these must have a means of charging refuse, both primary and secondary combustion chambers (with or without auxiliary fuel), a stack for the discharge of flue gases, and a means of ash removal. Various combinations are possible; proper selection depends on the nature and volume of the refuse to be burned. Unfortu-

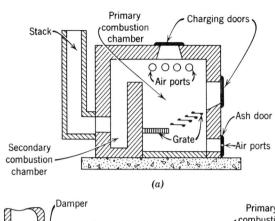

(a)

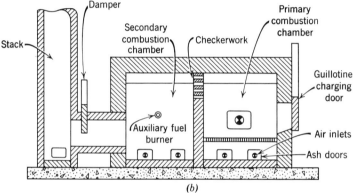

(b)

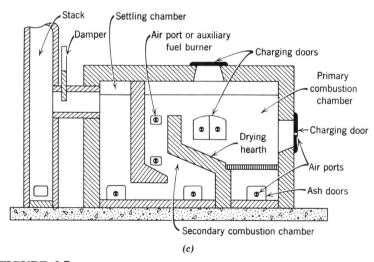

(c)

FIGURE 3.7.
Basic types of small and intermediate commercial and industrial incin-erators. (*Courtesy U.S. Bureau of Mines.*)

nately, air pollution considerations are seldom given sufficient weight in design and operation of the plant.

Two common methods of charging refuse are shown diagrammatically in Fig. 3.8. In a similar dumping method, the contents of the truck are

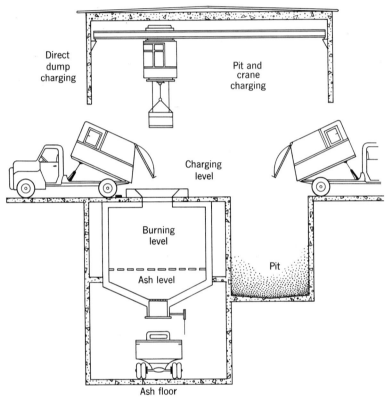

FIGURE 3.8.
Three-level construction for gravity flow of material.

dumped on a floor at the same level as the charging door and pushed into the incinerator by a bulldozer. Obviously, these methods lead to disturbance of the ash bed and distillation of volatiles from the raw refuse, so definite provision must be made in the rest of the unit to minimize smoke and fly-ash emissions. Another method of incinerator-charging makes use of continuous conveyers and mechanical rams. Effective use of this method requires that the refuse be uniformly cut or ground. Such material will burn more rapidly and more uniformly, but involves an obviously expensive and hazardous grinding step.

Combustion chamber design for municipal incinerators also varies widely. Differences are in size and shape of the primary chamber; type of grate (traveling, dumping, rabble-agitated, or no grate at all as in a rotary kiln); provision for auxiliary fuel; ratio of overfire and underfire air; shape and size of secondary chamber; relative position of bridge wall and baffles; method for heat recovery, if any; system for preheating air; etc. More complete details may be found in Ref. 2 and 13.

The principles involved in burning refuse are essentially the same as for burning any other solid fuel, except that special provisions must usually be made for preliminary drying of the charge. The low bulk-density and non-uniformity of the charge also presents problems. The most important design problem in municipal incinerators is the proper proportioning of equipment as to arch height, grate area, and type of grate.

Unfortunately, many municipal boards, in purchasing incinerators, place more emphasis on first cost than on proper design and end up with a king-size air pollution problem. Capital cost is an important factor in choosing an incinerator, inasmuch as costs range from $5,000 to $8,000 per ton daily capacity. To this must be added refuse collection costs (up to $10 per ton) and incinerator operating costs of $4 to $6 per ton.

A well-designed municipal incinerator will release as fly ash no more than 1% of the weight of the refuse burned, whereas a poor incinerator may release 2.5%. Both are far superior to open burning. For further discussion of fly-ash control, see Chapter 4.

Ash Removal

The amount of ash formed in municipal incinerators varies from 7 to 40% of the weight of refuse burned [1]. The ash falls into a pit below the grate and may be removed by gravity (see Fig. 3.8), or by a conveyor. A water-quenching system consisting of sprays or conveying the hot ashes through water is used to control incendiary cinders and to minimize dusting.

Stack System

Stack design follows well-known engineering principles and presents few unusual problems. When the other parts of the incinerator are poorly designed, means must be provided to reduce fly ash in the stack gases. The variety of methods available, settling chambers, centrifugal collectors, filters, wet collectors, and electrostatic precipitators, are discussed in the next chapter.

Special-purpose incinerators include those designed to burn wood waste, to remove combustible liquid and tar from used cans and drums (Fig. 3.9) [20], to remove organic coatings from scrap metal (Fig. 3.10), and to reclaim metal from scrapped auto bodies. The silo-type incinerator is

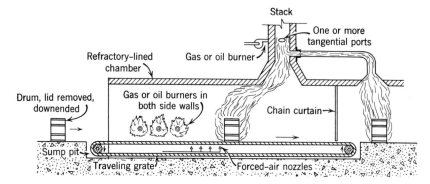

FIGURE 3.9.

Incinerator for removing organic residues from drums, etc. (*Courtesy U.S. Bureau of Mines.*)

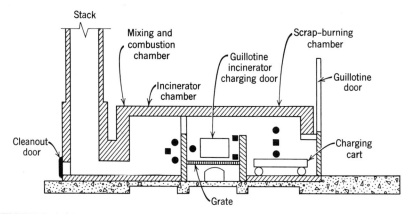

FIGURE 3.10.

Incinerator for removing organic coatings from scrap metal. (*Courtesy U.S. Bureau of Mines.*)

widely used for reducing wood waste (see Fig. 3.11). Usually these silo-type units are unsatisfactory from an air pollution standpoint [9].

The trench incinerator (see Fig. 3.12) [24] is a relatively new approach to the burning of chemical wastes. It is essentially an open rectangular pit lined with refractory brick. Waste is dumped or fed to the pit and ignited, after which all combustion air is admitted through closely spaced, high-velocity nozzles mounted above one wall of the pit at a 30-degree angle to the horizontal. The principal feature of the unit other than its open top is

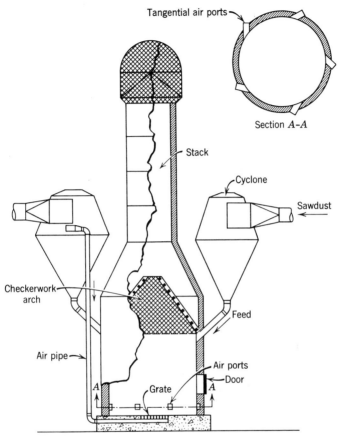

FIGURE 3.11.
Silo-type incinerator for wood waste. (*Courtesy U.S. Bureau of Mines.*)

the admission of all combustion air through the nozzles. Burning rates are said to be almost twice those obtainable with conventional designs. The air jets produce a cylindrical rolling flame that has adequate time to complete combustion. Flame temperatures are high, and mixing of the air and fuel are enhanced. Thus combustion is complete and thorough. The trench incinerator has been successfully used for reducing chemical wastes, plastics, rubber, film, paint, and heavy oils. Several units have been built to burn industrial trash. A steel mesh cage has been mounted on the top of the incinerator to retain any particulate matter that escapes the jets. Ash loss may well be greater than some air pollution control agencies will allow.

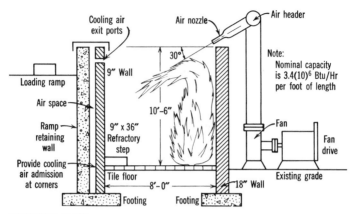

FIGURE 3.12.
Cross section of open pit incinerator.

Liquid Fuels

The combustion of liquid fuel falls into three categories, depending on the equipment in which they are burned. These are spark-ignited internal-combustion engines (gasoline), diesel engines (distillates), and furnaces and boilers (fuel oil). Only fuel-oil burning will be considered in this chapter. Although fuel oil is generally considered to be smokeless fuel, there are many cases where unsuitable burner design and poor maintenance have led to dense smoke emissions. The fundamental requirements for good combustion are (1) sufficient air, (2) intimate contact between fuel and air, and (3) a high temperature zone large enough to allow complete combustion. To insure good contact between fuel and air the fuel must be atomized. The heavier oils must be preheated before atomization to reduce viscosity (and thus promote good atomization and allow rapid ignition). Generally, only a slight excess of air (50%) is required. Care must be taken not to atomize the oil at too wide a spray angle or to allow droplets to impinge on solid surfaces (solid deposits of carbon will build up).

A variety of types of fuel-oil burners are available. Common types are (1) steam-atomizing, (2) high-pressure air-atomizing, (3) low-pressure air-atomizing, (4) mechanical-atomizing (spray nozzle), and (5) rotary mechanical atomizing. These are described in detail in engineering handbooks, e.g., Perry's *Chemical Engineers' Handbook* and Marks' *Mechanical Engineers' Handbook*. The particular type to use in a given service depends on many factors, which will not be discussed here.

The following list of possible causes of smoke from oil-burning equipment is taken from an excellent paper by Sambrook [28]:

Insufficient Air Supply
Inadequate chimney or deterioration of induced draught fan.
Accumulation of deposits (a self-aggravating effect).
Air infiltration due to leaky doors or boiler settings.
Inadequate ventilation of the boiler room.
Failure to open air registers or increase air pressure *before* increasing oil supply.

Air controls should be maintained in good order so that they can be set quickly and accurately.

Uncontrolled fluctuating in fuel supply due to fuel viscosity (i.e., temperature) variations and/or unsuitable fuel control valves.

Where fuel rate is controlled by a valve as is usual in steam and air atomizing burners, the valve should always be of the "metering" type and protected by a filter having openings smaller than the opening in the valve at the minimum operating position.

Poor Fuel and Air Mixing (Fuel atomization and air control)
Bad atomization by worn or damaged nozzles (pressure jet burners).
Fuel supplied at too high viscosity (too low temperature) either due to poor temperature control or large and/or unlagged branch lines between point of temperature regulation and burner.
Atomizing agent pressure too low (air or steam burners) or excessive pressure losses in supply pipes.
Unsuitable choice of atomizing nozzle (pressure jet burners) giving either too wide a spray resulting in fuel impingement on furnace walls or too narrow a spray giving coarse atomization and heavy fuel concentration in the centre of the spray.
Fuel pressure too low (pressure jet burners). The smallest nozzle operated at the highest pressure should be used to meet a given load.
Parts of air director damaged—air cones—swirler vanes burnt away or distorted. Adjustable controls jammed.
Refractory quarls (brickwork throats) damaged or displaced.

Flame Chilling
Insufficient distance between burner and cold surfaces (e.g., in water tube boilers) in relation to burner type and output.
Large excess of air resulting in low furnace temperature.

In extreme cases this can lead to white smoke (unburnt oil vapour). This sometimes occurs in marine boilers when manoeuvring due to cutting out burners without reducing air register openings or air pressure.

Indirect mixing air heaters, admission of diluting air into combustion chamber or before combustion of fuel is complete.
Insufficient refractory around root of the frame.

Although the majority of atomizing burners will give clean combustion in a water-cooled, unlined combustion chamber with only sufficient refractory to protect any uncooled surfaces, in some instances the provision of a small amount of refractory near the root of the flame materially assists rapid and clean combustion.

For certain high temperature furnaces, types of oil burners are used which do not produce stable flames in a cold environment. Attempts to use these oil burners before the furnace has been brought by other means to a temperature of perhaps 500°C can give rise to smoke formation until working temperature is attained.

Gaseous Fuels

Equipment for burning fuel gases embodies both atmospheric and pressure systems. Atmospheric burners run the gamut from the burners on kitchen ranges to large industrial units. The atmospheric system is simple, but the long heat-up time required often causes it to be rejected in favor of pressure systems. In the latter case, either the air or gas may be compressed before mixing, or the two may be premixed and then compressed. Descriptions of these burners may be found in the engineering handbooks by Perry and Marks. Smoke is usually no problem in gas burning.

One special situation where smoke is a problem is in the emergency flaring of fuel gases in petroleum refineries and other industrial plants. Several burners are on the market in which smokeless flaring is accomplished by mixing the waste fuel gases with steam [30].

Measurement of Smoke

The measurement of smoke is actually a special case of the measurement of particulate matter, which is described in the next chapter. Since smoke is generally considered to be only the black soot particles in combustion gases, one must resort to methods of estimating the degree of blackness of the smoky emission. The method used in a particular instance depends upon where the smoke density is being measured, i.e., in the stack, in the smoke emerging from the top of the stack, or in the open atmosphere.

One method commonly used *inside the stack* depends upon the diminution of light intensity when a beam of light passes through a cross section of the stack. Windows are cut in the side of the stack or chimney. A light beam from a constant source on one side of the chimney is directed across the stack to a photoelectric cell on the other side. When the smoke density, as measured by diminution of the light beam, exceeds a preset level, a visual or audible alarm system is activated and the event recorded. Several commercial units are on the market. One of the more annoying problems is keeping the glass slides on either side of the stack free from soot deposits.

The most suitable method for determining smoke concentration in the *open atmosphere* is based on filtration of a known volume of air through a white filter paper and estimation of the blackness of the resulting spot by optical means. One such method has been described by Hemeon [19]. In one automatic smoke filter sampler (see Fig. 3.13), a filter paper tape wound on a spool is moved at predetermined intervals by a time mechanism to expose a new section of filter paper to a flow of air. Approximately 0.25 cu ft of air per minute is drawn through a one-inch circle of filter paper for a given time period. Any smoke or colored particulate matter in the air stains the filter. The optical intensity of the stain by transmitted light is measured later in appropriate equipment. Results are expressed in Coh units per 1000 ft. The Coh unit is defined as the quantity of light-scattering solids capable of producing an optical density of 0.01 when the amount of light transmitted through the spot of dust collected on the tape of a smoke sampler (Fig. 3.13) is measured in a suitable densitometer. It has also been found convenient to reduce all of the results to multiples of 1000

FIGURE 3.13.
Automatic tape sampler with recorder. (*Courtesy Precision Scientific Company.*)

linear ft, so that the concentration of haze is expressed as so many Coh units per 1000 ft. In the city of Pittsburgh, the subjective impression of visibility was such that 1 Coh per 1000 ft represented a bright clear day, while that of 10 Cohs corresponded to one appreciably darkened by the presence of smoke in the atmosphere.

Ringelmann Chart.   Most ordinances regulating smoke emissions are based on estimation of the density of the smoke as it *emerges from the stack*. Of the several available methods, the one most commonly used is the Ringelmann Chart. The chart shows four shades of gray, as well as a pure-white and an all-black section. To overcome the difficulty of reproducing various shades of gray, the intermediate shades are built from black lines of various

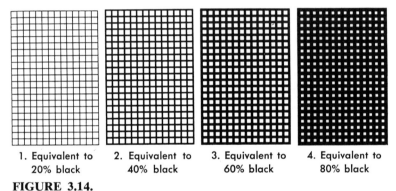

| 1. Equivalent to 20% black | 2. Equivalent to 40% black | 3. Equivalent to 60% black | 4. Equivalent to 80% black |

FIGURE 3.14.
Ringelmann's scale for grading the density of smoke.

widths (see Fig. 3.14). The four intermediate charts are printed by the United States Bureau of Mines on a single 26 in. x 10½ in. sheet. They may be reproduced as follows:

0. All white.
1. Black lines 1 mm thick, 10 mm apart, leaving white spaces 9 mm square.
2. Lines 2.3 mm thick, spaces 7.7 mm square.
3. Lines 3.7 mm thick, spaces 6.3 mm square.
4. Lines 5.5 mm thick, spaces 4.5 mm square.
5. All black.

In use the chart is set up at eye level in line with the stack at such distance (50 ft or more) that the sections appear to be different degrees of uniform gray shades. The appearance of the smoke at the top of the stack is matched against one of the shades on the card and reported as a specific "Ringelmann number" ranging from No. 0 (no smoke) to No. 5 (dense black smoke). With practice, an observer can estimate smoke density to half a number, particularly in the No. 2 to 4 range. Readings below No. 2 Ringelmann are subject to considerable error [11].

AIR QUALITY AND EMISSION STANDARDS

Air quality standards in the accepted sense are not applicable to smoke inasmuch as there is no suitable method for separating soot particles and fly ash from other atmospheric particulates. Some jurisdictions have adopted a "soiling index" based on Coh values as determined by a tape sampler, or automatic smoke sampler as it was originally called. Other jurisdictions have set Coh value standards as one measure of suspended particulate matter (usually in addition to weight per unit volume of air). Air quality standards based on Coh values have not been proposed by the Environmental Protection Agency.

In communities where coal smoke is emitted from low-level domestic and commercial chimneys, Coh values obtained by tape samplers may be useful for showing relative smoke density in the atmosphere. The subjective impression in Pittsburgh reported by Hemeon (loc. cit.) has already been mentioned. The state of New Jersey has prepared the following adjectival rating scale for suspended particulates as determined by a tape sampler.

> 0—0.9 Coh per 1000 lineal feet—light pollution
> 1.0—1.9 Coh per 1000 lineal feet—moderate
> 2.0—2.9 Coh per 1000 lineal feet—heavy
> 3.0—3.9 Coh per 1000 lineal feet—very heavy
> 4.0—4.9 Coh per 1000 lineal feet—extremely heavy

These figures probably reflect the subjective impression of air pollution intensity in New Jersey, but they probably would not fit many other localities. An example of the poor agreement between Coh values and other measurements of air pollution follows [17].

On a windy day in Las Vegas, Nevada, when the suspended particulate matter for a 24-hour period was 690 micrograms per cubic meter, the maximum Coh value measured was 2.0. On a day in Yakima, Washington, when the particulate loading was 111 micrograms per cubic meter, the Coh value was greater than 10. In the first case the atmospheric particulate matter was blowing desert dust; in the second case, considerable carbonaceous material was present because of orchard smudging in the vicinity.

The most common regulatory measure for control of smoke has been the Ringelmann number, so-called because it was first proposed as a method of measuring black smoke by Professor Maximilian Ringelmann

[26]. Smoke abatement officials long ago found that when coal was burned efficiently the plume appearance did not exceed a No. 2 Ringelmann, and that under these conditions soot did not settle from the plume and soil property. Accordingly, since the earliest days of smoke abatement legislation in the United States, smoke control regulations have restricted plume density to No. 2 Ringelmann. Exceptions are usually made for specified periods in the building of a new fire, cleaning a fire box, sootblowing, or making equipment changes. Most regulations also exempt plumes where violation of the regulation can be shown to be caused by the presence of water in the plume.

As long ago as 1937, Professor Lionel S. Marks of Harvard University pointed out the inadequacies of the Ringelmann Chart [23]. He showed both by experiment and by common sense that smoke showed a higher Ringelmann number if it was emitted from a wide-diameter stack, that it appeared darker on a clear day than on a hazy one, that the light passing through the plume was often not comparable with the reflectance from the Ringelmann Chart, and other inconsistencies. Despite these widely known problems, enforcement of a limitation of No. 2 Ringelmann on most smoke plumes has proved to be an acceptable means of smoke control. Even the gray plumes from large modern steam-electric plants have had to conform to an extension of the Ringelmann Chart known as equivalent opacity or vision obscuration. Inspectors are trained to estimate if a gray plume (the gray color reflects fly ash, not soot) obscures vision as much or more than does black smoke of Ringelmann shade No. 2. The method is patently a misapplication of the Ringelmann Chart, but most authorities deem the method reasonable because most combustion effluents can meet the test with available control equipment (usually electrostatic precipitators). Noteworthy exceptions are large power installations using some residual fuel oils or lignites.

In using Ringelmann numbers, and particularly the equivalent opacity concept, as a smoke control measure, it should be realized that the basic reason is an esthetic one rather than control of the amount of particulates emitted. Connor and Hodkinson [11] showed that the optical transmittance of a plume depends on size of particles rather than their mass (see Fig. 3.15). For instance, the figure shows that for a plume 3 meters in diameter, 60% optical transmittance (No. 2 Ringelmann) would be attained with only 0.005 grains per cubic foot of black particles 0.1 micron in diameter, whereas to meet the same transmittance would require 0.01 grains/ft^3 of 0.4-micron particles, 0.075 grains/ft^3 of 3.2-micron particles, and 0.15 grains/ft^3 of 6.4-micron particles. Thus a regulation based on a No. 2 Ringelmann plume might allow one operator to discharge 30 times more particles by weight than another even in the narrow particle-size range. Accordingly, one cannot relate the Ringelmann number of a plume to an air

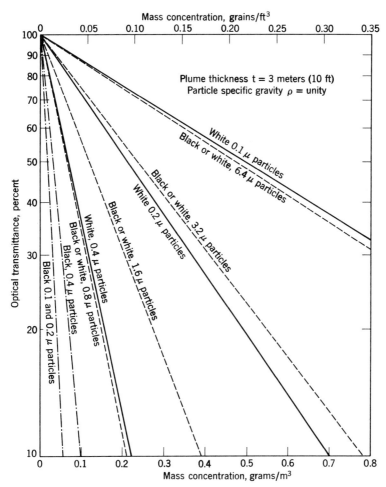

FIGURE 3.15.

Relationship of transmittance and mass concentration for plumes containing particles of various diameters and irregular shapes. (Ref. 11)

quality standard based on weight of particulate matter per unit volume of air.

A few communities have adopted or considered regulations restricting smoke emissions to No. 1 Ringelmann. Such actions are difficult to defend except in areas where no industrial operations are desired, or fuel use is restricted to gaseous fuels and distillate oils. Again, there is no relationship between plume appearance and the mass of the emission. A particularly difficult problem arises in reading a No. 1 Ringelmann sufficiently accurately for enforcement purposes. Conner and Hodkinson [11] report an experiment in which six highly trained smoke inspectors evaluated instrumentally calibrated black plumes of unknown (to them) Ringelmann numbers. Plumes in the No. 2 to No. 4 range were usually read within 0.25 of the calibrated Ringelmann number. When subjected to a plume of No. ½ (0.5) Ringelmann, five of the six inspectors read it greater than a No. 1 Ringelmann.

REFERENCES

1. American Public Works Association, Public Administration Service, *Refuse Collection Practice,* Third Edition (1966).
2. Anon., "Municipal Incineration," University of California, Institute of Engineering Research, Ser. 37, Tech. Bull. 5 (October 1951).
3. Anon., "Use of Wood for Fuel," United States Department of Agriculture Bull. 753.
4. ASTM D388-38.
5. ASTM Standards on Coal and Coke, *American Society for Testing Materials,* Philadelphia (1948).
6. Black, R. J., A. T. Muhich, A. J. Klee, H. L. Hickman, Jr., and R. D. Vaughn, "The National Solid Wastes Survey, an Interim Report." (Presented at the 1968 Annual Meeting of the Institute of Solid Wastes of the American Public Works Association, Miami Beach, Florida, October 24, 1968.)
7. Boubel, R. W., E. F. Darley, and E. A. Schuck, "Emissions from Burning Grass, Stubble and Straw," *J. Air Poll. Control Assoc.,* **19,** 497–500 (1969).
8. Carmen, E. P., and W. T. Reid, "Ignition Through Fuel Beds on Traveling-or-Chain-Grate Stokers," *Trans. Am. Soc. Mech. Engrs.,* **67,** 425–437 (1945).
9. Chass, Robert L., and Edward S. Feldman, "Incineration of Wood Waste in the Los Angeles Area," Los Angeles County Air Pollution Control District Publ. 24.
10. Christy, William G., "Stop Smoke and Save Money," *Building* (October 1951).
11. Conner, W. D., and Hodkinson, J. R., "Optical Properties and Visual Effects of Smoke-Stack Plumes," PHS Publication No. 999-AP-30, p. 28, U.S. Dept. of Health, Education, and Welfare, Cincinnati, Ohio (1967).
12. Ibid., p. 32.
13. Corey, Richard C., *Principles and Practices of Incineration,* Wiley-Interscience, New York (1969).
14. Darley, E. F., F. R. Burleson, E. H. Mateen, J. T. Middleton, and V. P. Osterli, "Contribution of Burning of Agricultural Wastes to Photochemical Air Pollution," *J. Air Poll. Control Assoc.,* **16,** 685–690 (1966).

15. Davis, J. D., and J. W. Greene, *Am. Gas Assoc. Proc.,* 1160–1164 (1926).

16. Essenhigh, R. H., "Incineration—A Practical and Scientific Approach," *J. Environmental Science and Technology,* **2,** 530 (1968).

17. Faith, W. L., "Inert Particulates—Nuisance Effects," *J. Occupational Medicine,* **10,** No. 9, 107–115 (1968).

18. Fox, L. L., "Some Measurements on the Smoke from Open Fires," *J. Inst. Fuel,* London (August 1954).

19. Hemeon, W. C. L., George F. Haines, Jr., and Harold M. Ide, "Determination of Haze and Smoke Concentrations by Filter Paper Samplers," *Air Repair,* **3,** 22 (1953).

20. Johnstone, H. F., et al, "Equipment and Processes for Abating Air Pollution," in *Air Pollution Handbook,* Paul L. Magill, Francis R. Holden, and Charles Ackley, eds., McGraw-Hill, New York, 1956.

21. Kreisinger, Henry, *Mech. Eng.,* **61,** No. 2, 115 (1939).

22. Landry, B. A., and R. A. Sherman, *Trans. Am. Soc. Mech. Engrs.,* **72,** No. 1, 9–17 (1950).

23. Marks, L. S., "Inadequacy of the Ringelmann Chart," *Mech. Engrg.,* 681–685 (1937).

24. Monroe, E. S., "New Developments in Industrial Incineration," Proc. 1966 ASME Intern. Incinerator Conf., pp. 226–230.

25. National Air Pollution Control Administration, "Control Techniques for Particulate Air Pollutants," NAPCA Publication No. AP-51, 3–11 (1969).

26. Ringelmann, M., "Method of Estimating Smoke Produced by Industrial Installations," *Rev. Technique,* 268 (1898).

27. Rosene, John, "General Examples of Industrial Furnaces and Fuels," Rept. to the city of Tacoma, Wash. (1952).

28. Sambrook, K. H., "The Efficient and Smokeless Combustion of Fuel Oils," Proc. Glasgow Conf., Natl. Smoke Abatement Soc. (London), 94 (1953).

29. Schalla, Rose L., Thomas P. Clark, and Glen E. McDonald, "Formation and Combustion of Smoke in Laminar Flames," *Natl. Advisory Comm. Aeronaut.* Rept. 1186, 20 (1954).

30. Smolen, W. H., "Smokeless Flare Stacks," *Petrol. Processing,* **6,** No. 9, 978–982 (1951).

31. Tebbens, B. D., J. F. Thomas, and Mitsugi Mukai, "Aromatic Hydrocarbon Production Related to Incomplete Combustion," *A.M.A. Arch. Ind. Health,* **14,** No. 5, 413–425 (1956).

32. Uicker, G. B., "Survey of Gas-Fired Domestic Incinerators," *J. Air Poll. Control Assoc.,* **5,** No. 4, 199–202 (1956).

33. U.S. Bur. of Mines reports and others sources.

34. Vandaveer, F. E., "The Domestic Gas-Fired Incinerator's Role in Air Pollution Control," *J. Air Poll. Control Assoc.,* **6,** No. 2, 90–97 (1956).

35. Van Winkle, Matthew, "Fuels," in *Handbook of Engineering Materials,* Douglas F. Miner and John B. Seastone, eds., Wiley, New York (1955).

4

DUSTS, FUMES, AND MISTS

Atmospheric pollutants may be classified broadly into two types—gases and particulate matter (liquid droplets and solid particles). Smoke, as described previously, consists of both types. Soot and fly ash are particulate matter; sulfur dioxide, carbon monoxide, etc., are gases. Aside from smoke, the most widespread air pollutants are the various types of particulate matter subclassified as dusts, fumes, and mists.

Dusts are solid particles, of natural or industrial origin, usually formed by disintegration processes. *Fumes* are solid particles generated by the condensation of vapors by sublimation, distillation, calcination, or chemical reaction processes. *Mists* are liquid particles, which may arise from vapor condensation, chemical reactions, or by atomization of a liquid. Hence, steam is a mist; fog is naturally occurring mist.

During the time particulate matter is suspended in the air it is known as an *aerosol*. After it has settled, either by virtue of its weight, by agglomeration, or by impact on a solid or liquid surface, the term no longer applies. Thus, particulate matter is an air pollutant only while it is an aerosol. It is a nuisance, however, both as an aerosol (visibility reduction) and as settled or deposited matter (soiling of surfaces, corrosion).

All air in the lower atmosphere contains some suspended matter. It is cleanest over the ocean, where salt nuclei and marine microorganisms are the major pollutants. Compared with ocean air, the average pollution of rural air is ten times greater; pollution in the air over small towns is 35 times greater; and the pollution over cities is 150 times greater [18]. Under unfavorable conditions the city values may reach 4000 times those found over the ocean.

The nature and magnitude of air pollution problems caused by particulate

100

matter depend on four factors: (1) the concentration of particles in the air, (2) size range of particles, (3) chemical composition, and (4) rate of settling or dustfall.

PARTICLE SIZE

The aerosols important in air pollution range from 0.01 μ to 100 μ in diameter. The size ranges for some common aerosols, dusts, and fumes are shown in Fig. 4.1. Below 1 μ are smoke from various sources, fumes from chemical and metallurgical industries, carbon black, and sulfuric acid mist. In this size range there is no tendency for particles to settle out, except by growth to larger particles or coalescence of several particles to produce much larger ones.

The other end of the spectrum, from 100 μ to 1000 μ, embraces large dust particles and raindrops, which settle rapidly. Between 1 μ and 100 μ are a wide variety of aerosols which, by virtue of settling rate, coalescence, and evaporation (in the case of natural fog, for instance), are removed from the air at varying, but still measurable, rates. Although the most effective size particles for visibility reduction (by light-scattering) are in the 0.3 to 0.6 μ range, the larger particles also reduce light by absorption.

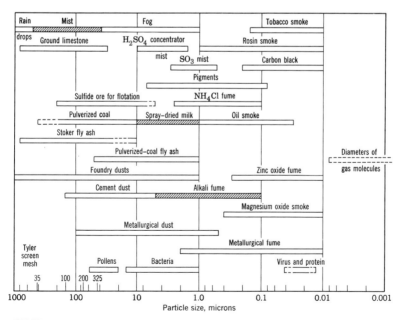

FIGURE 4.1.

Particle-size ranges for aerosols, dusts, and fumes (Ref. 23).

CHEMICAL COMPOSITION

The chemical composition of aerosols is quite complex. As pointed out previously, even soot particles, though largely carbon, contain adsorbed hydrocarbons. Particles of dust contain a variety of chemical compounds, depending largely on their origin. The importance of the chemical composition of dusts and fumes is related to the toxicity of some components, e.g., certain beryllium compounds, and to the catalytic effects of others. Manganese in fog droplets, for instance, is a very effective catalyst for the oxidation of sulfur dioxide to sulfur trioxide [13]. Another important factor is the hygroscopicity (tendency to take up water) of certain compounds. Salt nuclei are very effective in the formation of fog droplets.

Some of the more common elements found in urban air in the United States during the period 1957–1961 are shown in Table 4.1 [25]. Copper, iron, lead, manganese, and zinc were found in the largest amounts and were

TABLE 4.1

Geometric Mean and Maximum Concentrations of Selected Particulate Contaminants in U.S. Cities (1957–1961)

Pollutant	$\mu g/m^3$	
	Mean	Maximum
Suspended particles	104.0	1706.0
Benzene-soluble particles	7.6	123.9
Nitrates	1.7	24.8
Sulfates	9.6	94.0
Antimony	*	0.230
Bismuth	*	0.032
Cadmium	*	0.170
Chromium	0.020	0.998
Cobalt	*	<0.003
Copper	0.04	2.50
Iron	1.5	45.0
Lead	0.6	6.3
Manganese	0.04	2.60
Molybdenum	*	0.34
Nickel	0.028	0.830
Tin	0.03	1.00
Titanium	0.03	1.14
Vanadium	*	1.200
Zinc	0.01	8.4
Radioactivity	4.6†	5435.0†

* Less than minimum detectable quantity.
† Picocuries per cubic meter.
From Ref. 25.

universally distributed. Bismuth, cadmium, chromium, nickel, tin, titanium, and vanadium were widely distributed but in smaller quantities.

In addition to inorganic compounds, many aerosols contain various compounds in the liquid phase. In the case of fogs, the liquid is water.

The photochemical aerosol (see Chapter 8) contains an organic component. Analysis of the benzene-soluble fraction of atmospheric aerosols, and those produced synthetically by photochemical oxidation in the laboratory indicates the presence of carbonyl compounds, organic acids, halogenated compounds and "peroxy" compounds [20]. More detailed analyses by Cholak in Los Angeles [7] are shown in Tables 4.2 and 4.3.

TABLE 4.2

Composition of Benzene-Extractable Particulate Material

Height of inversion layer	0–600 ft				≥2000 ft	
Station	L. A.		Pasadena		L. A.	
Average daily particulate matter (mg/m³)	0.44		0.28		0.34	
Average daily benzene extractable (mg/m³)	0.09		0.05		0.07	
Per cent phenols	3.0		6.6		3.0	
Per cent organic bases	<1.0		<1.0		3.0	
Per cent organic acids and highly water-soluble compounds*	43.0		34.0		35.0	
Acids†		13.5		12.0		3.2
Dibasic acids						
Alcohols, etc.†		11.4		16.5		8.6
More polar organic*,†,‡		18.4		5.5		
Per cent neutral compounds	49.0		53.0		58.0	
Recovery	96.0		94.6		99.0	

* Determined by difference. Represent loss to water or not extractable by organics.
† These subdivisions overlap to a certain extent.
‡ May include organic bases which partition into aqueous $NaHCO_3$ rather than into benzene.

AEROSOL CONCENTRATION

The concentration of aerosols in any atmosphere varies widely.* It appears to be a function of meteorological conditions and the size of the community.

* The aerosol content of the air is usually expressed in terms of milligrams or micrograms per unit volume (cubic foot or cubic meter) of air, and sometimes in terms of millions of particles per unit volume. Dust loading in stack gases, however, is expressed in grains per cubic foot. Conversion factors for these and other units are listed in the Appendix. Dustfall is expressed in tons per square mile per month or per year and in milligrams per square centimeter per month.

TABLE 4.3

Chromatographic Analysis of Neutral Compounds*

Height of inversion layer	0–600 ft		≥2000 ft
Station	L. A.	Pasadena	L. A.
Saturates	44.5	33.3	38.8
Alkyl benzenes, etc.	6.4	4.2	5.6
Aromatics, 2 and 3 ring	5.5	4.2	3.7
Pyrenes and other hydrocarbons	1.5	<2.0	<2.0
Aromatics, 4 and 5 ring	4.7	4.2	5.6
Polynuclear aromatic hydrocarbons	11.7	10.4	9.2
Strongly held compounds, mostly nonhydrocarbon	27.5	42.0	38.8
Recovery	101.8	100.3	103.7

* Per cent of neutral compounds.

Since 1953, the United States Public Health Service has operated the National Air Sampling Network (NASN) to provide information on the nature and extent of air pollution. One of the air contaminants measured is "total suspended particulate matter." Currently the values obtained from the air sampling program are published annually under the title "Air Quality Data from the National Air Surveillance Network and Contributing State and Local Networks." Some of the data on average concentrations of suspended particulate matter are shown in Tables 4.4 and 4.5. The data indicate a positive relationship between size of cities and concentration of suspended particulate matter. Special studies of the data by USPHS also

TABLE 4.4

Distribution of Selected Cities by Population Class and Particle Concentration 1957–1967 (Ref. 25)

Population class	Average particle concentration ($\mu g/m^3$)										
	<40	40–59	60–79	80–99	100–119	120–139	140–159	160–179	180–199	>200	Total
>3 million							1		1		2
1–3 million							2	1			3
0.7–1 million			1		2		4				7
400–700,000				4	5	6	1	1	1		18
100–400,000		3	7	30	24	17	12	3	2	1	99
50–100,000		2	20	28	16	12	6	5	1	3	93
25–50,000		5	24	12	12	10	2	1	2	3	71
10–25,000		7	18	19	9	5	2	3	1		64
<10,000	1	5	7	15	11	2	1	2			44
TOTAL URBAN	1	22	77	108	79	52	31	16	8	7	401

TABLE 4.5

Distribution of Selected Nonurban Monitoring Sites by Category
of Urban Proximity, 1957–1967 (Ref. 25)

Category	Average particle concentrations (μg/m³)				Total
	<20	20–39	40–59	60–79	
Near urban*		1	3	1	5
Intermediate†		5	6		11
Remote‡	4	5			9
TOTAL NONURBAN	4	11	9	1	25

* Near urban—although located in unsettled areas, pollutant levels at these stations clearly indicate influence from nearby urban areas. All of these stations are located near the northeast coast "population corridor."
† Intermediate—distant from large urban centers, some agricultural activity, pollutant levels suggest that some influence from human activity is possible.
‡ Remote—minimum of human activity, negligible agriculture, sites are frequently in state or national forest preserve or park areas.

show that unusually high values are associated with either atmospheric stagnation or strong, dry winds [39].

Detailed data for selected cities for the year 1966 are shown in Table 4.6. The wide range of values at a single site reflects both meteorological variaables and differences in human activities from one season of the year to another, e.g., home heating in the colder months. Both arithmetic and geometric means of the samples are given in the referenced document, although the geometric mean is most commonly used to describe the atmospheric dust regime in a given locality.

One of the limitations of the NASN data is the sampling of the air at only one location, usually the downtown area, in each urban community. The variation in suspended particulate matter concentrations from one part of a city to another may show a greater range than is shown in the biweekly samples from a single location. In a 12-month survey conducted in the greater St. Louis area [41] the annual geometric means for suspended particulate matter ranged from 61 micrograms per cubic meter in a light commercial area to 222 micrograms per cubic meter in a site surrounded by industry and burning dumps. The geometric mean for all 17 stations was 108 micrograms per cubic meter.

DUSTFALL

A common measurement of air pollution severity in cities is the amount of dust which settles per square mile in a given time. The amount will vary widely in different parts of a given city, depending on nearby emissions. Values may be expected to reflect dust particles greater than 30 microns in diameter [9].

TABLE 4.6

Concentration of Suspended Particulates in Selected
U.S. Cities 1966 (Ref. 24)

City	Annual geometric mean ($\mu g/m^3$)	Maximum $\mu g/m^3$
Birmingham, Ala.	128	329
Anchorage, Alaska	81	349
Phoenix, Arizona	128	297
Bakersfield, Calif.	161	293
Los Angeles, Calif.	113	235
Denver, Colo.	117	230
Washington, D.C.	72	216
Miami, Fla.	45	100
Atlanta, Ga.	89	189
Honolulu, Hawaii	33	74
Chicago, Ill.	114	273
Kansas City, Kansas	83	148
Louisville, Ky.	132	332
New Orleans, La.	82	401
Baltimore, Md.	133	296
Detroit, Mich.	143	323
Jackson, Miss.	62	113
St. Louis, Mo.	135	255
Albuquerque, N.M.	106	302
Buffalo, N.Y.	117	1321
New York City, N.Y.	124	252
Columbus, Ohio	114	253
Portland, Ore.	66	174
Philadelphia, Pa.	148	261
Columbia, S.C.	70	142
Chattanooga, Tenn.	131	347
Dallas, Texas	89	390
Salt Lake City, Utah	86	172
Seattle, Wash.	72	181
Charleston, W. Va.	174	684
Cheyenne, Wyo.	35	379

Although dustfall values are usually given in tons per square mile per month, there is a trend toward reporting dustfall in milligrams per square centimeter per month. The change is reasonable because, in a sense, stating dustfall in tons per square mile per month is misleading, inasmuch as the area of the mouth of a dustfall jar is less than one two-hundred millionth of a square mile. Uniform deposition over a large area does not occur.

Variations in dustfall from one month to another and from one location to another are even greater than the variations in suspended particulate matter. In a dustfall survey at twenty stations in San Bernardino County, California in 1961, monthly dustfall values ranged from 11 to 1014 tons per square mile per month. Minimum monthly values ranged from 11 to 50, the low value in a desert recreational area, the high value in a windy, industrial location. The maxima for the same two stations were 269 and 1014 tons per square mile per month, respectively, both values reflecting periods of high winds. Review of a large mass of dustfall data leads one to the conclusion that soil and surface dust re-entrained by the wind has a greater effect on dustfall than any other factor.

In most heavily polluted areas deposited matter is generally high in water-soluble sulfates and is strongly acidic in solution. This reflects the high concentrations of fly ash and grit from the burning of coal. Occasionally, natural phenomena, e.g., dust storms and volcanic eruptions, are responsible for very high rates of dustfall. It is estimated that air currents carried 10 million tons of red dust from northwest Africa and deposited it in England in February, 1903. During the dust-snow storm in the Minneapolis-St. Paul area on March 12, 1954, 128.8 tons per square mile of dust settled on the Twin Cities [31].

Dustfall values are often reported in terms of both total solids and soluble solids, but the trend is to report only the former. A disadvantage of the dustfall jar, either with or without water, is coalescence of the collected particles so that particle-size measurements cannot be made. If particle-size determinations are desired, a large plastic sheet may be used to collect falling dust [9].

SOURCES OF PARTICULATE MATTER

The sources of aerosols and heavier dust particles in the atmosphere may be classified as industrial, natural (including agriculture), and domestic. Under each class there will be tremendous variations in amount, chemical composition, particle size, and density. There are even large variations in identical sources, depending on the degree of control exercised. Accordingly, quantitative data may be misleading. The only true method of determining emissions in a given situation is actual collection and measurement of emitted particulate matter over a sufficient period of time to cover all operating variables. Nevertheless, average data which have been published can serve as a general guide.

Industrial dust sources are classified in Table 4.7.

Extensive estimates of particulate emissions from uncontrolled sources have been published by the National Air Pollution Control Administration [27]. Selected values from the publication are shown in Table 4.8.

TABLE 4.7

Sources of Atmospheric Dust

Combustion	Materials handling and processing	Earth moving	Miscellaneous
Fuel burning	Loading and unloading	Construction	House cleaning
Incineration	(sand, gravel, ores,	(roads, dams,	Sand blasting
Open fires	coal, lime, bulk	buildings, site	Crop spraying
Burning dumps	chemicals)	clearance)	Poultry feeding
Forest fires	Mixing and packaging	Mining (blasting,	Rubber-tire
	(fertilizers, chem-	sorting, refuse	abrasion
	icals, feed)	disposal)	Engine exhaust
	Crushing and grinding	Agricultural oper-	
	(ores, gravel, chem-	ations (land	
	icals, cement)	preparation,	
	Food processing (mill-	soil tilling)	
	ing, e.g., flour, corn-	Natural (winds)	
	starch; drying; han-		
	dling grain)		
	Cutting and forming		
	(sawmills, wallboard,		
	plastics, etc.)		
	Manufacturing and		
	processing solids		
	(cement, chemicals,		
	carbon black)		
	Metallurgical (smelters,		
	blast furnaces,		
	foundries)		

Very few data on particulate emissions are published in sufficient detail to show the effect of operating variables on the amount and character of the emissions. Nevertheless, this type of information must be available for the design of adequate controls.

SAMPLING AND ANALYSIS

The determination of the amount and nature of particulate matter and aerosols contributing to air pollution falls into three categories:

1. Measurement of particulate matter in ducts and stacks.
2. Measurement of settled dust (dustfall).
3. Measurement of atmospheric aerosols.

TABLE 4.8

Particulate Emission Factors for Selected Uncontrolled Sources (Ref. 27)

Source	Rate
Fuel combustion	
General (pulverized)	(16 × % ash in coal)/ton of coal burned
Solid waste disposal	
Open burning dump	16 lb/ton of refuse burned
Municipal incinerator	17 lb/ton of refuse burned
Single-chamber incinerator	10 lb/ton of refuse burned
Multiple-chamber incinerator	3 lb/ton of refuse burned
Flue-fed incinerator	28 lb/ton of refuse burned
Chemical industry	
Sulfuric acid mfr.	0.3–7.5 lb/ton of acid produced
Food and agricultural industries	
Direct-fired coffee roaster	7.6 lb/ton of green beans
Cotton-ginning	11.7 lb/500-lb bale of cotton
Starch flash drier	8 lb/ton of starch
Primary metal industry	
Iron and steel manufacture	
Sinter plant gases	20 lb/ton of sinter
Open-hearth furnace	
Oxygen lance	22 lb/ton of steel
No oxygen lance	14 lb/ton of steel
Blast furnace	
Ore charging	110 lb/ton of iron
Agglomerate charging	40 lb/ton of iron
Coking	2 lb/ton of coal
Secondary metal industry	
Brass and bronze smelting	
Reverberatory furnace	26.3 lb/ton of metal charged
Gray iron foundry	
Cupola	17.4 lb/ton of metal charged
Lead smelting	
Cupola	300 lb/ton of metal charged
Mineral product industry	
Asphalt batch plant drier	5.0 lb/ton of mix
Cement manufacture	
Dry process kiln	46 lb/barrel of cement
Wet process kiln	38 lb/barrel of cement
Concrete batch plant	0.2 lb/yard of concrete
Lime production	
Rotary kiln	200 lb/ton of lime
Vertical kiln	20 lb/ton of lime

TABLE 4.8 (*continued*)

Source	Rate
Rock, gravel, and sand production	
Crushing	20 lb/ton of product
Conveying, screening, shaking	1.7 lb/ton of product
Petroleum industry	
Fluid catalytic cracker	0.1–0.2 lb/ton of catalyst circulated
Kraft pulp industry	
Smelt tank (uncontrolled)	20 lb/ton of dry pulp
Lime kiln	94 lb/ton of dry pulp
Recovery furnace with primary	
stack gas scrubber	150 lb/ton of dry pulp

Each of these categories requires specific techniques, some of which may be adapted with modifications to all three cases. The discussion which follows is limited to the measurement of the amount of particulate matter and determination of particle size range. Reference is made to simple methods of chemical analysis. Detailed methods of determining chemical composition are reserved for the professional chemist.

Sampling in Ducts and Stacks

No determination of the components of a flowing gas stream can be accurate unless the sample obtained for analysis is representative and reproducible. This is particularly true in the determination of particulate matter where concentration differences exist in different portions of the stream. Accordingly, a suitable sampling device must meet the following requirements:

1. The sampler must be made of noncorrosive material.
2. Temperature in the sampling line must be held above the dew point of normally gaseous components.
3. The sample must be obtained isokinetically (rate and direction of flow of gases into the sampler must be the same as in the gas stream being analyzed).
4. The long axis of the sampling head must be parallel to the direction of gas flow.
5. The collector must be at least 95% efficient. Occasionally, lower efficiencies can be accepted if reproducibility of samples is high and representative of the total gas stream.
6. The sample must be of sufficient size for required analysis.
7. The total gas flow must be known. The volume of gas flow is usually determined by traversing the cross section of the duct or stack with a Pitot tube (see Fig. 4.2) to obtain gas velocity. Calculated velocities

based on the Pitot readings are averaged and multiplied by the cross sectional area to give the gas volume. Temperature and static pressure of the gas stream must also be determined so that volume may be referred to a standard condition, e.g., 68°F and 29.92 in. Hg.

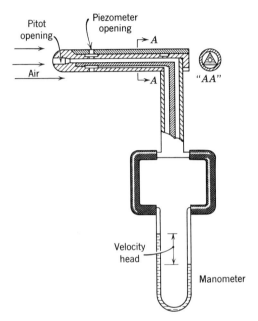

FIGURE 4.2.
Pitot-static tube.

The most difficult criterion of adequate sampling met in practice is maintenance of isokinetic sampling conditions. After the duct or stack has been traversed * the average stack velocity may be calculated. The suction device that will draw the gas sample through the sampling apparatus is then set so that velocity at the probe nozzle will be the same as that of the main gas stream at the sampling point. A sampling nozzle is then chosen for a suitable gas flow. Kanter [15] suggests 0.5 to 1 cfm for combustion effluents at the selected velocity.

Besides selecting a proper sample nozzle, care must be taken in choosing the other components of the sampling train, i.e., the trap for particulate matter, the meter, and the suction device. A suitable arrangement for the apparatus is shown in Fig. 4.3.

* Ref. 15 is an excellent article on sampling and contains a discussion on the selection of traverse points.

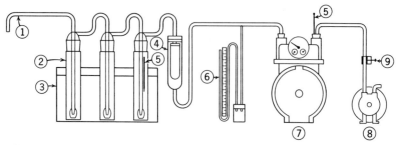

FIGURE 4.3.

Sampling apparatus for stack gases. (1) Sample probe. (2) Impinger (dust concentration sampler). (3) Ice bath container. (4) Dry filter. (5) Thermometer. (6) Mercury manometer. (7) Sprague dry gas meter (Zephyr No. 1A). (8) Vacuum pump. (9) Hose clamp to control gas flow rate. (*Courtesy Los Angeles Air Pollution Control District.*)

A wide variety of meters may be used, including wet test meters, dry gas meters, capillary flowmeters, and rotameters. The important factor is that they must be accurately calibrated. Commonly used suction devices include vacuum pumps and water or steam ejectors.

The biggest variation in particulate matter samplers is in the trap that removes the solid matter from the air stream. Common techniques involve the use of filtration, impaction (impingement), centrifugal separation, liquid scrubbing, thermal precipitation, electrostatic precipitation, visual methods, and sedimentation. A tabular comparison of the various types of devices, as prepared by Silverman [35], is shown in Table 4.9.

Filtration.   Many types of filters are available for removal of particulate matter from gas streams. Chief variations are in the filter media material and in the shape of the membrane. One common type is the thimble filter shown diagrammatically in Fig. 4.4. The filter-paper thimble is filled with well-fluffed cotton and is operated at a sampling rate of 2 cfm. By determining the change in weight of the dried filter after exposure, dust concentration in the gas stream may be determined. The paper thimble may not be used with high temperature gases; an alundum thimble is useful in this case.

A very popular type of filter is the *molecular* or membrane filter. This filter consists of a porous membrane disk supported in a filter holder (see Fig. 4.5). The membrane is composed of a mixture of cellulose nitrate and cellulose acetate. Filtration is by screening rather than by impaction, so collected particles are held on the surface. Accordingly, the collected particles are visible under a microscope when incident light is used. The mem-

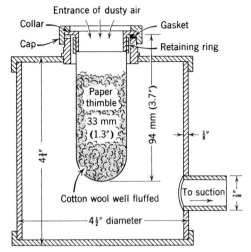

FIGURE 4.4.
Thimble filter.

brane is soluble in several organic solvents, so it may be dissolved and the size distribution of the collected particles may be determined by sedimentation. The particles may also be examined under an electron microscope after the membrane has been dissolved. Obviously, the membrane filter is most useful in the determination of extremely fine particles. Molecular membranes are currently available in three ranges of pore sizes: type hydrosol assay, 0.1 to 0.3 μ; aerosol assay, 0.5 to 0.7 μ; and aerosol protective, 1.2 to 1.6 μ.

Impaction. The operation of impactors and impingers is based on the differential momentum of the gas and particulate phases in a flowing gas stream. By changing the direction of an aerosol stream at a solid surface, the aerosol particles are driven toward the surface and cling to it.

A diagram of a cascade jet impactor is shown in Fig. 4.6. A photograph of a similar instrument is shown in Fig. 4.7. The device is used to collect particles and classify them into several size ranges. In the three stages shown, the jets are progressively smaller in size, i.e., 8, 3, and 1.5 mils. The jets direct the gas stream to small collodion-coated microscope slides upon which the particles are collected. The largest particles are collected on the first slide, where the gas velocity is lowest, and so on. The slides may be removed and the particles examined under a microscope. Because of the extremely high velocities through the jets, efficiency is very high (100% for particles as small as 0.6 μ).

Several commercial instruments that use the *impinger* or *impactor* (the

TABLE 4.9*

A Comparison of Various Devices Commonly Used for Sampling Particulate Matter in Gases

Principle of method	Instrument	Application	Sample volume	Sampling rate	Per cent efficiency	Particle size range (microns)
Sedimentation	Setting pan	General atmospheres and stacks (low concentrations)	440 ml	Unknown	?	High-low 10-mesh-variable
	Sedimentation cell		50 ml small	Instantaneous	100	10-mesh-0.2
Visual or photometric	Ringelmann chart	Stacks and high concentrations			Direct reading	
	Tyndallometer			Variable	Direct reading	10-0.1
	Photoelectric densitometer			Variable	Direct reading	10-0.1
Inertial and centrifugal	Labyrinth	Stacks and process effluents	Large	Variable	?	10-mesh-1
	Midget cyclone		Large	25–50 cfm	98	10-mesh-5
	Aerotec tube Design 2		Large	35 cfm	94	10-mesh-5
Impingement (impaction and jet condensation)	Owens	General atmospheres and after gas-cleaning devices	50–100 ml small	Instantaneous	99	10-0.2
	Konimeter		2.5–5 ml small	Instantaneous	99	10-0.2
	Bausch and Lomb		0.01 cu ft	Instantaneous	99	10-0.2
	Impinger		Medium	1 cfm	93–96	25-0.5
	Cascade impactor		Small	17 lpm	95	100-0.2

Filtration	Paper thimble		Small, medium, or large	1–3 cfm	95	10-mesh–0.2
	Paper disks or pleated paper		Small, medium, or large	1–3 cfm	99	10-mesh–0.2
	Cotton, wool, or asbestos bags	Stack and general atmospheres	Large	40–60 cfm	99	10-mesh–1
	Glass fibers, cloth, or alundum thimbles		Small, medium, or large	10–50 cfm	90–95	10-mesh–1
	Wire screen or gauge		Large	1–10 cfm	50–70	10-mesh–10
	Volatile or soluble crystals		Small, medium, or large	1–30 cfm	35–90	10-mesh–0.2
Washing or scrubbing	Simple bubbling devices	Stack and general atmospheres	Small	0.5–1 cfm	90–95	10-mesh–10
	Sintered glass absorbers		Small	0.1–0.5 cfm	95	10–0.2
	Wetted packed towers (bead columns, etc.)		Medium	1–3 cfm	90–95	10-mesh–10
	Venturi scrubber		Large	30–40 cfm	90–95	10-mesh–0.2
Precipitation	Thermal precipitator	General atmospheres and after cleaning devices	Small	5–10 ml/min	99	5 μ–0.2
	Electrostatic precipitator	Stack and general atmospheres	Small, medium, or large	1–3 cfm	99	80 μ–0.2

* From L. Silverman [35].

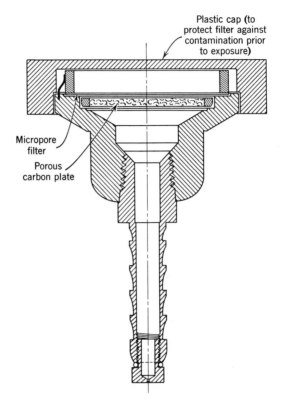

Plastic cap (to
protect filter against
contamination prior
to exposure)

Micropore
filter

Porous
carbon plate

FIGURE 4.5.
Molecular filter holder. (*Courtesy Alexander Goetz.*)

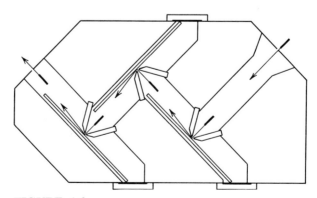

FIGURE 4.6.
Plastic jet impactor.

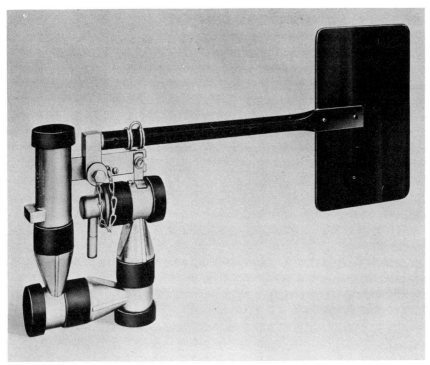

FIGURE 4.7.
Casella Cascade Impactor. (*Courtesy Mine Safety Appliances Co.*)

terms are interchangeable) principle are available. One of the most useful is the Greenburg-Smith impinger (see Fig. 4.8), which employs a high velocity air jet that impinges on a glass plate immersed in water or alcohol. The particles are collected in the surrounding fluid. Another is a midget-type impinger, which uses a sampling rate of 0.1 cfm and may be operated with a hand pump. Efficiency is high for particles of 1 μ or larger.

Centrifugal Separation. The most common device using centrifugal force to separate solid particles from a gas stream is the cyclone sampler, which is smaller but similar in design to cyclone collectors used for dust abatement (Fig. 4.9). Particle-laden gases enter a cylindrical chamber tangentially, and the solid particles are thrown toward the wall where they collect and fall into a flask at the bottom. Denuded gas leaves through a pipe extending through the top to the center of the cyclone. Although collection efficiency is poor for particles less than 5 μ, the device is well adapted to the sampling of heavily loaded gases in stacks and ducts.

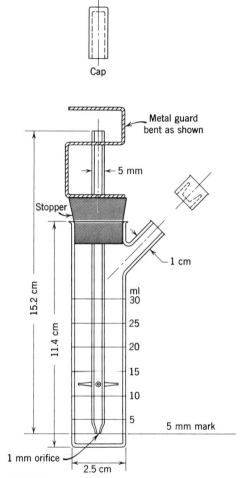

FIGURE 4.8.
Greenburg-Smith Impinger.

Thermal Precipitation.   When an aerosol passes between a hot wire and a cold plate, particles in the stream are repelled by the hot wire and deposited on the cold plate. Advantage is taken of this phenomenon in the thermal precipitator, which has proved to be highly efficient for fine particle collection. Unfortunately, capacity is quite low. Figure 4.10 shows a thermal precipitator of the type supplied by Casella Company, Ltd., London, England.

Electrostatic Precipitation.   In the electrostatic precipitator, dust-laden air is passed between two surfaces carrying a high electric potential, and under the force of the electric field, particles are driven to a collecting electrode and precipitated. As the collecting plate is small, the collected particles can be weighed accurately. The precipitation is highly efficient (approaches 100%); air flows are high and pressure drops small. As little as 0.1 mg per 10 cu m of air can be determined by this technique. However, because of sparking, the precipitator is not recommended for explosive atmospheres.

Operating characteristics and limitations of the various particle collectors described are listed in Table 4.9.

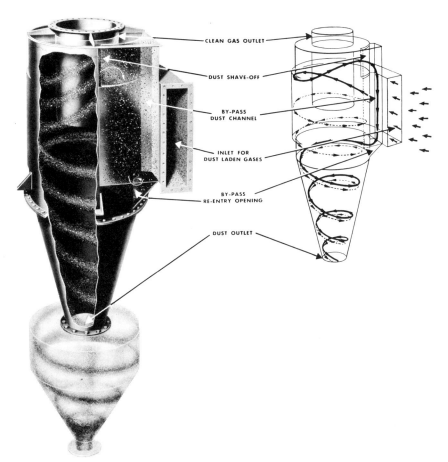

FIGURE 4.9.
Van Tongeren Cyclone. (*Courtesy Van Tongeren, Heemstede, Holland*)

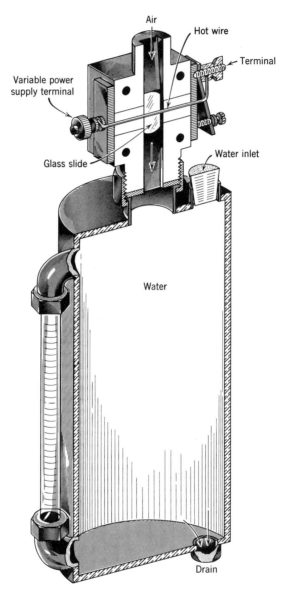

FIGURE 4.10.
Thermal precipitator.

Measurement of Settled Dust

Two methods commonly used for the determination of the nature and amount of solid particles settling from air are dust jar collection and greased plate collection. The dust jar method consists only of placing an open jar or bucket at a location of interest and determining the amount of dust collected in a known period of time. Obviously only the heavier particles in the air are collected. Sometimes water is placed in the bottom of the jar to retain the samples.

A recommended standard method for a continuing dustfall survey has been published by the Air Pollution Control Association [12].

The greased plate method involves exposure under arbitrary conditions of a 1 in. x 3 in. glass slide coated with petrolatum. This method is usualy employed to determine the pollen count in the atmosphere.

Atmospheric Sampling

Most of the methods used for sampling flues and stacks may be adapted to open-air sampling. Of course, concentrations are usually lower, as is temperature. Some of the devices suitable for open air sampling are indicated in Table 4.9.

The most commonly used particulate matter sampler for atmospheric use is the high-volume sampler, such as is used in the National Air Surveillance Network. Various designs are available commercially, but all involve passing a large volume of air through suitable filter media. One specific unit (see Fig. 4.11) makes use of a specially fired glass fiber filter web (8 in. x 10 in.) mounted on a rectangular aluminum filter holder. Capacity of the sampler is 60 cfm, so that 85,000 cu ft of air may be filtered in 24 hours. The collected particulate matter may be dried and weighed, then removed by washing for subsequent size separation and composition analysis.

In the NASN program, the samplers are installed in a vertically-oriented shelter. Air entering the shelter must pass upward around the outside of the sampler, then down through the filter media and out the discharge tube. Upward flow within the shelter prevents particles above 100 μ from reaching the filter.

The method used by the NASN for dividing the sample for detailed analysis is shown in Fig. 4.12 [40].

Determination of Particle Size

An important characteristic of any sample of particulate matter removed from a stack or the open atmosphere is the range or distribution of particles by size. The simplest method of measurement is by screen analysis obtained by separating the particulate matter into size groups by means of a

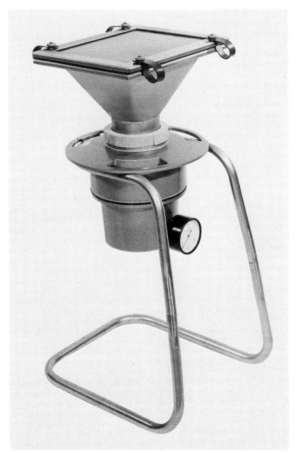

FIGURE 4.11.
High-volume sampler. (*Courtesy Precision Scientific Co.*)

graduated series of metal-cloth screens. The method is limited to particles larger than 40 μ. Separation of dust particles finer than 40 μ into various size ranges is usually done by air elutriation. The Roller particle-size analyzer (see Fig. 4.13) utilizes this method and permits the separation of a 25 g dust sample into six size ranges (0–5 μ, 5–10 μ, 10–20 μ, 20–40 μ, 40–80 μ, and >80 μ) in a few hours. Particle counts and size determination may also be made microscopically. A most powerful tool for the examination of aerosols is the electron microscope, which has a resolving power of the order of 0.02 μ. The usefulness of the electron microscope in con-

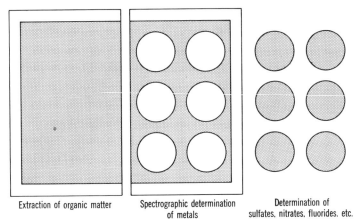

| Extraction of organic matter | Spectrographic determination of metals | Determination of sulfates, nitrates, fluorides, etc. |

FIGURE 4.12.

Division of high-volume particulate sample for analysis.

junction with a thermal precipitator has been described by Froula, Bush, and Bowler [11].

Other aerosol sampling equipment, such as cascade impactors and electrostatic precipitators, may also be used to collect aerosols for microscopic examination.

The Coulter Counter has also proved to be useful for determining the number and size of particles in a sample of dust. Particles suspended in a suitable liquid are pumped through a current-carrying orifice. In passing through the orifice, a particle alters the resistance of the circuit. The change in the resistance produces a voltage pulse having a magnitude proportional to the volume of the particle. By electronically counting the pulses above a certain threshold voltage, the instrument produces a count of all particles above a selected size. Comparatively small samples may be used, but the particles must be insoluble in the liquid carrier and contain but few fibrous particles.

A widely used unit involving centrifugal classification is the Bahco microparticle classifier, which is a combination air centrifuge-classifier. It is specified as the standard instrument for particle-size measurement by the ASME Power Test Code [4]. A dust sample of about 10 grams is required. Calibration of each specific unit is a necessity.

Other Analytical Methods

Aerosols that are collected on filters or precipitator tubes can be removed for chemical, petrographic, spectrographic, or X-ray analysis. Thus a

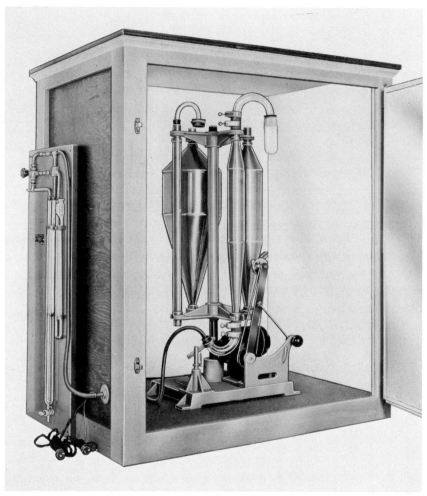

FIGURE 4.13.
Roller particle-size analyzer. (*Courtesy American Instrument Co.*)

qualitative and quantitative analysis may be made. X-ray diffraction methods will indicate crystalline materials. Spectrography will show the elements present in the sample. A wide variety of chemical methods are available for determination of specific components.

Optical Methods of Measuring Aerosols

Both visual and photometric methods have been used to determine the concentration of aerosols in the atmosphere. Actually, the Ringelmann chart described in Chapter 3 is a visual method of estimating the concentration of black particulate matter in a smoke plume.

For the estimation of the number of small particles in the atmosphere, or even in a dusty room, the forward-scattering Tyndallometer [36] is extremely sensitive. Several commercial modifications of the instrument are available.

Another useful device is the photoelectric particle counter described by O'Konski and Doyle [29]. The counter has ten channels so that it will record the size distribution in ten intervals, as well as the total number of particles and the number of particles larger than the largest falling into a presented channel. In order to avoid having more than one particle in the scattering beam at a time, the ambient air is diluted before passing through the counter.

Another type of optical instrument is the visibility meter. These are photometers which measure the attenuation of a light beam by scattering and absorption by aerosols in the path of the beam.

All instruments based on light-scattering and absorption can be adapted to continuous observations.

Several aerosol measuring instruments based on light scattering techniques are available commercially [2]. Other promising techniques for automatic continuous monitoring of particulate emissions include beta radiation attenuation and a piezoelectric microbalance technique [34].

CONTROL OF AEROSOL EMISSIONS

Three basic means of controlling or abating emissions that may contaminate the air are recognized: reduction at the source, dilution, and allocation of air space. By far the most effective method is reduction at the source. This involves two approaches: elimination of the effluent by process control, or treatment of the effluent to remove or alter potential pollutants. If air pollution problems are properly considered when an industrial unit or other source of pollution is designed and built, real economy can be effected. But in most instances air pollution control is an afterthought and means must be devised to treat noxious effluents. Even in the best-planned industrial plant, some potential air pollutants will be evolved, so here, too, the effluent must be treated.

The following discussion will deal only with the removal of particulate matter from gas streams. This is an engineering problem. To attack this problem, certain basic data must be available:

1. The volume of gas to be treated and its variation with time.
2. Temperature and pressure of the gas stream.

3. The nature and concentration of the particulate matter to be removed.

4. The nature of the gas phase (for solubility and corrosive effects).

5. The desired condition of the treated effluent.

With the information available, the engineer is then in a position to select a method for removal of particulate matter from the gas stream and to design equipment to do the job. Actual design should be done by engineers qualified in the particular specialty. Sometimes, suitable basic designs are available and need only be modified to meet a specific application. This is normally done by the equipment manufacturer.

A list of common types of collection equipment for solid and liquid aerosols is shown in Table 4.10 [16]. Usually the choice of a unit for a specific application is fairly obvious, or limited to two or three alternates. A final selection can be made by a detailed design and economic analysis of the alternate propositions. Approximate costs are shown in Table 4.11.

Lapple [19] suggests the following guide for preliminary selection of equipment:

> In over 80% of the dust collection problems, the solution will take the following form:
>
> 1. If the dust concentration is under 10 grains/1000 cu ft, a unit of the air filter class will be applicable.
>
> 2. If the dust concentration is over 1 grain/cu ft, the following will be most applicable in the order specified: (*a*) cyclone separator; (*b*) cloth collector; (*c*) scrubber.
>
> Applications where the dust concentration is between 10 grains/100 cu ft and 1 grain/cu ft are not too frequent, although, when they do occur, may be the most troublesome to handle because they are usually associated with very small particle size and large gas volumes.
>
> The selection of equipment for removal of liquid particles can be resolved to an even greater extent:
>
> 1. If the concentration is under 1 grain/cu ft, the particles are probably present as a mist (i.e., a "condensed" dispersoid with particle size under 10 microns). The equipment applicable in this case will probably be a packed bed, a scrubber, or an electrical precipitator, with inertial separators a possibility if only a mediocre collection efficiency is required.
>
> 2. If the concentration is over 1 grain/cu ft, the particles are probably present as a spray (i.e., a "mechanical" dispersoid with a particle size predominantly over 50 microns). An inertial-type separator or a coarse packed bed will usually be adequate for this case.

Brief discussions of the theoretical aspects of various types of collectors, their usual application, and pertinent performance data follow.

Settling Chambers

The simplest type of dust collection equipment is the settling chamber. When the velocity of a dust-laden gas stream is reduced sufficiently, some of the particulate matter will settle out. Thus, some of the suspended particles in a gas stream in a pipe or duct may be removed by simply enlarging the duct so that gas velocity is reduced. If a series of shelves or plates is inserted in the chamber, dispersed particles will not have to settle very far until they are removed from the gas stream. Of course, gas velocity must be sufficiently low (less than 10 fps) to prevent re-entrainment of the settled particles; less than 1 fps for best results.

Settling chambers are sometimes used in the process industries, particularly the food and metallurgical industries, as a first step in dust and fume recovery. Because the gas velocity is low, pressure drop is also low (0.1 to 0.5 in. water) and consists mostly of entrance and exit losses. Only comparatively large particles (certainly not less than 10 μ if very dense and 40 μ if of low density) may be collected, so some other type of collector must follow the chamber if a high degree of cleanup is required.

Cyclone Separators

This type of separator is the most common of a general group of separators that are classified as centrifugal or inertial separators. The operation of all devices in this class depends upon the tendency of suspended particles to move in a straight line when the direction of the gas stream is altered.

The simple cyclone collector (see Fig. 4.9) consists of a cylinder with a tangential inlet and an inverted cone attached to the base. Gas enters the cyclone through the tangential inlet, which imparts a whirling motion to the gas. Suspended particles are thrown toward the wall on which they collect and slide down into the conical collector. Near the bottom of the cone, the gas turns abruptly upward and forms an inner spiral, which leaves through the pipe or duct extending into the center of the cyclone body.

Many different designs of cyclone separators are available commercially. Some have vanes located in the inlet to impart a spiral motion to the gases; others have helical plates in the cyclone body to guide the gases through the collector in specific channels.

One widely used modification of the cyclone is the Multiclone (see Fig. 4.14). This device is a nest of individual cyclones in parallel, with a single header and a single dust hopper. The individual cylinder diameter varies from 6 to 24 in. Thus, the advantage of the short path for the "settling"

TABLE 4.10

Collection Equipment—Solid and Liquid Aerosols*

Type	Dust characteristics			Pressure drop, in. H_2O	Advantages	Disadvantages	% efficiency (wt. basis)
	Type	Particle size, μ	Sp. gr.				
CENTRIFUGAL COLLECTORS							
Simple cyclone	Wood dust	50–1000	0.4–0.7	0.5–2.0	Simple in construction	Low efficiency	70–90
	Grain dust	10–200	0.9–1.1				60–80
	Mineral dust	10–500	2.0–3.0				70–90
	Pulverized chemicals	10–500	1.5–3.0				70–90
High-efficiency cyclone	Catalyst dust	2–80	1.5–3.5	2.0–6.0	Relatively high efficiency	Subject to abrasion damage	65–80
	Fly ash	0.1–100	0.4–1.5				50–70
	Other fine dust	5–200	1.0–3.0				85–98
Impeller	Foundry dust	10–300	2.5–4.0	Acts as own fan	Low space requirement	Impeller abrasion, causing unbalance	70–90
ELECTROSTATIC PRECIPITATORS							
Single-stage	Gray iron Cupola fume	0.5–50	3–6	0.25–0.5	High efficiency under severe conditions	High initial cost, operating difficulties	90–97
	Electric steel Furnace fume	0.1–20	5–7				90–97
	Open hearth steel† Furnace fume	0.1–3	5–7				96–99
Two-stage	Catalyst dust	2–80	1.5–3.5	0.25–0.5	High efficiency for low dust loading, safe in operation	Limited in use	85–98
	Oil mist	10–400	~1				85–99
	Air conditioning	0.2–10	—	—			95–99

				CLOTH FILTERS			
Tubular	Metallurgical fume Nonferrous	0.03–1.0	~5	0.5–6.0	High efficiency over wide particle size range	Caking from moisture	98–99.5
	Ferrous	0.10–50.0	3–6				97–99.5
Screen or frame	Ceramic dusts	1–50	1–3	0.5–4.0	Somewhat self-cleaning	Higher stresses on filter mediums	95–99.0
	Metallurgical fume	0.1–50	3–6				94–97
Reverse flow (standard cloths)	Same as screen type shown above			0.5–3.0	Higher dust loadings possible	—	—
Reverse jet (felt mediums)	Carbon black	0.1–10	1.5	1.0–6.0	High filter ratios possible	Bag wear	99.5
	Flour dust	5–100	0.9				99.9
				WET COLLECTORS			
Spray chamber	Rock dust	40–500	2–3	0.5–1.0	Low pressure drop	High nozzle pressure required for good collection	60–75
	Asphalt mist	10–400	~1				70–80
	Acid mist	20–500	1.1–1.3				70–90
Inertial	Al and Mg Grinding dust	50–1000	1.7–2.8	2.0–4.0	No nozzle maintenance	Higher pressure drop	80–95
Centrifugal spray	Foundry dust	10–300	2.5–4.0	1.0–4.0	Combined scrubbing and centrifugal action	Abrasion	70–90
	Rock and sand Dust	20–500	2–3				75–95
Venturi scrubber	Acid and caustic mist	20–500	1.1–1.3	10.0–15.0	High efficiency, low water rate	High power consumption	75–95
	Sulfuric acid	2–10	~1.5				85–95
	Mist from concentrator Chemical fume	0.1–50	1.5–3.5				60–85

* Ref. 16.
† Cold metal furnaces.

TABLE 4.11

Cost of Dust Collectors (100,000 cfm)

Type	Equipment (¢/cfm)	Installation (add to equip)	Operating (¢/cfm/yr)	Maintenance (¢/cfm/yr)
Multiclones	12–18	50%	10–12	1.5
Scrubbers	18–30	100–200%	30–100	4
Bag filters	50–90	75%	20–40	5
Electrostatic precipitators	65–95	70%	20–30	2

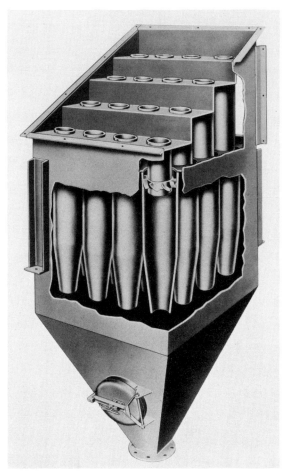

FIGURE 4.14.

Multiclone. (*Courtesy Western Precipitation Corp.*)

particle, inherent in cyclones of small diameter, is provided in a single unit. Several collectors of this type are available commercially.

Cyclone collectors are commonly used to separate dust from disintegration operations, such as rock crushing, ore handling, woodworking, and sand pulverizing. They are also widely used to collect dust from dryers in the chemical and food industries, in the recovery of catalyst dusts in the petroleum industry, and in the reduction of fly-ash emissions.

Commercial units handle gas volumes ranging from 30 to 25,000 cfm at very wide temperature ranges. Efficiencies vary from 50 to 90%, depending on particle size and density. They are most generally adapted to remove particles in the 10 to 200 μ range. High efficiency cyclones can remove particles down to 2 μ efficiently. The lower limit for dust concentration in the feed gases is about 1 grain per cu ft. Pressure drop is low (0.5 to 2.0 in. water for simple cyclones and 2.0 to 6.0 in. water for high efficiency cyclones). Maintenance requirements are nominal, although abrasion is a problem where hard particles are being collected.

Other devices based on applying inertial force to the separation of particles from a gas stream are mechanical centrifugal separators and impingement separators. In *mechanical centrifugal separators,* rotational force is applied to the gas by a fan or impeller, rather than by motion of the gas alone. The Roto-Clone (see Fig. 4.15) is a typical commercial unit. The blades of the fan are specially shaped to direct separated dust into an annular slot leading to a collection hopper. The clean gas continues to the scroll. The chief advantage of this type of equipment over simple cyclones is compactness with attendant lower space requirements. Efficiencies are comparable to high-efficiency cyclones, but power requirements are higher. If solids accumulate on the rotor, the unit must be cleaned.

Impingement separators depend on inertial deposition of particles when a gas stream meets an obstruction. The simplest form is a baffled chamber similar to the entrainment separators used on evaporators and distillation equipment. Others contain slotted plates, staggered channels, and similar mechanical obstructions. Packed beds of granular solids or metal ribbon serve the same purpose. These separators are most commonly used to separate mists from gases.

Filters

Filtration is one of the oldest methods of removing particulate matter from gases. Two general types of filters are recognized: fibrous or deep-bed filters, and cloth filters. In the *deep-bed filter,* a fibrous medium acts as the separator and the collection takes place in the interstices of the bed. Dry fibrous filters, mats of wool, asbestos, cellulose, glass, or iron fibers are extensively used in air-conditioning units, in hot-air furnaces, and as aftercleaners fol-

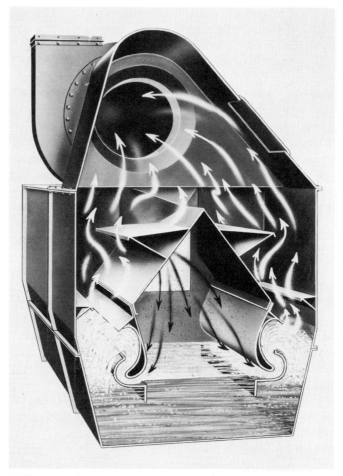

FIGURE 4.15.
Type N Roto-Clone. (*Courtesy American Air Filter Co.*)

lowing other collection devices. They are most useful for light dust loads (0.0002 to 0.01 grain per cu ft). Air velocities must be kept low; pressure drop increases as the entrained dust layer builds up. Because of the requirement of low pressure drop (not over ⅜ in. water), efficiencies are seldom greater than 50%.

High efficiency dry fibrous filters have been developed for special applications, such as the removal of radioactive or toxic particles or the cleaning of air in plants manufacturing photographic film. Loosely packed pads of glass fibers have been found particularly applicable for such cases. A filter bed of glass fibers used at Hanford Atomic Products Operation had an efficiency of 99.99% for submicron particles of radioactive dust at a pressure drop of 4 in. of water [6].

The efficiency of fibrous filters may be improved by coating the fibers with a viscous liquid, such as a high flash point, low volatile oil. The resulting unit is called a *viscous filter*. A common filter of this type is the air cleaner on the modern automobile. Automatic viscous filters (see Fig. 4.16) are well adapted for handling large quantities of atmospheric air containing high concentrations of dust (above 2 grains per 1000 cu ft). The fiber is formed into an endless belt, which continually moves through an oil bath at the bottom of the housing. Air velocities of the order of 300 to 500 fpm may be handled.

Cloth filters are used in the form of tubular bags or as cloth envelopes pulled over a wire screen frame like a pillowcase. The tubular bags are a large-scale version of the common dust bag used on household vacuum cleaners. A baghouse or bag filter consists of numerous vertical bags, 6 to 30 in. in diameter and 6 to 30 ft long (see Fig. 4.17). They are suspended so that their upper ends are closed; the open lower ends are attached to an inlet manifold, which also serves as a receiving hopper for the dust. The dust-laden gas flows into the bottom of the bags and deposits particles on the inside of the bags as it passes through the tubes. Periodically, the tubes are shaken mechanically to dislodge the dust and cause it to fall into the collecting hopper. Some bag filters have manifolds both at the top and bottom, in which case the bags are open on both ends.

Leaf- or envelope-type filters are also multiple units housed in a metal shell. Here, the gases pass from the outside of the envelope into the frame unit and thus deposit dust on the outside of the cloth. The dust is removed by either a shaking or rapping device. In both tubular- and screen-type filters, the filter is operated for four or five hours; the gas flow is then turned off and the filters are shaken for five or ten minutes. A damper arrangement may allow air to be drawn through the filter in the reverse direction during the cleaning operation. In continuous equipment, one section of the filter unit is cleaned while the other sections are operating normally.

Another type of bag filter is the self-cleaning, reverse jet filter (see Fig. 4.18). A slotted ring, to which compressed air is connected, travels slowly up and down the outside of the bags and continuously dislodges dust from the inside. Gas flows as high as 40 fpm may be used. This is 5 to 10 times that used on most cloth filters.

A wide variety of filter cloths are available commercially. Cotton fabrics are used extensively where gas temperature is below 180°F and acid gases are not present. Wool fabrics, which are more resistant to acids than cotton, are used in the collection of metallurgical fumes and for fine and abrasive dusts such as cement and diatomaceous earth. Many synthetic filter fabrics,

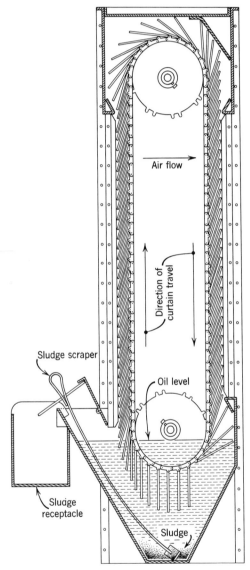

FIGURE 4.16.
Automatic viscous filter. (*Courtesy American Air Filter Co.*)

FIGURE 4.17.
Multi-bag filter.

e.g., Orlon, nylon, and Vinyon, are used in special applications where higher priced cloths can be used.

The greatest problem inherent in cloth filters is rupture of the cloth, which results from shaking. It is often difficult to locate ruptures, and when they are found the maintenance time is often excessive.

Cloth collectors are used for the high efficiency removal of solid particles when an appreciable fraction has a particle size smaller than 10 μ. Efficiency is usually quite high (up to 99+%), but flow rates must be kept low (about 3 cfm per sq ft cloth area). Pressure drop varies from 1 to 6 in. of water. Considerable performance data on cloth filters have been published (see Ref. 30).

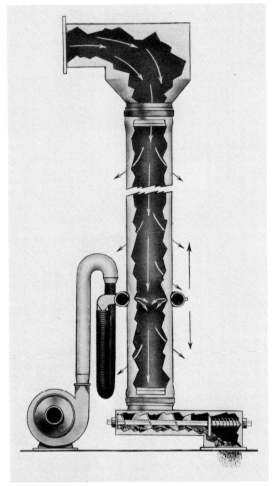

FIGURE 4.18.

Schematic diagram of reverse air jet filter. (*Courtesy Carter Day Co.*)

The most extensive use of cloth filters is in the metallurgical industry. They are also widely used in the food and chemical process industries in connection with grinding and drying operations.

Large, modern baghouses (see Fig. 4.19) such as are used for handling hot gases from cement or lime kilns or other calcining operations are a specialized adaption of the cloth filter. The filter media, usually silicone-treated glass-fiber fabric, handles gases at sustained temperatures up to 500°F and will withstand surges up to 550°F. A graphite-coated fiberglass fabric is also becoming popular. Mechanical shakers are seldom used. Periodic reversal of the air flow allows dirty bags to collapse and the collected filter cake to fall into the hopper. The high dew point of kiln gases (up to 180°F) makes it in many cases necessary for the baghouse to be closed, welded, and fully insulated to prevent corrosion of metal parts and shortened bag life. Below 135°F, an open baghouse, i.e., one which allows infiltration of outside air into the housing around the bags, may be used. Internal partitions may be required in closed baghouses to allow isolation of a com-

FIGURE 4.19.
Cutaway model of fabric filter (baghouse). (*Courtesy of Joy Manufacturing Co.*)

partment for maintenance. In an open baghouse, maintenance men may enter the unit during operation to tie off or replace a ruptured bag. Generally, the welded, completely insulated baghouse costs about twice as much as the open type per square foot of filter cloth.

Besides the limitation of handling wet gases from which moisture may condense in the baghouse, there are also limitations in handling dusts which may abrade, corrode, or blind the cloths. Baghouse problems usually arise from misapplication where these factors were not adequately considered.

Wet Collectors

Scrubbers and washers make use of a liquid medium to increase the size of the aerosol particle to facilitate its removal from the gas stream. The liquid also collects particles by impact and continuously removes the collected material from the impact surfaces. Many types of equipment are available to disperse the liquid phase for good contact between the liquid and the aerosol particle. Wet collectors are often divided into four classes: spray chambers, atomizing scrubbers, deflector washers, and mechanical scrubbers. Other classifications have been proposed, but this one is adequate.

A *spray chamber* is the simplest type of gas scrubber. Gas is passed countercurrent to falling drops of liquid (usually water) from a bank of spray nozzles. Only moderate contact between phases is effected, so units of this type are used only for removing coarse dusts when high efficiency is not necessary. A typical application is the removal of fly ash and cinders from flue gases.

Various means have been used to increase contact between the liquid droplets and the aerosol particles. One widely used unit is the Pease-Anthony cyclone scrubber (see Fig. 4.20). Gas enters the unit tangentially at the bottom and cuts through jets of fine spray from a centered spray manifold. The theory of this operation and typical performance data have been published by Kleinschmidt and Anthony [17]. Table 4.12 is taken from their article. The scrubber has been used commercially on gas volumes from 500 to 175,000 cfm in a single unit. For high efficiencies, dust particle size should be at least 1 μ.

In *atomizing scrubbers,* water is introduced at the throat of a venturi through which the aerosol is passing at high velocity. The theoretical and practical aspects of one unit of this type (Pease-Anthony venturi scrubber; see Fig. 4.21) have been published [14]. Units of this type remove particles in the 0.1 to 1.5 μ range with efficiencies of 90% or more. Pressure drop is high (13 to 20 in. water). Venturi scrubbers are used for removing mists and dusts from gases from Kraft mill furnaces (salt-cake fume), various metallurgical furnaces, and sulfuric acid concentrators. They are also used for removal of coarser particles from flue gases and blast-furnace gases, in which cases lower velocities and water rates may be used with consequent lower pressure losses.

Several types of *deflector washers* are on the market in which deflector plates are used to catch and to help disperse the spray droplets. A typical unit is shown in Fig. 4.15. These collectors are usually designed to remove particles larger than 1 to 5 μ.

Mechanical scrubbers utilize rotating blades or disks to disperse the liquid phase.

Several disadvantages are inherent in all wet collectors. Corrosion and caking both offer problems.

One must also be wary of the deposition of organics in the scrubber and

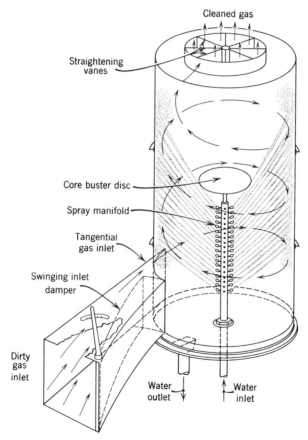

FIGURE 4.20.

Pease-Anthony cyclonic-spray scrubber.

TABLE 4.12

Performance Data of the P-A Cyclonic Scrubber; Dust

Source of gas	Dust	Particle size range, μ	Dust loadings, grains/cu ft		Removal efficiency, %
			Inlet	Exit	
Boiler flue gas	Fly ash from pulverized coal	2–5+	0.49–2.58	0.02–0.046	88–98.8
Blast furnace (iron)	Iron ore, coke	0.5–20	3–24	0.03–0.08	99
Lime kiln (kraft mud)	Lime	1–25	7.7	0.25	97
Lime kiln (raw stone)	Lime	2–40	9.2	0.08	99
Reverberatory lead furnace	Lead compounds	0.5–2+	0.5–2	0.023–0.04	95–98
Rotary dryer	Ammonium nitrate	Large but unstable	. . .	. . .	99+

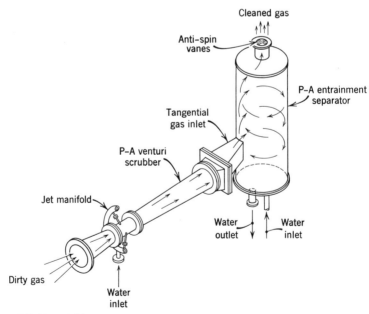

FIGURE 4.21.

Pease-Anthony venturi scrubber.

subsequent growth of micro-organisms. Complete blockage of scrubbers and even total collapse caused by heavy growths are not unknown. Mist often accompanies the gaseous effluents from scrubbers, so "mist eliminators" may be required.

The chief advantage of wet collectors or scrubbers is low capital cost, but this may often be overweighed by high energy consumption. It has been shown [33] that, regardless of the type of wet collector used, particle removal efficiency is dependent only upon the power per unit volumetric gas flow rate that is dissipated in the gas-liquid contact process. Thus high efficiency requires high power expenditure.

Electrostatic Precipitators

When gas containing an aerosol or dispersed phase is passed between two electrodes that are electrically insulated from each other and between which there is a considerable difference in electric potential, aerosol particles precipitate on the low-potential electrode. The high-voltage electrode usually has a small cross section and some curvature, e.g., a wire. The other electrode may be a plate or a surface of only slight curvature. The high voltage on the electrode of small cross section ionizes the gas and aerosol particles, which are then attracted to the larger collecting electrode. When the aerosols reach the collecting electrode they precipitate.

Basically, then, an electrostatic precipitator has four principal parts: (1) a source of high voltage, (2) high voltage ionizing electrodes and collecting electrodes, (3) a means for disposal of the collected material, and (4) an outer housing to form an enclosure around the electrode. A diagram of a simple horizontal flow precipitator is shown in Fig. 4.22. This is a single-stage precipitator in which ionization and precipitation take place simultaneously. In two-stage precipitators, the pre-ionizing section is separated from a second nonionizing collecting section. Usually, the two-stage precipitator is used for lightly loaded gases and the single-stage, for more heavily loaded industrial gas streams.

There are also two general types of precipitators, depending on physical arrangement of the electrodes. In pipe-type precipitators, the high-voltage ionizing electrodes hang axially in collecting electrode pipes. Collected material is removed from the inside of the pipes by a continuous flow of liquid down the pipes, or by periodic rapping of the pipes. The plate-type precipitator may have either a horizontal or vertical gas flow arrangement. The collecting electrodes are in the form of curtains, which act as ducts with high-voltage electrodes equidistantly located in the ducts between the collecting electrode curtains. A cutaway view of a plate-type precipitator is shown in Fig. 4.23. The critical design factors of electrostatic precipitators have been discussed in detail by White [44].

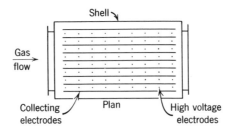

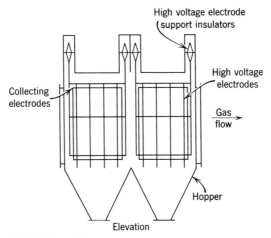

FIGURE 4.22.
Diagram of horizontal-flow precipitator. (*Courtesy Western Precipitation Corp.*)

The important applications of electrostatic precipitators are:

Cement Plants:
   1. Cleaning kiln gases.
Paper Mills:
   1. Recovering salt cake from the flue gases of Kraft mill recovery boilers.
Steel Plants:
   1. Cleaning blast furnace gases to permit use as a fuel.
   2. Removing tars from coke oven gases.
   3. Cleaning open hearth gases.
Utility Plants:
   1. Collection of fly ash.
Oil Industry:
   1. Collection of catalyst in the fluid catalyst plants.

Carbon Black Industry:
  1. Agglomeration and collection of furnace carbon black.
Nonferrous Metals Industry:
  1. Recovering valuable material from flue gases.
  2. Collecting acid mist.
  3. Cleaning gas streams to acid plants.
Chemical Industry:
  1. Collection of sulfuric and phosphoric acid mist.
  2. Cleaning various types of gas, such as hydrogen, CO_2, and SO_2.
  3. Removing the dust from elemental phosphorus in the vapor state.

Commercial precipitators have been built in a wide range of sizes, from those handling as little as 100 or 200 cu ft gas per minute to at least one

FIGURE 4.23.
Cutaway view of a plate-type precipitator. (*Courtesy Western Precipitation Corp.*)

installation that will handle 4,000,000 cfm. Sometimes the electrostatic precipitator is used in conjunction with simpler devices, such as the cyclone collector.

Generally, the collection efficiency of the electrostatic precipitator is high, approaching 100%. Many installations operate at 98 and 99% efficiency. Of course some materials ionize more readily than others and thus are more adapted to removal by electrostatic precipitation. Acid mists and catalyst recovery units often have efficiencies in excess of 99%; carbon black, because of its agglomerating tendency, has a normal collection efficiency of less than 35%. However, by proper combination of an electrostatic precipitator with a cyclonic collector, high efficiencies may be realized in collecting carbon black.

Electrostatic precipitation is best adapted to particles with an electrical resistivity below 10^{10} ohm-cm. At higher resistivities, particles are too difficult to charge; if the resistivity is too low ($<10^5$), particles accept a charge easily but dissipate it so quickly at the collector electrode that the particles are re-entrained in the gas stream. At times, particles of high resistivity may be conditioned with moisture to bring them into the acceptable range.

Precipitators may be operated above, below, and at atmospheric pressure, and under a wide range of gas temperatures. Because flow rates are low (2–15 fps), pressure drop across the unit is low (0.10 to 0.50 in. of water). The dust loading that can be handled in dry precipitators may be as high as 25 grains per cu ft; at high dust loadings, mechanical separators should precede the precipitator. At very low dust loadings, 0.2 to 0.4 grain per cu ft, wet precipitation is advisable.

Operating conditions of electrostatic precipitators are indicated in Table 4.13 [44].

TABLE 4.13

Range of Precipitator Operating Conditions*

Gas flow	1 to over 3,000,000 cfm
Gas temperature	to 1200°F
Gas pressure	to 150 psi
Gas velocity	3 to 15 fps for most applications; 25 to 50 fps for a few special air cleaning units
Draft loss	0.1 to 0.5 in w.g.
Particle size	0.1 to 200+ μ
Particle concentration	0.0001 to 100 grains/ft³
Particle composition	no basic limit; solid, liquid, corrosive chemicals
Treatment time	1 to 10 sec for most applications; as low as 0.1 for a few special cases
Efficiency	most applications 80 to 99%; some 99.9+%

* From H. J. White [44].

There has been a tendency in the United States to overextend the application of electrostatic precipitators or at least to overestimate their applicability to air pollution problems. Variations in particle resistivity caused by raw material or process changes are sometimes difficult to accommodate. Inlet ducts of too great a cross-section often lead to stratification and overloading of the precipitator. But when the unit is designed for a specific use by an expert, it does an excellent job. Most manufacturers now insist on model studies to determine suitable design to prevent stratification and to maintain laminar flow through the precipitator.

Ultrasonic Agglomerators

A development of considerable interest is the use of high-frequency sound waves to clean industrial gases. However, only a few commercial installations have been made. A typical sonic collection system has been described by Neumann [28].

THE FLY-ASH PROBLEM.

The control of fly-ash emissions is often considered an entirely different problem from the control of dusts, mists, and fumes. Basically, it is not different, and the same principles apply that govern the control of other types of particulate matter in stack emissions.

Fly ash is a problem when any ash-containing fuel is burned; it is particularly troublesome in the case of pulverized coal. Everything else being equal, the quantity of fly ash emitted varies with the ash content of the coal. For this reason, coal cleaning or washing for ash reduction may be considered a means of air pollution abatement. Even though the ash of coal used in the United States has been reduced an average of 50% since 1900, the quantities of fly ash produced at central power stations is still great. It is not unusual for a utility system to produce a million tons or more per year of fly ash. Accordingly, disposal is a problem. Considerable research has been carried out to find uses for the material. Small quantities have been used as a constituent of Portland cement mixes.

The ease of recovery of fly ash depends on many factors that affect the shape, size, density, and composition of the ash. Type of coal, boiler design, and method of feeding coal to the furnace all affect the nature of the particle. Size of the particles varies from 0.01 μ to over 300 μ. Some are hollow spheres; others vary from slivers to sponge- or lace-like particles. Carbon content may be as high as 40%. In some pulverized-coal installations, 85% of the ash may appear as fly ash; in others, loading may be as low as 25%. Electrical properties vary markedly and, accordingly, affect the efficiency of electrostatic precipitators. White [44] has discussed many of these factors in detail.

Recovery is usually effected by cyclonic collectors or electrostatic precipitators, or a combination of the two. Selection of proper equipment presents many difficult engineering problems of both technical and economic nature.

The trend toward extremely large steam-electrical power plants has led to a demand for a very high-efficiency fly-ash collector. The new ASME "Recommended Guide for the Control of Dust Emission—Combustion for Indirect Heat Exchangers" [5] recommends maximum fly-ash emissions of 0.8 lb per million Btu input, which for high-ash coal (ash > 15%) would mean 95% collector efficiency even with a high stack. Much more restrictive performance standards were published by EPA in 1971 (see Table 7.1).

The cost of an electrostatic precipitator increases markedly as collection efficiency increases (see Table 4.14). Usually an economic balance must be drawn between collector efficiency and stack height.

TABLE 4.14

Relative Cost of Electrostatic Precipitators for Different Collector Efficiencies

Collection efficiency (%)	Relative installed cost for electrostatic precipitators
90	1
95	1.3
97	1.5
98	1.7
99	2.0
99.75	2.6
99.9	4.7

The problem has been complicated somewhat by use of low-sulfur fuels. The presence of sulfur oxides in boiler effluent gases is an aid to collector efficiency. As the sulfur oxide content is decreased, collector efficiency decreases, and precipitator cost is increased. Studies made by the Tennessee Valley Authority indicate that precipitator efficiency drops markedly when coal contains less than 2% sulfur [32].

One recently installed system locates the electrostatic precipitator ahead of, instead of after, the air heater on a large steam generator [37]. The precipitator is rated at 99% efficiency and handles gases at 600° to 700°F, in which temperature range resistivity of the fly-ash particles is favorable for collection. The greater volume of gas which must be handled requires a larger precipitator and this is more expensive. Long-term performance evaluation will be required to determine suitability of the unit.

HAZARDOUS PARTICLES

The Clean Air Act of 1970 directed the Environmental Protection Agency to publish a list of hazardous air pollutants which in its judgment might cause or contribute to an increase in mortality or an increase in serious, irreversible, or incapacitating illnesses; and to establish national emission standards for source categories known to emit these hazardous pollutants.

Proposed emission standards were published for asbestos, beryllium, and mercury on December 7, 1971 (36 F.R. 23239). Asbestos standards covered mines, mills, and other facilities handling asbestos. Source operations are required to install fabric filters or their equivalent; visible emissions are prohibited.

Beryllium emissions from machine shops, ceramic plants, propellant plants, foundries, extraction plants, and incinerators were restricted to 10 grams in a 24-hr day and to an out-plant air concentration of $0.01/\mu g/m^3$ (30-day average). Special standards were published for beryllium-rocket motor firing.

Mercury emissions were restricted to 5 lb per 24-hr from both ore-processing facilities and mercury cell chlor-alkali plants.

Standards for other hazardous substances are being developed.

RADIOACTIVE PARTICLES

Rapid developments in the field of atomic energy have brought a new and very serious type of pollution to the atmosphere: radioactive particles. Some of the radioactive materials formed in nuclear reactions are the most potent poisons known, from several million to several billion times as toxic as chlorine. Moreover, these materials cannot be detected by taste, smell, or the other human senses. Further, since there is no way of neutralizing radioactivity (except by time), it is imperative that such material be kept out of the air.

The half-lives of some common radionuclides are shown on Table 4.15.

A United States Weather Bureau document [43] lists the possible sources of radioactive emissions to the atmosphere as follows:

1. Mining and handling uranium ores.
2. Chemical production of brown oxide of uranium (UO_2).
3. Machining radioactive and toxic metals (e.g., uranium and beryllium).
4. Atomic laboratories.
5. Particle accelerators.

TABLE 4.15

Half-Lives of Important Radionuclides Produced by Nuclear Reactions

Radionuclide	Half-life	Radionuclide	Half-life
^{137}I	30 years	^{132}I	8.05 days
^{137}Cs	30 years	^{133}Xe	5.27 days
^{90}Sr	28 years	^{133}I	20 hours
^{3}H	12 years	^{135}Xe	9.2 hours
^{85}Kr	10.4 years	^{88}Kr	2.8 hours
^{144}Ce	285 days	^{87}Kr	1.3 hours
^{95}Zr	65 days	^{41}A	112 min.
^{89}Sr	50 days	^{138}Xe	17 min.
^{140}Ba	12.8 days	^{135m}Xe	15 min.
^{131}I	8.1 days		

6. Nuclear reactors.
7. Chemical processing plants for reactor fuels.
8. Waste disposal.

The tremendous growth of the nuclear power industry in the last several years has exceeded previous forecasts considerably. It is expected that by 1973 over 50,000 MWe (megawatts, electrical) of nuclear power facilities will be in commercial operation in the United States. It has also been estimated that by 1980, 80% of the annual added electrical-generating facilities will be nuclear. One by-product of this technological change will be less pollution from combustion of fossil fuels. Counteracting this, of course, will be some increase in radioactivity release to the environment.

Fortunately, plants handling radioactive materials are well aware of the hazards involved and go to extreme methods to prevent emission of radio-active material to the atmosphere.

It is general practice in the nuclear industry to treat routine off-gases to reduce radioactive aerosols and gases to a very low level before discharge to the atmosphere, and to provide backup devices to minimize the proba-bility of release in the event of an accident.

The most important likely sources for the emission of radioactive aerosols and gases are uranium fuel-element fabrication plants, nuclear power reactors, and fuel recovery plants. In each case, routine off-gases (if any) and contaminated ventilation air are passed through high-efficiency particulate air (HEPA) filters (see Fig. 4.24) capable of removing at least 99.97% of the 0.3-micron particles in the gases. To prevent clogging of the HEPA filters, roughing filters are usually installed up-stream to remove large particles.

In the United States the test aerosol is prepared by atomization of di-octyl phthalate (DOP) [3]. The United States Atomic Energy Commission has two Quality Assurance Stations (filter testing facilities) for verifying the efficiency of new filters. HEPA filters are delicate and should be shipped, handled, and installed with care. Because they are easily damaged, filters must also be tested in place at regular intervals. If there is any likelihood of the filter being subjected to wet steam or fog they are usually preceded by moisture separators (demisters).

In addition to the release of radionuclides in reactor off-gases and in

FIGURE 4.24.
High-efficiency particulate filter (HEPA). (*Courtesy American Air Filter Co.*)

ventilation air, there is also the possibility of release of massive quantities in case of accident. This possibility is minimized by requiring nuclear power reactors to be fitted with a variety of engineered safeguards and other safety features. Among these are (1) the ready availability of backup facilities in case of the malfunction of a component that could result in the release of fission products, (2) reactor siting and construction to forestall any credible natural or man-made force, e.g., earthquakes, flooding, hurricanes, aircraft or other missiles, etc., and (3) sufficient isolation to protect nearby populations from excessive dosages of radioactivity in case a fission product release penetrates the multiple barriers installed. One important feature of the safeguard system is a pressure resistant containment structure to receive and contain any radioactive gases that might be released in case of accident. Among other equipment in the containment structure are prefilters and HEPA filters to remove radioactive particulates during the post-accident period. Other features of the safety air-cleaning system are discussed in Chapter 5 (see Radioactive Gases).

AIR QUALITY AND EMISSION STANDARDS

Air quality standards for dusts, fumes, and mists are usually concerned with two general categories of particulate matter, i.e., suspended particulate matter and dustfall. In those cases where a specific material is toxic, e.g., beryllium particles or asbestos fiber, specific standards are written.

In reality *dustfall* does not describe the quality of ambient air, so the standard is often directed toward *settleable particulate matter*. In practice the latter is measured as dustfall so the difference is academic.

The standards are nearly always stated in terms of monthly dustfall in tons per square mile per month, or, more meaningfully, in milligrams per square centimeter per month. Accordingly, dustfall values are only very rough estimates of the fallout in a community. Each value is really only that which falls through a specific 15-square-inch opening over a period of 30 days. Theoretically, all of the dust could fall in a period of a few minutes or it could fall uniformly over a full monthly period. The actual mechanism, in practice, falls somewhere between the two extremes. It has been shown [9] that dustfall jars set side by side for a month seldom give values that check within $\pm 15\%$.

Selection of standards for settled dust is highly subjective and appears to be based on a blend of opinion, desire, and what the traffic will bear. Differentiation is usually made between residential and industrial areas on the basis that dust particles fall comparatively close to their source and that nonresidential areas (particularly heavy industrial) have no need for low values nor can they be obtained.

The standards for several communities are shown in Table 4.16. Most values either contain an allowance for background (settled dust from natural sources) or specify that the standards are in addition to background. In times of high winds, background values of both settled and suspended dust can be quite high [1].

TABLE 4.16

Typical Dustfall Standards (30-day basis)

State	Standard
Pennsylvania	1.5 mg/cm² maximum
New York	0.3–0.8 mg/cm²/mo. (50% of values less than shown depending on area of state).
	0.45–1.20 mg/cm²/mo. (85% of values less than shown depending on area of state).
Minnesota	25 tons/mi²/mo. for zoned heavy industrial areas.
	10 tons/mi²/mo. for all other areas.
	(*Note:* In addition to specified background of 5 tons/mi²/mo. in both cases.)
Louisiana	15–35 tons/mi²/mo. maximum, depending on area of state.
Washington	3½–10 g/m²/mo., depending on type of particulate and area of state.

Ambient air quality standards for suspended particulate matter are usually stated in micrograms per cubic meter as measured by a high-volume sampler over a 24-hr period. Some jurisdictions also state the standard in Cohs per 1,000 lineal feet as determined by a tape sampler (see p. 95) but there appears to be little justification for this type of standard [10].

Standards stated in terms of micrograms per cubic meter (24-hr basis) are most commonly stated both as a mean 24-hr value over a period of 3 months, 6 months, or a year, and as a maximum 24-hr value not to be exceeded more than 1.0% of the days in a given period. This type of standard is obviously a goal or a guideline rather than a legal limit in that violations can be determined only after the fact. Nevertheless, the standard adopted by the Environmental Protection Agency (see Chapter 1) is stated as a 24-hr maximum and an annual geometric mean.

Just what the basis of a suspended particulate matter regulation should be is moot inasmuch as there are few sound criteria upon which to base a standard. There have been many attempts to relate particulate matter concentration to visibility but no suitable relationship has been found. The state of California evaded the issue when it adopted standards in 1960 and in 1969 by listing as a standard for suspended particulate matter "(the amount) sufficient to reduce visibility to less than 3 miles when relative humidity is less than 70 per cent." No numbers were listed.

There have also been suggestions that standards be set on the basis of public opinion surveys, in which residents of a community have been asked to decide whether or not the atmosphere on a given day was acceptable or not. The answers were then correlated with the measured concentration of suspended particulates on that day. A variety of numbers has resulted from such polls [10].

Emission standards for particulate matter cover a far greater range of values and units than do air quality standards.

The more common standards are those stated in terms of concentration in effluent gases, discharge rate in weight per unit time, and plume opacity (visual obscuration), and those stated in terms of some design parameter such as stack height or degree of collection.

Effluent concentration standards are usually stated in grains per standard cubic foot, grains per cubic foot at stack conditions, or pounds per 1,000 pounds of effluent gas. Note that the first and last mentioned parameters are independent of temperature. The concentration measured at stack conditions gives credit to the heat content of the effluent, which is justified on the basis that heat adds buoyancy to the plume, carries it higher into the atmosphere and thus gives lower ground-level concentration of particulates. Where concentration per standard cubic foot or per unit weight of gas is specified, care must be taken in defining a standard cubic foot, particularly as to whether the gas is assumed to be saturated with water vapor at standard conditions or bone dry. To date there has been no standardization of the definition either with respect to presence of water vapor or to temperature ($60\,°F$, $70\,°F$, and $32\,°F$ are commonly used) and pressure.

Standards defined in terms of grains per cubic foot or pounds per 1,000 pounds of gases are usually related to technological capability of a given industry rather than a specific effect on a receptor or an air quality standard. The most common general values used are 0.3 and 0.2 grains per standard cubic foot. Occasionally one will find 0.1 gr/scf or 0.4 gr/scf applied to specific industries. The 1967 Illinois standards allowed 0.75 gr/scf for corn wet-milling process dusts under certain conditions, but only 0.1 gr/scf for new blast furnaces. Technological capability was the determining factor in both cases.

When the concentration standard is applied to combustion effluents, the standard cubic foot is usually adjusted to a specified percentage of excess air (often 50%) or to a specific carbon dioxide percentage (often 12%).

Process-Weight Regulations. Standards expressed in terms of weight discharge per unit time are most often related to a process-weight chart which relates allowable emissions in pounds per hour to the weight of material processed in tons per hour. The first regulation of this type and the first process-weight chart was developed by the Los Angeles County Air Pollution Control District in 1948 [21, 22]. The chart (Table 4.17) was developed originally to control emissions from foundries and secondary smelters which met the No. 2 Ringelmann restriction yet caused an unacceptable amount of fallout on nearby property. Thus the regulation was developed to solve a fallout problem. Basis for the regulation in addition to reducing emissions was that the required degree of control should be within the economic reach of industry. A corollary to this principle was that large units could afford more restrictive control than small units. To prevent compliance by switching from solid fuels to oil or gas, it was determined that solid fuel could be included in the weight of material processed, but that liquid or gaseous fuels could not. The degree of control found to be within the economic range of the affected industries ranged from 80% for small units to 90% for the largest plant tested (20,000 lb per hour process weight). The figures developed were extended to 60,000 pounds per hour process weight for which 98% collection efficiency was deemed feasible. Although the regulation was developed specifically for metallurgical operations it was found later to be applicable to other industries in Los Angeles County. It should be noted that there are no large process industries such as cement and lime manufacture in the county.

Several years later the Bay Area APCD developed another process-weight chart (see Table 4.18) to suit the conditions in that jurisdiction. Because industry was much larger there, the Bay Area curve was extended to 6,000,000 lb per hour process weight, whereas the Los Angeles curve stopped at 60,000 lb. The Bay Area regulation also contained a proviso that exempted from the process-weight regulation those industries which could meet a very restrictive effluent concentration.

Several other jurisdictions have adopted the Bay Area process-weight regulation or a modification of it. Because of the different nature of industry in other locations, some exemptions have been adopted along with the process-weight chart. In Illinois, a process-weight table based on the Bay Area chart was the basic regulation for control of particulates from most process operations. A special table (see Table 4.19) was developed for small foundry cupolas and foundry open hearths. Certain other specified industries are controlled by concentration limitations.

The Indiana law also incorporates a process-weight chart similar to the Bay Area chart, a special chart for foundries similar to the Illinois chart, and a special formula for existing cement plants, to wit:

TABLE 4.17

Allowable Particulate Emissions Based on Process Weight
(Los Angeles County APCD, as of 1971)

Process wt/hr (lb)	Maximum weight disch/hr (lb)	Process wt/hr (lb)	Maximum weight disch/hr (lb)
50	.24	3400	5.44
100	.46	3500	5.52
150	.66	3600	5.61
200	.85	3700	5.69
250	1.03	3800	5.77
300	1.20	3900	5.85
350	1.35	4000	5.93
400	1.50	4100	6.01
450	1.63	4200	6.08
500	1.77	4300	6.15
550	1.89	4400	6.22
600	2.01	4500	6.30
650	2.12	4600	6.37
700	2.24	4700	6.45
750	2.34	4800	6.52
800	2.43	4900	6.60
850	2.53	5000	6.67
900	2.62	5500	7.03
950	2.72	6000	7.37
1000	2.80	6500	7.71
1100	2.97	7000	8.05
1200	3.12	7500	8.39
1300	3.26	8000	8.71
1400	3.40	8500	9.03
1500	3.54	9000	9.36
1600	3.66	9500	9.67
1700	3.79	10000	10.0
1800	3.91	11000	10.63
1900	4.03	12000	11.28
2000	4.14	13000	11.89
2100	4.24	14000	12.50
2200	4.34	15000	13.13
2300	4.44	16000	13.74
2400	4.55	17000	14.36
2500	4.64	18000	14.97
2600	4.74	19000	15.58
2700	4.84	20000	16.19
2800	4.92	30000	22.22

TABLE 4.17 (*continued*)

Process wt/hr (lb)	Maximum weight disch/hr (lb)	Process wt/hr (lb)	Maximum weight disch/hr (lb)
2900	5.02	40000	28.3
3000	5.10	50000	34.3
3100	5.18	60000	40.0
3200	5.27	or	
3300	5.36	more	

TABLE 4.18
Allowable Rate of Particulate Emission Based on Process Weight Rate*
(Bay Area APCD, 1961–1970)

Process weight rate lb/hr	Process weight rate tons/hr	Rate of emission, lb/hr	Process weight rate lb/hr	Process weight rate tons/hr	Rate of emission, lb/hr
100	0.05	0.551	14,000	7.00	15.1
200	0.10	0.877	16,000	8.00	16.5
400	0.20	1.40	18,000	9.00	17.9
600	0.30	1.83	20,000	10.	19.2
800	0.40	2.22	30,000	15.	25.2
1,000	0.50	2.58	40,000	20.	30.5
1,500	0.75	3.38	50,000	25.	35.4
2,000	1.00	4.10	60,000	30.	40.0
2,500	1.25	4.76	70,000	35.	41.3
3,000	1.50	5.38	80,000	40.	42.5
3,500	1.75	5.96	90,000	45.	43.6
4,000	2.00	6.52	100,000	50.	44.6
5,000	2.50	7.58	120,000	60.	46.3
6,000	3.00	8.56	140,000	70.	47.8
7,000	3.50	9.49	160,000	80.	49.0
8,000	4.00	10.4	200,000	100.	51.2
9,000	4.50	11.2	1,000,000	500.	69.0
10,000	5.00	12.0	2,000,000	1,000.	77.6
12,000	6.00	13.6	6,000,000	3,000.	92.7

* Interpolation of the data in this table for process weight rates up to 60,000 lb/hr shall be accomplished by use of the equation $E = 4.10\ P^{0.67}$, and interpolation and extrapolation of the data for process weight rates in excess of 60,000 lb/hr shall be accomplished by use of the equation;

$$E = 55.0\ P^{0.11} - 40, \text{ where } E = \text{rate of emission in lb/hr and}$$
$$P = \text{process weight rate in tons/hr.}$$

TABLE 4.19

Allowable Particulate Emissions from Small Foundry Cupolas and Foundry Open Hearths (Illinois)

Process weight rate lbs/hr	Allowable emission lbs/hr
1,000	3.05
2,000	4.70
3,000	6.35
4,000	8.00
5,000	9.58
6,000	11.30
7,000	12.90
8,000	14.30
9,000	15.50
10,000	16.65
12,000	18.70
16,000	21.60
18,000	23.40
20,000	25.10

Existing cement manufacturing operations equipped with electrostatic precipitators, bag filters, or equivalent gas-cleaning devices shall be allowed to discharge concentrations of particulate matter in accordance with $E = 8.6P^{0.67}$ below 30 tons per hour of process weight.

There is also a general exemption for plants where the process weight exceeds 200 tons per hour if the loading in the discharge gases is less than 0.10 pounds per 1,000 pounds of gases.

The basis for process-weight regulations, as with concentration restrictions, is technological capability and local conditions both with respect to air pollution and to the economy. There is no relationship to any air quality standards that may have been adopted. It is apparent that no one regulation or type of regulation is adaptable to all industry. Accordingly, care should be taken in developing local regulations for control of particulate matter. The guiding principle should be protection of the public without undue and unnecessary damage to the economy.

In this light it is difficult to understand a process-weight regulation published as an example by EPA in its instructions to the various states in preparation of their implementation plans (Federal Register v.36, No. 58, 1971). No basis has ever been given for the EPA table, and there are some industrial operations which could not comply even with the most modern control equipment. Further EPA departed from the individual unit basis for which process-weight regulations historically had been developed. For example: if a plant had 3 kilns or 3 furnaces each performing like the other, the process-weight rule has in the past been applied to each unit separately. EPA suggests the table apply to "mass emission limitations on the basis of similar units at a plant—in order to avoid unequal application of this type of limitation to plants with the same total emission potential but different size units." The suggestion is patently arbitrary and discriminates against modern economic-sized units.

Where baghouse collectors are used, no sliding-scale rule is appropriate inasmuch as collection efficiency does not vary with size. EPA has recognized this fact in its performance standards for cement kilns (Title 40 CFR Part 60).

Plume Opacity Restrictions. Although the Ringelmann Chart was developed originally to measure black smoke (see Chapter 3), Los Angeles authorities extended its use to nonsmoke plumes as early as 1945 by means of an equivalent opacity concept. This concept, in which inspectors judge whether a plume obscures vision through the plume to the same extent as a black plume of No. 2 Ringelmann numbers, is undoubtedly a misapplication of the Ringelmann Chart but it has worked fairly well in Los Angeles County. Where it could not be met, as with steam power plants using residual fuel oil, variances were granted. In other areas of California, enforcement has been spotty.

More problems have arisen in other areas with different types of industry when attempts have been made to extend the equivalent opacity concept to dust and other nonsmoke plumes. There can be little doubt that the concept is invalid. Conner and Hodkinson, who have studied the problem to a greater extent than others, say "if the smoke is not black, then evaluation by comparing its luminance with filters or Ringelmann charts becomes unrealistic" [8]. They also point out the lack of correlation between light transmission through plumes and the weight of particulate matter in the plume. As a result, control equipment manufacturers who will guarantee collection efficiencies and exit grain loadings will not make guarantees with respect to plume opacity [45].

In spite of the California precedent and vigorous promotion of equivalent opacity by some agencies, the concept lacks a sound scientific basis. Further it is susceptible to considerable error when measured by sight instead of instrumentally [8].

Design Standards. The most widely used design standards are those directed toward the control of particulate emissions from combustion processes, particularly the combustion of coal. Most regulations are based on the Recommended Guide for the Control of Dust Emission—Combustion for Indirect Heat Exchangers" developed by the American Society of

Mechanical Engineers [5]. By means of the guide one can select a stack height and degree of collection efficiency to limit maximum 24-hour ground level concentration of particulate matter to 17 or 34 micrograms per cubic meter, depending on which of two charts are used. The two concentrations shown are equivalent to 3–15 minute maxima of 100 and 200 micrograms per cubic meter. The more restrictive chart of the two is shown in Fig. 4.25.

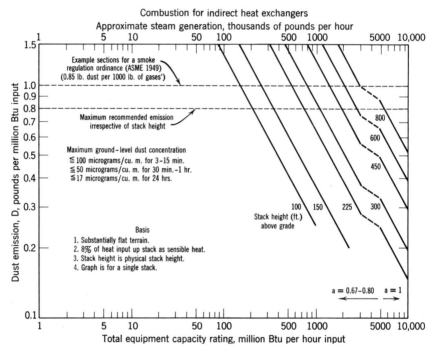

FIGURE 4.25.
ASME Standard No. APS-1, second edition, November 1968 (Figure 2).

The guide is very restrictive, and in some cases will call for precipitator efficiencies above 99%. The guide includes a maximum recommended emission irrespective of stack height of 0.8 pounds per million Btu input. Large units must meet much more restrictive limits than small ones.

A related type of standard is one in which the allowable emission of a process dust is limited to that amount which would not violate ambient air quality standards. The allowable amount may be calculated according to an approved plume dispersion formula or a chart based on the formula. Separate formulas are usually given for dustfall and for suspended particulate matter. States with regulations of this type include Pennsylvania, South

Carolina, Texas, and New Jersey. An appealing aspect of this type of regulation as well as the ASME type is its direct relatability to air quality standards.

Another type of design standard used by some jurisdictions is specification of collector efficiency either for a broad class of operations or for a specific type of equipment. The Indiana regulation, for instance, requires existing petroleum catalytic cracking units to be equipped with systems that will "recover 99.97% or more of the circulating catalyst or total gas-borne particulate." Other regulations would exempt operations using electrostatic precipitators of a stated efficiency from other stated restrictions on emissions of particulate matter.

Regulations for Wind-Blown Dust. In many communities in arid portions of the western United States, and sometimes in other areas, there has been a demand for regulations directed toward control of blowing dust from construction sites, stockpiles, unpaved parking lots and school yards, mining sites, and similar operations. Several regulations have been enacted but whether or not they are enforced is uncertain. Some of the regulations specify a permit from the air pollution control officer who can require the ground to be kept wet or the stockpile covered as a condition of obtaining the permit. Other regulations are more general and require the operator to "take reasonable precautions to prevent particulate matter from becoming airborne." Many of these regulations appear to be window dressing.

REFERENCES

1. Adley, F. E., and W. E. Gill, "Atmospheric Particulate Background in a Rural Environs," *Am. Ind. Hygiene Assoc. J.,* **19,** 271–275 (1958).
2. American Conference of Governmental Industrial Hygienists, "Air Sampling Instruments," pp. B-8-73 to B-8-77, 2nd Edition (1962).
3. American Industrial Hygiene Association, "Air Pollution Manual, Part II," 56–61 (1968).
4. American Society of Mechanical Engineers, "Determining the Properties of Fine Particulate Matter," ASME Power Test Code PTC 28-1965, New York (1965).
5. American Society of Mechanical Engineers, "Recommended Guide for the Control of Dust Emission—Combustion for Indirect Heat Exchangers," ASME Standard No. APS-1, 2nd Edition (1968).
6. Blasewitz, A. G., and B. J. Judson, "Filtration of Radioactive Aerosols by Glass Fibers," *Air Repair,* **4,** No. 4, 223–229 (1955).
7. Cholak, J., L. J. Schafer, D. W. Yaeger, and R. A. Kehoe, "The Nature of the Suspended Matter," *in* "An Aerometric Survey of the Los Angeles Basin,

August-November, 1954," N. A. Renzetti, ed., Air Pollution Foundation (San Marino, Calif.) Rept. 9 (1955).

8. Conner, W. D., and J. R. Hodkinson, "Optical Properties and Visual Effects of Smoke-Stack Plumes," PHS Publication No. 999-AP-30, U.S. Dept. of Health, Education, and Welfare, Cincinnati, Ohio (1967).

9. Fairweather, John H., A. F. Sidlow, and W. L. Faith, "Particle Size Distribution of Settled Dust," *J. Air Poll. Control Assoc.*, **15**, 345–347 (1965).

10. Faith, W. L., "Inert Particulates—Nuisance Effects," *J. Occupational Medicine*, **10**, No. 9, 107–112 (1968).

11. Froula, H., A. F. Bush, and E. S. C. Bowler, "Use of Thermal Precipitator and Electron Microscope for Evaluation of Air-Borne Particles," *Proc. Natl. Air Pollution Symposium, 3rd Symposium, Pasadena, Calif., 1955*, 102–111.

12. Herrick, R. A., "Recommended Standard Method for Continuing Dustfall Survey (APM-1, Revision 1)," *J. Air Poll. Control Assoc.*, **16**, 372–377 (1966).

13. Johnstone, H. F., and D. R. Coughanowr, "Absorption of Sulfur Dioxide from Air and Oxidation in Drops Containing Dissolved Catalysts," *Ind. Eng. Chem.*, **50**, 1169–1172 (1958).

14. Jones, W. P., and A. W. Anthony, Jr., "Pease-Anthony Venturi Scrubbers," in *Air Pollution*, Louis C. McCabe, McGraw-Hill, New York (1952).

15. Kanter, C. V., R. G. Lunche, and A. P. Fudurich, "Techniques of Testing for Air Contaminants from Combustion Sources," *J. Air Poll. Control Assoc.*, **6**, No. 4, 191–199 (1957).

16. Kirk, Raymond E., and Donald F. Othmer, eds., *Encyclopedia of Chemical Technology*, Vol. 12, Interscience Publishers, Inc., New York, pp. 566–567 (1954).

17. Kleinschmidt, R. V., and A. W. Anthony, Jr., "Pease-Anthony Cyclonic Spray Scrubbers," in *Air Pollution*, Louis C. McCabe, McGraw-Hill, New York (1952).

18. Landsberg, H. E., "Climatology and Its Part in Pollution," *Meteorol. Monographs*, **1**, No. 4, 7–8 (1951).

19. Lapple, C. E., "Dust and Mist Collection," in *Air Pollution Abatement Manual*, Manufacturing Chemists' Association, Washington, D.C., Chapter 9 (1951).

20. Mader, P. P., R. D. Macphee, R. T. Lofberg, and G. P. Larson, "Composition of Organic Portion of Atmospheric Aerosols in the Los Angeles Area," *Ind. Eng. Chem.*, **44**, 1352–1355 (1952).

21. McCabe, L. C., "Can One State's Code Solve Another State's Problems?" *Environ. Sci. and Technology*, **4**, 210–213 (1970).

22. McCabe, L. C., A. H. Rose, W. J. Hamming, and F. H. Viets, "Dust and Fume Standards," *Ind. Eng. Chem.*, **41**, 2388–2390 (1949).

23. Munger, H. P., "The Spectrum of Particle Size and Its Relation to Air

Pollution," in *Air Pollution,* Louis McCabe, McGraw-Hill, New York, 160 (1952).

24. National Air Pollution Control Administration, "Air Quality Data from the National Air Sampling Networks, 1966," Durham, N.C. (1968).

25. National Air Pollution Control Administration, "Air Quality Criteria for Particulate Matter," National Air Pollution Control Administration, Washington, D.C. (1968).

26. National Air Pollution Control Administration, "Control Techniques for Particulate Air Pollutants," NAPCA Publ. No. AP-51, pp. 4-66 to 4-76 (1969).

27. National Air Pollution Control Administration, "Control Techniques for Particulate Air Pollutants," pp. 5-4 to 5-11 (1969).

28. Neumann, E. P., C. R. Soderberg, Jr., and A. A. Fowle, "Design, Application, Performance, and Limitations of Sonic Flocculators and Collectors," in *Air Pollution,* Louis McCabe, McGraw-Hill, New York, 373–381 (1952).

29. O'Konski, C. T., and G. J. Doyle, "Light Scattering Studies in Aerosols with a New Counter-Photometer," *Anal. Chem.,* **27,** No. 5, 694–701 (1955).

30. Perry, J. H., "Chemical Engineers' Handbook," Fourth Edition, McGraw-Hill, New York (1963).

31. Prokopovich, N., "Dust-Snow Storm in the Minneapolis-St. Paul Area on March 12, 1954," *Science,* **120,** 230–231 (1954).

32. Reese, J. T., and Joseph Greco, "Experience with Electrostatic Fly-Ash Collection Equipment Serving Steam-Electric Generating Plants," ASME Preprint 67-WA/APC-3, American Society of Mechanical Engineers, New York (1967).

33. Semrau, K. T., "Dust Scrubber Design," *J. Air Poll. Control Assoc.,* **13,** 587–594 (1963).

34. Sem, G. J., J. A. Borgos, and J. G. Olin, "Monitoring Particulate Emissions," *Chem. Eng. Progress,* **67,** No. 10, 83–88 (October 1971).

35. Silverman, L., "Sampling of Industrial Stacks and Effluents for Atmospheric Pollution Control," *Proc. Natl. Air Pollution Symposium, 1st Symposium, Pasadena, Calif., 1949,* 55–60.

36. Sinclair, David, and V. K. LaMer, "Light Scattering as a Measure of Particle Size in Aerosols," *Chem. Revs.,* **44,** 245–267 (1949).

37. Smith, M. C., and A. A. Salerno, "Engineering for Low Sulfur Fuels," ASME Preprint 68-WA/APC-1, American Society of Mechanical Engineers (1968).

38. U.S. Dept. of Health, Education, and Welfare (Public Health Service, Division of Air Pollution, Cincinnati, Ohio), "Air Pollution Measurements of the National Air Sampling Network, 1957–1961," 6–8 (1962).

39. U.S. Dept. of Health, Education, and Welfare, Public Health Service,

Division of Air Pollution, "Air Pollution Measurements of the National Air Sampling Network—Analyses of Suspended Particulates, 1963," 8 (1965).

40. U.S. Dept. of Health, Education, and Welfare, Public Health Service, "Air Pollution Measurements of the National Air Sampling Network," Robert A. Taft Sanitary Engineering Center, Cincinnati, Ohio, 249 (1958).

41. U.S. Dept. of Health, Education, and Welfare, Public Health Service, National Center for Air Pollution Control, "Interstate Air Pollution Study—III, Air Quality Measurements" (1966).

42. U.S. Dept. of Health, Education, and Welfare, Public Health Service, National Center for Air Pollution Control, "National Air Surveillance Networks, Air Quality Data, 1967," Cincinnati, Ohio (1968).

43. United States Weather Bureau, "Meteorology and Atomic Energy," Report for United States Atomic Energy Commission, July, 1955, Supt. of Documents, U.S. Govt. Printing Office, Washington, D.C. (for second edition, see Ref. 31, Chapter 2).

44. White, Harry J., "Industrial Electrostatic Precipitation," Addison-Wesley Publishing Co., Inc., Reading, Mass., 376 pp. (1963).

45. Wilson, Earl L., Testimony before Subcommittee on Air and Water Pollution of the U.S. Senate Committee on Public Works (May 18, 1967).

5

GASES

Gaseous contaminants in the atmosphere arise from two general sources: combustion of fuels and the handling and processing of chemicals. The latter category includes not only chemical manufacture but related activities, such as petroleum refining, smelting of ores, and various solvent-handling activities. Almost every gaseous material known escapes into the air at one time or another, but by far the most common are sulfur dioxide, carbon monoxide, hydrogen sulfide, nitrogen oxides, hydrocarbons and their simple oxidation products, and halogens and their derivatives (chiefly chlorine, bromine, hydrogen fluoride, and chlorinated solvents). Sulfur dioxide, carbon monoxide, and nitrogen oxides occur chiefly in combustion products. The others, with the exception of hydrocarbons, come chiefly from industrial operations. Hydrocarbons, in most areas, occur largely in the form of unburned or partially burned fuel from internal-combustion engines. Smaller, but significant, quantities are lost by evaporation in the storage, refining, marketing, and use of gasoline.

SULFUR DIOXIDE

Sulfur dioxide, SO_2, is a colorless gas with a suffocating odor. Concentrations of 6 to 12 ppm cause instantaneous irritation of the nose and throat. Most people can detect concentrations as low as 0.3 to 1 ppm by taste and about 3 ppm by smell. On the other hand, the maximum allowable concentration in air for 8 hours of exposure in working areas is 10 ppm.

A great deal of research work has been done and is also underway in an attempt to relate various atmospheric concentrations and dosages (concentration multiplied by time) of sulfur dioxide with effects of the gas on people and their property. Most of the pertinent literature has been compiled into a single volume by the National Air Pollution Control Administration [54]. To draw valid conclusions from these studies is an exasperating experience because of the mitigating effects of other pollutants and even normal air components which occur along with sulfur dioxide in the ambient atmosphere.

Laboratory studies under closely controlled conditions indicate that most people will experience mild chronic respiratory irritation at concentrations above 5 ppm. Sensitive individuals will notice 1 to 2 ppm, and have severe bronchospasms at 5 to 10 ppm. Epidemiological studies, on the other hand, have brought forth correlations of much lower atmospheric concentrations of sulfur dioxide with a wide variety of pathological symptoms, incidence of disease, and even death.

Sulfur dioxide has been suspect in several air pollution disasters—notably Donora, the Meuse Valley, and the several episodes in London. Although the concentrations found in these areas were higher than normal, they were still less than those often found elsewhere, so sulfur dioxide can neither be blamed nor exonerated for the severity of the attacks. This lack of correlation may well be due to the high reactivity of sulfur dioxide. It is soluble in water, e.g., fog droplets, and is oxidized fairly readily in solution to sulfur trioxide, which in turn may be hydrated to sulfuric acid, particularly if catalytic nuclei are present [54]. So in a foggy atmosphere, any sulfur dioxide present will usually be accompanied by sulfur trioxide and sulfuric acid mist. The latter material will eventually react with metal oxides in dust to form sulfates.

Considerable study has been directed toward the rate of oxidation of sulfur dioxide in air and the effect of the resulting sulfur trioxide on health. Johnstone and his co-workers [34] have shown that SO_2 is oxidized slowly in the gas phase in the atmosphere, but that after absorption on liquid droplets it will react much more rapidly, especially if manganese compounds are present. Of course, some sulfur trioxide (about 1 to 5% of total sulfur) usually accompanies sulfur dioxide in flue gases. The blue haze seen at or near the top of stacks is believed to be hydrated sulfur trioxide (sulfuric acid). Considerably more work remains to be done before the atmospheric reactions of SO_2 are well understood [87].

Some investigators have attributed the Meuse Valley and Donora episodes to a combination of sulfur dioxide, sulfuric acid mist, and certain metallic ions; others disagree. So as far as health is concerned, sulfur dioxide and its derivatives are still suspect but not convicted. The real meaning of reported findings still awaits clarification [8]. On the other hand, the threshold concentrations of sulfur dioxide and sulfuric acid aerosols for vegetation damage are much better known

It is generally agreed that acute damage to vegetation (certain types of leaf injury) occurs seldom, if ever, at concentrations below 0.4 ppm. Chronic symptoms (chlorosis) and excessive leaf drop may occur at long-term exposures between 0.25 to 0.50 ppm.

Structural damage caused by corrosion by sulfuric acid, to which all atmospheric sulfur dioxide eventually degrades, is widely noticeable. Some studies have indicated considerable corrosion at annual average sulfur dioxide concentrations below 0.05 ppm [54].

Poor visibility has also been related to atmospheric sulfur dioxide through oxidation to a sulfuric acid aerosol. Laboratory investigations have shown dramatically the formation of an aerosol of sulfuric acid from sulfur dioxide in an atmosphere where a rapid photochemical oxidation of olefinic hydrocarbons was occurring [26].

Atmospheric Concentrations of Sulfur Dioxide

The actual amount of sulfur dioxide found in the air varies from almost nil in some rural areas to about 3 ppm in heavily industrialized areas. The greatest volume of data available is that gathered by the Continuous Air Monitoring Program (CAMP network) of the U.S. Public Health Service; currently by the Environmental Protection Agency. Data from eight cities for the years 1962 to 1967 are shown in Table 5.1.

TABLE 5.1
CAMP Data on Sulfur Dioxide Concentrations (1962–1967)

City	5 minute		1 hour		24 hour		Annual	
	Max.	Mean	Max.	Mean	Max.	Mean	Max.	Mean
Chicago	1.94	0.104	1.69	0.111	0.79	0.121	0.18	0.142
Cincinnati	1.15	0.016	0.57	0.018	0.18	0.021	0.04	0.029
Denver	0.96	0.013	0.36	0.014	0.06	0.015	0.02	0.018
Los Angeles	0.68	0.013	0.29	0.014	0.10	0.015	0.02	0.018
Philadelphia	1.25	0.055	1.03	0.060	0.46	0.067	0.10	0.081
St. Louis	1.42	0.028	0.96	0.031	0.26	0.036	0.06	0.047
San Francisco	0.33	0.005	0.26	0.006	0.08	0.007	0.02	0.010
Washington	0.87	0.039	0.62	0.042	0.25	0.046	0.05	0.050

Data are also collected by the National Air Surveillance Network (NASN). Twenty-six random 24-hr samples are taken at each station annually. Data for 46 cities were reported for 1966 [55]. Highest values were obtained in New York City where 24-hr concentrations varied from 114–720 micrograms per cubic meter (0.043–0.27 ppm) with a geometric mean of 297 micrograms per cubic meter (0.111 ppm). Lowest mean value was 3 $\mu g/m^3$ (Guayanilla, Puerto Rico).

Sources of Sulfur Dioxide

Almost all fuels, with the exception of wood, contain a small amount of sulfur. The sulfur content of bituminous coal ranges from 0.3 to 5.0% and higher—most commonly, from 0.5 to 2.5%. Most crude petroleum contains less than 1% sulfur; a few crudes contain up to 5%. Refining processes tend to concentrate sulfur compounds in the heavier fraction, with the result that gasoline seldom contains more than 0.25%. Light fuel oils have a maximum sulfur specification of 0.5 or 0.75%; heavy fuel oils (No. 6 or Bunker C) vary from 0.3 to 3.0% sulfur. Fuel gases, on the other hand, are much lower in sulfur content because they are treated during manufacture or production to remove sulfur, which is nearly always present as hydrogen sulfide. As a consequence, the sulfur specification for natural gas is 0.1 grain per cu ft maximum; most commercial natural gas contains considerably less sulfur.

Nearly all raw fuels are treated prior to use to reduce their sulfur content. With coal, this is usually done to increase calorific value and to decrease corrosion by flue gases. Various methods of reducing the sulfur content of fuels are outlined in Ref. 59.

About 80% of the sulfur in coal and nearly all that in liquid and gaseous fuels appears in flue gases in the form of sulfur dioxide. The remaining sulfur in coal is that which is present as inorganic sulfur and thus remains in the ash. The concentration of sulfur dioxide in flue gases depends, therefore, on the sulfur content of the fuel (in the case of coal, the organic sulfur) and the percentage of excess air. Generally, the concentration in flue gases ranges from 0.05 to 0.25%, occasionally as high as 0.4%.

Another common source of sulfur dioxide in the atmosphere is metallurgical operations. Many ores, e.g., zinc and copper, are primarily sulfides. During the smelting of these ores, sulfur dioxide is evolved in stack concentrations of 5 to 10% SO_2. As will be seen later, this is usually recovered as sulfuric acid.

Among the more important miscellaneous operations releasing sulfur dioxide into the air are sulfonating operations and sulfuric acid plants. Volumes are usually low and thus easily amenable to control measures (see Chapter 7). Generalized factors for calculating emissions of sulfur dioxide from various operations are shown in Table 5.2.

Sampling and Analysis

The sampling of gases in stacks, ducts, and vents for sulfur dioxide and sulfur trioxide must follow the same principles of isokinetic sampling described in Chapter 4.

Kanter [36] described one satisfactory method of analyzing stack gases:

TABLE 5.2

Generalized Emission Factors for Sulfur Dioxide from Selected Sources*

Source	Emission factor
Combustion of coal	(38 times %S by wt) lb SO_2/ton of coal
Combustion of fuel oil	(158.8 times %S by wt) lb SO_2/1000 gal of oil
Open-burning dumps and municipal incinerators	1.2–2.0 lb SO_2/ton of refuse
Sulfuric acid manufacture	20–70 lb SO_2/ton of 100% acid
Copper smelting (primary)	1400 lb SO_2/ton of conc. ore
Lead smelting (primary)	660 lb SO_2/ton of conc. ore
Lead smelting (secondary cupola)	64 lb SO_2/ton of metal charged
Zinc smelting (primary)	1090 lb SO_2/ton of conc. ore
Kraft mill recovery furnace	2.4–13.4 lb SO_2/ton of air-dried pulp
Sulfite mill recovery furnace (assuming 90% recovery)	40 lb SO_2/ton of air-dried pulp

* Data from Ref. 59.

The sampling train for sulfur dioxide and sulfur trioxide consists of a paper thimble *maintained just above the dew point of the stack gases* followed by 3 series-connected impingers immersed in an ice bath, a dry gas meter and a pump. The thimble acts as a collector for the sulfuric acid aerosol formed from the sulfur trioxide. The first 2 impingers contain 100 ml each of approximately 5% sodium hydroxide solution. The third impinger is dry. The sulfur dioxide gas passes through the thimble and is collected in the impingers. The thimble is extracted with hot water and the solution is titrated with standard sodium hydroxide solution to determine the sulfur trioxide. The sulfur dioxide collected in the impingers is determined by oxidation with bromine, acidification, and precipitation as barium sulfate.

Colorimetric methods are widely used for determining sulfur dioxide in the atmosphere. A particularly sensitive and simple method has been developed by West and Gaeke [89]. A red-violet color is developed in the absorbing solution and may be compared with standard solutions to estimate sulfur dioxide concentrations. The method is sensitive to 0.005 ppm.

Details of the method recommended by the Environmental Protection Agency have been published [16].

A widely used method for approximate measurement of sulfur oxide pollution involves the use of the lead peroxide candle, sometimes called a "sulfur candle." A cylindrical glass or porcelain tube is wrapped with a cotton gauze supporting a paste made of lead peroxide and gum tragacanth. The candle is mounted in a louvered shelter where it is exposed to the ambient air. Any sulfur compound impinging on or diffusing to the surface of the candle is converted to lead sulfate. At the end of thirty days the candle is removed and taken to the laboratory for determination of lead sulfate. Sulfation rates are reported as milligrams of sulfur trioxide per square decimeter of exposed candle surface per day. Obviously the method is empirical and sensitive to many variables. It has been widely used in England as an inexpensive method to measure long-term variations in sulfur oxide pollution. Sulfation rates in some American cities have been said to range from a few tenths of a milligram to 9 mg $SO_3/dm^2/day$ [80].

Automatic Atmospheric Instruments

A variety of automatic instruments are available for monitoring sulfur dioxide in the open atmosphere. Most of them operate on one of two methods, the colorimetric method or the conductiometric method.

Typical of the *colorimetric method* is the West-Gaeke method. Sulfur dioxide is absorbed in 0.1 normal aqueous sodium tetrachloromercurate solution to form the volatile sulfitomercurate ion. This ion then reacts with formaldehyde and bleached pararosaniline methylsulfonic acid. The color intensity of the resulting red-violet dye, which is proportional to the concentration of sulfur dioxide, is measured at a wavelength of 560 mμ. To eliminate interference by nitrogen dioxide, *o*-toluidine or sulfamic acid is added to the collected sample. Oxidation of the disulfitomercurate, which would result in low sulfur dioxide values, is prevented by addition of ethylene diamine tetraacetic acid to the absorbing reagent to sequester metallic ions. The method is specific for sulfur dioxide and will determine concentrations in the air ranging from 0.002 ppm to 5 ppm. A flow diagram for an instrument utilizing the West-Gaeke procedure is shown in Fig. 5.1.

The *conductiometric method* (Fig. 5.2) utilizes the oxidation of sulfur dioxide to sulfuric acid by acidic aqueous hydrogen peroxide and subsequent measurement of the increased electrical conductivity of the solution. Precautions must be taken either not to operate in atmospheres containing large quantities of other acid or alkaline gases, specifically hydrogen chloride or ammonia, which may increase or decrease the conductivity of the solution. Weakly acidic gases, e.g., hydrogen sulfide and nitrogen dioxide, do not interfere because of poor absorption. The limitations of continuous instruments for sulfur dioxide have been discussed in detail [60, 65].

Several spectroscopic techniques have been developed recently which may be useful for remote monitoring of stacks [54].

Abatement Methods

Specific equipment and methods used for abating sulfur dioxide emissions depend largely on the concentration of sulfur dioxide in the gas stream and the volume of the gases. Where the concentration is of the order of

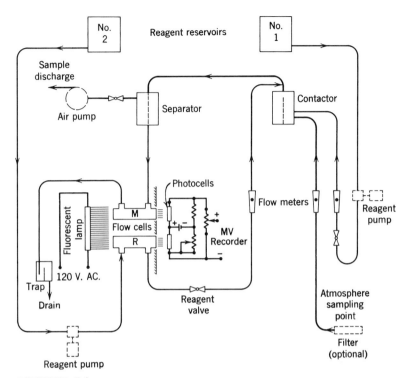

FIGURE 5.1.

Schematic flow diagram for a continuous sulfur dioxide analyzer using West-Gaeke method.

5 to 10% sulfur dioxide, as in smelter gases, the sulfur content may be recovered economically. The usual method at smelters is to use the sulfur dioxide stream as the raw material for the manufacture of sulfuric acid. For a discussion of sulfuric acid manufacturing processes, see *Industrial Chemicals,* by Faith, Keyes, and Clark [20].

If sulfur dioxide concentration is high and sulfuric acid manufacture is not feasible, the sulfur dioxide may be converted to elemental sulfur by reaction with hydrogen sulfide:

$$2SO_2 + 4H_2S \rightarrow 4H_2O + 3S_2$$

This process is normally used for the recovery of sulfur from sour natural or sour refinery gases [21].

The most vexing problem of sulfur dioxide abatement arises where large volumes of gas containing only comparatively low concentrations of sulfur dioxide are evolved, as in power plant flue gases. These gases have been

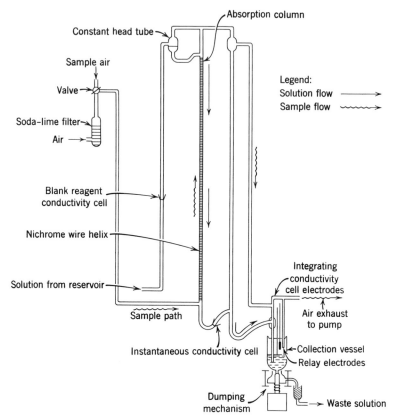

FIGURE 5.2.
Autometer (Ref. 17).

the subject of a tremendous amount of research and development activity, but no economic means of recovering the sulfur dioxide from flue gases has been found.

In England three power stations used the "Battersea" process (Fig. 5.3) for years, in which flue gas was washed with alkaline river water from the Thames to which a chalk slurry was added. Crude manganese sulfate was added to the scrubber effluent to allow dissolved sulfite to oxidize to calcium sulfate in an aeration tank. Prior to oxidation, the solution passed through a settler to remove excess solids. It was finally pumped back to the river. Such a process can be used only in a river already so heavily polluted that fishing and other riparian rights are not interfered with. It has now been abandoned.

Many other processes for recovery of sulfur dioxide from power plant stack gases [59] have been proposed and several are in the large pilot plant or demonstration plant stage. Which, if any, will prove to be feasible, is moot. The following are typical of the various approaches.

Limestone injection [72] is under test in several commercial power plants. Powdered limestone is introduced into the boiler where it calcines and reacts with some of the sulfur oxides. The stack gas is then led to a wet scrubber where it meets a slurry of calcined limestone which combines

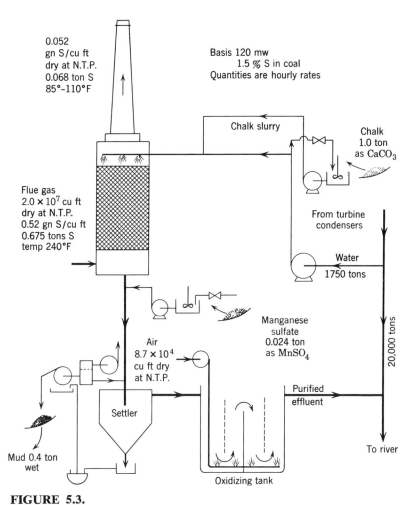

FIGURE 5.3.

Typical arrangement of modified Battersea effluent process. (*Courtesy Central Electricity Generating Board of England and Wales;* **Ref. 48, pp. 143–154.**)

with the remaining sulfur dioxide to produce sulfates and sulfites. Sulfur removal efficiency of 83 to 99% is claimed. The process is also said to require a lower capital investment than other proposed methods of sulfur removal.

A dry process which also makes use of either limestone or dolomite injection is also under test. The calcium or magnesium sulfates formed are removed along with fly ash in dry collectors. Only 50% or less removal of sulfur oxides is claimed but lower capital and operating costs are claimed.

A catalytic oxidation process [78] is also offered commercially. The process is based on the catalytic oxidation of SO_2 to SO_3 and subsequent conversion to dilute sulfuric acid. The first stage of the system is an electrostatic precipitator which removes 99% of the fly ash in the stack gases. The cleaned gases then pass over a vanadium pentoxide catalyst to convert up to 90% of the SO_2 to SO_3. Waste heat is recovered and the gases are sent to a packed absorption tower where they meet a stream of cool sulfuric acid where the SO_3 reacts further to form sulfuric acid. The effluent gases then pass through a mist eliminator and are discharged to the atmosphere. A capital cost of $20 to $30 per kilowatt of installed capacity is claimed. Economics of the process depend upon disposal of the sulfuric acid.

A *chemical absorption* process is also offered commercially. Complete details are not available, but the process is said to be based on removal of SO_2 by absorption in an aqueous solution of potassium sulfite to form potassium bisulfite. Prior to absorption, fly ash and SO_3 are removed from the stack gases by wet scrubbing. The potassium bisulfite solution formed during the absorption step is cooled to allow a portion of the bisulfite to separate as potassium pyrosulfite crystals, which are subsequently heated to regenerate potassium sulfate and release sulfur dioxide. The SO_2 is recovered by steam stripping [49, 69]. Capital costs of $6 to $9 per installed kilowatt are claimed. It is also claimed that the process can be operated at a profit.

Three other processes in various stages of development have been discussed widely. The U.S. Bureau of Mines *alkalized alumina* process [46] entails contact of stack gases with granular alkalized alumina to form sodium oxide and hydrogen sulfide. The regenerated alumina is recycled and the hydrogen sulfide may be converted to elemental sulfur by the Claus process [21]. Several activated carbon adsorption processes have been proposed. In one of these, the *Reinluft* process, sulfur dioxide is adsorbed on charcoal where it reacts with oxygen and moisture to produce sulfuric acid *in situ*. The adsorbent is removed to a regenerator, heated to 700°F so that the acid may be reduced by the carbon to sulfur dioxide which is evolved along with carbon dioxide. The gas mixture is then converted to sulfuric acid in a conventional acid plant.

A *molten carbonate absorption* process [39] has engendered interest. Sulfur dioxide in stack gases contacts a molten mixture of lithium, sodium, and potassium carbonates to convert the SO_2 to mixed sulfites and sulfates, which remain dissolved in the molten carbonate mixture. The salt mixture is recirculated to a regeneration system in which the sulfites and sulfates are reacted with carbon monoxide and hydrogen to produce hydrogen sulfide and regenerate the carbonates. In addition to the processes mentioned, a variety of others are under study.

Considerable study has also been given to removal or reduction of the sulfur content of fuels. Numerous processes for the *desulfurization of fuel oil* have been developed and evaluated [9]. Most of these processes involve hydrogenation processes and shifting of product balance at the refinery. Costs are highly variable and are generally regarded by petroleum technologists as highly dependent on the cost of hydrogen.

Desulfurization of coal appears to be even further from reality than does desulfurization of oil. Suitable processes probably involve either complete gasification or complete liquefaction of coal. Considerable research over a period of years has failed to develop a currently feasible process. A solvent extraction process developed by Gulf Oil Corporation has received some attention [33]. Substantial reduction of inorganic sulfur can be achieved with some coals by several physical separation processes [92].

In the absence of suitable processes for removal of sulfur either from fuel or from stack gases, large fuel users have resorted either to burning naturally occurring low sulfur fuels (natural gas, low-sulfur oils, and low-sulfur coals), or to the use of *tall stacks*. The availability of low-sulfur fuels is limited so this approach can be only of a temporary nature. Tall stacks, although expensive, present a means by which high-sulfur fuels may be utilized without creating an air pollution problem [74]. Obviously the stacks must be properly designed and located so that adequate atmospheric dilution may be accomplished.

Air Quality and Emission Standards

Air quality standards for sulfur dioxide have probably been the subject of more controversy in the United States than any other facet of air pollution. This state of affairs has been occasioned largely by the relatively high cost of sulfur dioxide abatement and lack of agreement on sulfur dioxide criteria. Nevertheless, standards have been adopted by several governmental jurisdictions, including the Environmental Protection Agency (see Table 1.3). The specific values chosen will probably continue to be controversial for some time.

Closely related to air quality standards for sulfur dioxide are standards for "suspended sulfates," "sulfuric acid mist," and "sulfation." There may well be a need for an air quality standard for sulfuric acid mist in those areas where specific sources (e.g., a sulfuric acid plant) may emit a mist, but the need for the other two is difficult to substantiate. A likely reason for their being suggested is the low cost of monitoring equipment. "Suspended sulfates" may be determined by chemical analysis of particulate matter collected by a high-volume sampler; "sulfation" is determined by the relatively inexpensive lead peroxide (sulfur) candle. It is almost impossible to relate the values obtained to a specific source or to an effect on people or their property.

Emission standards for sulfur dioxide are stated in various manners. Stack concentration limitations have been in use for many years. Thus Los Angeles County as early as 1947, limited the sulfur dioxide in stack gases to 0.2% by volume, except from scavenger plants (those recovering sulfur that would otherwise be emitted to the atmosphere from many sources).

More recently adopted regulations relate emissions to an allowable ground-level concentration downwind from the source. The Texas regulation lets allowable ground-level concentrations vary according to land use.

The American Society of Mechanical Engineers has suggested a regulation for combustion effluents in which the allowable weight of sulfur dioxide emitted varies with heat input and height of stack in such a way that a predetermined ground-level concentration will not be exceeded [4]. It may well be that when suitable abatement methods for power plant stack effluents are developed that emission standards will be stated in terms of concentration or weight per unit time.

In some communities, emission standards have been abandoned, at least temporarily, in favor of restrictions on the sulfur content of fuels. In some areas, the effect of such regulations is to force the use of natural gas, in others, the use of solid or liquid fuels with as low a sulfur content as can be obtained. Unless feasible desulfurization processes can be developed, this type of regulation can be only temporary because of limited supplies of suitable fuels. There probably is a place for restrictions on type of fuel used, particularly for small users or those with emission points relatively near the ground. Stack-gas treatment will probably be limited to comparatively large users. Until suitable processes are developed, sound natural-resource conservation practices would favor the use of high-sulfur fuels and high stacks for large power plants.

Some areas, e.g., Los Angeles, have considered a ban on use of all fossil fuels for large steam-electric plants on the assumption that power could be generated either by nuclear energy or in comparatively uninhabited areas from which electric power could be transmitted by high-voltage lines. This has been accomplished in effect by Los Angeles' controversial Rule 67 which limits emission from new power plants to 200 lb SO_2 per hour, 140 lb NO_x per hour, and 10 lb combustion particulates per hour.

HYDROGEN SULFIDE

Of the various sulfur-bearing gases, hydrogen sulfide ranks second to sulfur dioxide as an air pollutant. Hydrogen sulfide is an extremely toxic, evil-smelling gas that can readily be detected by smell in concentrations as low as 0.0005 ppm [43]. However, the nose soon becomes accustomed to the odor, so this method is not adequate when the concentration is high or persistent. Hydrogen sulfide may readily be burned to sulfur dioxide; in fact, large quantities of formerly wasted gases are now burned for use in the manufacture of sulfuric acid [20]. Besides having an objectionable odor, hydrogen sulfide reacts with lead pigments and thus causes darkening of some painted surfaces after prolonged exposure in humid atmospheres. This is particularly noticeable with some white paints, which become brown or even black. Under favorable conditions, darkening will occur at 0.1 ppm for 1 hour. Eventually, hydrogen sulfide in the air is converted to sulfuric acid and thus is a potential corrosion agent.

The maximum allowable concentration for an 8-hour exposure in working areas is 20 ppm, a value far above that found in the open atmosphere. Even in the vicinity of viscose rayon plants, where hydrogen sulfide is produced during spinning, hydrogen sulfide concentration is rarely as high as 0.1 ppm. Cholak [15] analyzed Cincinnati air over a period of five years and rarely found hydrogen sulfide to exceed 0.01 ppm. A survey carried out in Houston, Texas, in 1957 showed average values of 0.02 ppm in the most highly polluted section of the city; the highest value found was 0.28 ppm [75]. On the other hand, Katz [38] found mean concentrations of 0.1 ppm and a maximum of 0.6 ppm in the Windsor, Ontario, area.

The only air pollution disaster laid to hydrogen sulfide is the Poza Rica, Mexico, incident in 1950 [45]. This was an industrial accident in which large quantities of hydrogen sulfide were released into the still early morning air.

Sources

The chief potential sources of hydrogen sulfide in the atmosphere are industrial operations, although vegetable matter, volcanos, and natural springs are also among the contributors. The industrial effluents of greatest importance come from petroleum refineries, coke-oven plants, viscose rayon plants, and some chemical operations.

In the processing and refining of crude oil, a considerable portion of the sulfur in the crude is converted to hydrogen sulfide. The sulfide normally appears as a contaminant of gaseous hydrocarbon streams used as fuel in the refinery. On burning the fuel, the hydrogen sulfide is converted to sulfur dioxide and emitted as such. Because of the magnitude of the air pollution problem and the value of recovered sulfur, the hydrogen sulfide is now commonly recovered and converted either to elemental sulfur or to sulfuric acid [20]. Regenerable solutions used to absorb the hydrogen sulfide are sodium phenolate, potassium phosphate, and the ethanolamines [42, 63].

Coke-oven gas and natural gas are other sources of hydrogen sulfide, but these gases are seldom pollutants, inasmuch as the sulfide must be removed to prevent corrosion in distribution lines and burners.

In the chemical process industry, the chief source is viscose rayon manufacture where hydrogen sulfide is liberated during spinning. Effluent gases also contain carbon disulfide vapors. Because the spinning room air must be kept below 20 ppm hydrogen sulfide, sufficient sulfide-laden air is exhausted from the room to hold the hydrogen sulfide concentration below that level.

Analytical Methods

Hydrogen sulfide in gas streams in the atmosphere may be collected by passing a gas sample through a bubbler or impinger containing a 1% alkaline zinc acetate solution [67]. Upon addition of p-aminodimethylaniline and ferric chloride, a methylene blue color is developed in the presence of the sulfide ion.

The California Air Resources Board has adopted a method wherein H_2S-containing air is scrubbed with an absorbing slurry of cadmium hydroxide containing arabino-galactan (STRactan) in a midget impinger. The cadmium hydroxide reacts with the H_2S to form cadmium sulfide. The STRactan is used to prevent decomposition of the cadmium sulfide. At the end of the sampling period, solutions of N,N-dimethyl-p-phenylene diamine and ferric chloride are added to form methylene blue which is measured spectrophotometrically. For an evaluation of the method, see Ref. 7.

For monitoring hydrogen sulfide in the atmosphere, the Autometer, described in the section on sulfur dioxide, may be used after slight modifications. One of the most convenient methods for monitoring hydrogen sulfide has been described by Sensenbaugh [70]. The instrument is a modification of the AISI Smoke Sampler. A paper tape, impregnated with lead acetate solution, is utilized as the sensing element.

Control Methods

The recovery of hydrogen sulfide from *petroleum* and *natural gases* and *coke-oven gas* is usually dictated by economic necessity and may be considered an air pollution control method only incidentally. On the other hand, the effluents from *chemical operations* are controlled primarily to prevent an air pollution problem. Any type of liquid-gas contactor may be used, with the final selection based on economic factors. Both spray towers and packed towers are used for treating high volumes of gases. Jet and cyclone scrubbers are used for smaller volumes. Pratt and Rutherford [62] have described the design and operation of a spray scrubber used to reduce the hydrogen sulfide in the effluent from a viscose rayon plant from 300 ppm to 10 to 20 ppm. Low volume emissions may be flared.

Air Quality and Emission Standards

Ambient air quality standards range from 0.1 ppm for 1 hour (Pennsylvania) down to 0.03 ppm for ½ hour, not to be exceeded more than once in 5 days (Montana).

Emission standards are generally stated in terms of allowable ground-level concentrations, downwind from a source.

HYDROGEN FLUORIDE

In the atmosphere, the only toxic gas that contains fluorine is hydrogen fluoride. Fluorine itself is far too reactive to exist as such very long, or even to be emitted in large quantities from chemical processes. The fluorocarbons used in refrigeration and as an aerosol propellant are non-toxic.

Hydrogen fluoride is an extremely corrosive, poisonous liquid, but because of its low boiling point (19.4°C) it may readily be emitted into the atmosphere as a gas. Although the maximum allowable concentration in air is 3 ppm, it is practically never found in the air except in very small concentrations. Cholak [14] reported analyses for fluorides in six American cities. Values ranged from 0.000 to 0.08 ppm, calculated as HF; the mean values for the six cities ranged from 0.003 to 0.018 ppm. How much of these loadings represented hydrogen fluoride and how much solid fluorides was not recorded. This is an important consideration, because laboratory experiments have shown that many plants are susceptible to hydrogen fluoride in concentrations as low as 0.02 to 0.05 ppm. Actually the effect of fluorides on vegetation covers a very wide range of concentrations [83] depending on the particular species. Inorganic fluorides apparently have little effect, although one investigator believed the Meuse Valley disaster was caused by fluorides [50]. Actually, fluorine is a cumulative poison even under conditions of prolonged exposure in subacute concentrations. Therein lies its danger.

The most important effect of fluorides appears to be fluorosis of cattle, caused by ingestion of vegetable matter that has collected fluorine-containing dusts. Even here, the solubility of the fluoride appears to be important, with sodium fluoride twice as effective as the calcium salt. According to Phillips and colleagues, the safe level of fluoride for animal rations is as shown in Table 1.2.

Sources of Fluorides

The chief sources of fluoride pollution of the atmosphere are gases and dusts from certain high-temperature metallurgical plants, the ceramic industry, and the superphosphate industry [76]. Electrolytic aluminum plants have a very particular problem in that cryolite (sodium aluminum fluoride) is added to the electrolytic bath and some hydrogen fluoride is evolved. Meetham [50] reports fluoride concentrations in the air near an aluminum plant of 0.02 to 0.1 mg fluorine per cu m. Agate [3] quotes values of 0.02 to 0.22 mg per cu m, calculated as fluorine, near an aluminum plant in Scotland. The gaseous fluorine ranged from 15 to 81% of the total. Fluorine compounds are also emitted as incidental contaminants from other metallurgical operations, such as zinc foundries and open-hearth steel furnaces. Schrenk [68] estimated that fluorides in the form of particulate matter from open hearth stacks of the Donora steel plant amounted to 39 lb per day, calculated as fluorine. In some areas, superphosphate plants are the principal source of fluorine compounds, either as hydrogen fluoride, silicon tetrafluoride, or fluosilicic acid. The smaller quantities in many urban communities, e.g., up to 0.025 ppm in Cincinnati, probably came from the burning of coal, which normally contains about 0.01% fluorine [14].

Analytical Methods

Because of the low concentrations of fluoride in the atmosphere, large quantities of air must be sampled and sensitive analytical methods must be used. Fortunately, several colorimetric methods are available [13, 22]. A procedure based on the use of limed filter paper has been described by Hendrickson [28]. An automatic recorder for stack monitoring has been developed by Adams [1].

Abatement Methods

The chief factor affecting the choice of equipment and processes to control emissions of fluorine compounds is the form of the compound, i.e., whether gaseous or particulate. In either case, only very low concentrations should be present in the effluent stream after treatment. Fortunately, the affinity of water and alkaline water solution for both gaseous hydrogen fluoride and soluble fluorides, such as sodium fluoride, is very high. Therefore, any form of gas-liquid contact apparatus may be used for abatement. Both packed towers and spray towers are effective [81]. Most units operate at 97 to 99% efficiency.

When elemental fluorine is present in gases, water scrubbing is not practiced because of the potential explosion hazard. A caustic soda solution is effective. The presence of silicon tetrafluoride is no problem, since it hydrolyzes readily to hydrogen fluoride and silicon dioxide. When insoluble fluoride dusts, such as calcium fluoride, must be controlled, the several types of dust-recovery equipment described in Chapter 4 may be considered.

Air Quality and Emission Standards

Ambient air quality standards for fluorides are conspicuous by their nonexistence. The California Department of Public Health studied the problem in 1959, but did not arrive at a consensus. Accordingly the only references to fluorides in the California air quality standards are in footnotes, i.e.:

> *Footnote 11.* Hydrogen fluoride and other airborne fluorides settle upon and are absorbed into vegetation. When forage crops containing 30–50 ppm of fluoride measured on a dry basis are regularly consumed over a long period, the teeth and bones may show changes, depending upon age, nutritional factors, and the form of fluoride ingested. Fluorides at these levels do not necessarily cause injury to the forage plants themselves. However, injury may be produced in certain species of vegetation upon long exposure to low levels of atmospheric fluorides.
>
> *Footnote 12.* The irritating properties of hydrogen fluoride in experimental human exposure have been manifested by desquamation of the skin at concentrations of 2–5 ppm. Mucous membrane irritation also occurs from hydrogen fluoride but quantitative data are not adequate to support a standard.

The Fluoride Air Pollution Control Commission has adopted several rules directed toward the control of atmospheric fluorides. The Commission assumes there is fluoride pollution if grasses in the vicinity of a fluoride emitter contain an average of 40 ppm fluorine, or if gladioli contain, in the terminal six inches of their foliage 35 ppm fluorides, both on a dry weight basis.

Emission restrictions are directed specifically to phosphoric acid manufacturing and the calcination and defluorination of phosphate rock. Maximum allowable emissions are 0.6 and 0.4 lb fluoride (as F) per ton of P_2O_5, or equivalent, produced in a 24-hr period, the specific value depending on the age of the facility.

Presumptive limits of the Alkali Inspectorates of England and Wales and for Scotland are:

Fertilizer Plants (Superphosphate).   The acidity of gases escaping to the atmosphere should be below the equivalent of 0.1 grain SO_3/cu ft, or there should be greater than 99% efficiency of condensation or scrubbing of the gases.

Hydrofluoric Acid Works.   HF in exit gases should not exceed the equivalent of 0.1 grain SO_3/cu ft.

There are, of course, many regulations governing emissions of dusts and particulate matter, but few specify fluorides as such. The New York State Air Contaminant Emission Guides divide particulates into six classes. Greatest degree of control is required for Class A materials, e.g., beryllium, least for Class F, e.g., fly ash. Fluorides are listed in Class D.

Air quality criteria for fluorides have been discussed in detail by Hodge and Smith [30]. Pertinent reviews have been published by McCune [47] and Shupe [71].

CHLORINE AND HALOGENATED COMPOUNDS

Chlorine is found in polluted atmospheres as the element itself, chlorine; as hydrogen chloride; as chlorine-containing organic compounds, such as perchloroethylene; and as inorganic chlorides. The last-mentioned compounds are solids, hence found in particulate form; the other materials mentioned are present as gases. Occasionally, organic bromine compounds and inorganic bromides from auto exhaust are present in the air.

The concentrations of chlorine and its compounds in the air are usually quite low. Cholak [14] reported the average chloride concentration in several American communities in the range of 0.016 to 0.078 ppm, calculated as Cl^-. A similar compilation by Katz [37] showed average values in five cities in the range of 0.033 to 0.095 ppm. A few analyses have been made for organic halides. The values generally found are less than 0.1 ppm.

The principal effects of chlorine and its compounds are corrosion, by hydrogen chloride and salt nuclei; respiratory irritation, from chlorine [25]; possibly more deep-seated respiratory effects from complex ammonium chlorides [27]; and damage to vegetation from chlorine [91].

The most common acute effects, however, are the result of industrial accidents rather than air pollution episodes. One highly publicized accident occurred in Brooklyn, N.Y., in 1947 when a faulty tank valve failed while chlorine was being transported by truck. Thirty-three persons were hospitalized for one to two weeks. Most of the casualties were in a subway where the heavy gas (sp. gr. 2.49) accumulated. Comparatively high concentrations are required for such effects. The MAC for chlorine is 3 ppm; for hydrogen chloride, 5 to 10 ppm; and for chlorinated organic compounds, 100 to 200 ppm. Certain specific compounds are troublesome, even at very low concentrations. Examples are tear gases (chloroacetophenone), poison gases (phosgene), and weed killers (2,4-dichlorophenoxy acetic acid).

The lachrymatory properties of some halogenated componds are so intense that concentrations in the parts-per-billion range are noticeable. One episode was reported in Michigan in which small amounts of bromine or chlorine passed over a sewer outfall containing very small amounts of styrene [2]. The resulting halogenated styrene caused severe eye irritation downwind from the sewer outfall.

Sources of Chlorine and Its Compounds

The most common sources of chlorine in the atmosphere are from operations in which it is manufactured or used to produce other chemicals. In the electrolytic manufacture of chlorine, a waste gas containing 40% chlorine in air (snift gas) is sometimes exhausted to the atmosphere. In rural areas this may be no problem, but in urban areas control is necessary. Inasmuch as chlorine is used in water-purification plants, in sewage plants, and in swimming pools, equipment failure sometimes leads to inadvertent losses from these operations.

Hydrogen chloride is evolved in numerous industrial chemical processes, but it is so easy to recover that little reaches the atmosphere.

As mentioned previously, most of the chlorine in the air is found in particulate form, as inorganic chlorides. This may arise from the reaction of hydrogen chloride with the metallic constituents of dusts and fumes, from salt particles that winds have picked up from ocean spray, or from lead compounds emitted by automobile exhaust systems. In the last-mentioned case, the source of the chloride is the antiknock fluid in the gasoline. This fluid is a mixture of tetraethyllead, ethylene chloride, and ethylene bromide. When subjected to the high-temperature oxidizing conditions in the engine cylinders, it is converted to a complex mixture of lead and ammonium chlorides and bromides [29].

The principal organic halides found in the atmosphere are the so-called chlorinated solvents, chiefly perchloroethylene and trichloroethylene. These solvents are widely used in dry cleaning and metal degreasing. For further discussion, see p. 192.

Analytical Methods

The quantities of chlorine and water-soluble chlorides in effluent gas streams or in the atmosphere may be determined by passing a known volume of gas through an impinger containing dilute sodium hydroxide solution. The chloride content may then be determined by either gravimetric or titrimetric methods [24]. Free chlorine in the atmosphere may be determined colorimetrically by absorption in o-tolidine solution [13].

Peterson [61] has suggested that halogenated hydrocarbons in air may be absorbed on silica gel, then desorbed by heat and aeration. The resulting halogen and halogen acid may then be determined by conventional means.

Abatement Methods

The methods used to remove chloride-containing particulates from effluent streams are similar to those described in Chapter 4. Chlorine, hydrogen chloride, and halogenated organic compounds may be removed from gas streams in scrubbers or spray towers. The solvent used depends, of course, on the particular contaminant to be recovered. Hydrogen chloride is very highly soluble in water and thus comparatively simple to recover. By recirculation of dilute acid, and use of several stages, hydrochloric acid may be recovered in concentrated form. Low concentrations of halogenated hydrocarbons can be removed from air streams by adsorption on activated charcoal or similar adsorbent. High-boiling compounds may be adequately controlled by refrigeration. Chlorine may be absorbed effectively by milk of lime.

OXIDES OF NITROGEN

Seven oxides of nitrogen (N_2O, NO, NO_2, NO_3, N_2O_3, N_2O_4, N_2O_5) and two hydrated oxides (HNO_2 and HNO_3) can theoretically exist in the atmosphere. The only ones present in noticeable amounts, however, are nitrous oxide (N_2O), nitric oxide (NO), nitrogen dioxide (NO_2), and possibly nitrogen pentoxide (N_2O_5). Nitrous oxide is a normal constituent of the atmosphere and found in concentrations of about 0.5 ppm. It is probably formed in the upper atmosphere by the reaction of nitrogen and atomic oxygen or ozone. It is, however, extremely unreactive and not considered an atmospheric pollutant.

On the other hand, nitric oxide and nitrogen dioxide arise from many human activities and are classified as pollutants. In atmospheric analyses they are usually reported as "total oxides of nitrogen," or "NO_x."

Data from the Continuous Air Monitoring Program of the National Air Pollution Control Administration for the year 1964 indicate average annual concentrations in six large cities varied from 0.04 to 0.15 ppm (76 to 288 $\mu g/m^3$). Hourly maxima for the year varied from 0.43 ppm to 1.41 ppm (825 to 2700 $\mu g/m^3$). Somewhat higher values have been reported by the Los Angeles County Air Pollution Control District, which has operated at least six continuous monitoring stations in Los Angeles County for thirteen years. The arbitrary value of 0.25 ppm for one hour has been exceeded in Los Angeles on more than 800 days in the past nine years. An instantaneous peak value of 3.93 ppm was recorded on January 13, 1961 [6].

In all probability, nitrogen oxide enters the atmosphere as nitric oxide and is oxidized first to NO_2 and then to N_2O_5, which, in the presence of water vapor, forms nitric acid (HNO_3). This in turn reacts with various metal salts and winds up in the particulate phase as a metal nitrate. National Air Sampling Network data for the years 1962 through 1964 show average concentrations of nitrates in suspended particulates in 96 cities to be 2.6 $\mu g/m^3$. The maximum value measured was 39.7 $\mu g/m^3$.

The deleterious effects of nitrogen oxides appear to be of four types: biological, phytotoxic, coloration of the atmosphere, and formation of photochemical smog (see Chapter 8). The limited *biological* data available indicate that, for transient exposures, 3.5 ppm NO_2 for 1 hour is the lowest value that affects man. The nature of long-term chronic effects is unknown. With respect to phytotoxicity, scientists at the University of California at Riverside have reported damage to some types of vegetation for one hour exposures to 0.5 ppm NO_2. So far as both people and vegetation are concerned, nitric oxide is far less toxic than nitrogen dioxide and thus is usually disregarded in this respect.

On the other hand, concentration of nitrogen oxides (either as NO or NO_2) in concentrations as low as 0.1 ppm readily contribute to the nuisance effects of photochemical smog (see Chapter 8) under unfavorable meteorological conditions.

The *atmospheric coloration* effects of nitrogen dioxide are based on the absorption of light by NO_2 in both the ultraviolet and visible spectrums. Thus at some atmospheric concentration of NO_2, light absorption will cause white objects on the horizon to appear to be yellowish-brown. Based on calculations by Hodkinson [11], the California Department of Public Health concluded that at 0.25 ppm NO_2, on a day when visibility is 20 miles or more, the color effect on objects 10 miles distant would be objectionable to the public. The day of 20-mile visibility was chosen because atmospheric aerosols tend to cancel the effect.

Sources of NO and NO_2

The most important sources of nitrogen oxides in the atmosphere are combustion of fuels and certain chemical manufacturing operations. When fuel is burned with air as the oxidant, some of the nitrogen in the air is oxidized to nitric oxide (NO). The amount in the combustion effluent depends on the flame temperature and the rate of cooling or quenching of the combustion products. Flame temperature is important because the equilibrium concentration of nitric oxide in air increases with temperature (Table 5.3). The

TABLE 5.3
Equilibrium Concentrations of Nitric Oxide

Temperature, °C	Concentration NO, ppm
20	0.001
427	0.3
527	2.0
1,538	3,700.
2,200	25,000.

rate of cooling is important because the rate of decomposition of nitric oxide to elemental nitrogen and oxygen decreases rapidly as temperature falls.

As a general rule the highest concentrations of nitrogen oxides in gaseous emissions occur in effluents from manufacturing facilities where nitric acid is being produced or used in chemical reactions. The next highest concentration is in automobile exhaust, then effluents from large power plants and thence progressively down to very small sources like range-top burners.

A study made by the U.S. Public Health Service on emissions from twelve nitric acid manufacturing plants indicated emissions of nitrogen oxides ranged from 0.1 to 0.69 volume percent, with an average of 0.37. About one-third to one-half of the gases was nitrogen dioxide, the remainder was nitric oxide. Emissions are roughly 55 lb per ton of acid produced.

Concentrations as high as 6000 ppm have been reported for motor vehicle exhaust during acceleration; up to 3000 ppm during cruise. Very little NO is found during deceleration and idling. Average emissions will depend on the type of operation in the various modes and the percentage of time the vehicle is operated in each mode. For the average urban cycle as defined by the U.S. Department of Health, Education, and Welfare, average emissions for 1966 model vehicles (those without hydrocarbon and carbon monoxide controls) are about 1500 ppm or 3.75 grams per mile. New models (1968 et seq.) equipped with exhaust control systems average close to 2000 ppm NO_x (5.0 grams per mile).

The concentration in diesel exhaust is only half the average concentration in exhaust from gasoline engines, but high gas flows make the diesel contribution significant. Aircraft turbine engines yield concentrations in their exhaust of 100 to 400 ppm NO_x. Gas flows are extremely high.

Low heat burners and furnaces yield 10 to 100 ppm in combustion gases, whereas large power plants are in the range of 200 to 1500 ppm [52, 58].

Measurements as high as 1750 ppm have been made. There are considerable variations in the data, with identical boilers operating in the same manner giving widely divergent results. Emission factors suggested for various stationary combustion sources are listed in Table 5.4 [58].

Analytical Methods

Methods for the analysis of stack gases, engine exhaust, and the open atmosphere leave a great deal to be desired with respect to reproducibility and specificity. The most commonly used method in air pollution work is based on the use of the Griess-Ilosvay reagent as modified by Saltzman [66]. The method entails the bubbling of air or a stack gas sample through the reagent to produce a pink color, the intensity of which is dependent on the amount of nitrogen dioxide in the sample. The color may be measured photometrically. If the nitric oxide content of the gas sample is also desired, the usual situation, the NO may be converted to NO_2 by bubbling the sample through potassium permanganate solution [85], then analyzed by the Saltzman procedure. The Saltzman method has been adapted to a continuous instrument (see Fig. 5.4).

Another commonly used analytical procedure for total oxides of nitrogen determination is the phenoldisulfonic acid method [32] which measures total combined nitrogen (except nitrous oxide), whether it be nitric oxide, nitrogen dioxide, nitric acid, an organic nitrogen compound, or an inorganic nitrate.

A recent development of promise for the determination of nitric oxide (NO) both in the atmosphere and in stack gases is a dry, electrochemical sensor. Several commercial units are available.

Nitric oxide may be determined in some stack gases by infrared spectrometry. Unfortunately, water vapor absorption bands interfere in some cases. A chemiluminescent method for either NO or NO_x determination is said to be promising [19a]. It has not been widely tested.

Abatement Methods

The time-honored method for recovering oxides of nitrogen in chemical process streams is absorption by water in bubble towers, spray towers, venturi scrubbers, and the like. Nitrogen dioxide is absorbed by water easily, but if any nitric oxide is present, air or oxygen must be added and space must be provided for the gas-phase oxidation of nitric oxide to nitrogen dioxide.

For economic reasons, the method is not adapted to combustion effluents and other streams with comparatively lower nitrogen oxide concentrations. Similarly, adsorption of nitrogen dioxide by silica gel [23] is not economical at low concentrations.

TABLE 5.4

Emission Factors for Nitrogen Oxides during Combustion of Fuels and Other Materials

Source	Average emission factor
Fuels	
Coal	
Household and commercial	8 lb/ton
Industry	20 lb/ton
Utility	20 lb/ton
Fuel oil	
Household and commercial	12–72 lb/10^3 gal
Industry	72 lb/10^3 gal
Utility	104 lb/10^3 gal
Natural gas	
Household and commercial	116 lb/10^6 ft^3
Industry	214 lb/10^6 ft^3
Utility	390 lb/10^6 ft^3
Wood	11 lb/ton
Combustion sources	
Gas engines	
Oil and gas production	770 lb/10^6 ft^3
Gas plant	4,300 lb/10^6 ft^3
Pipeline	7,300 lb/10^6 ft^3
Refinery	4,400 lb/10^6 ft^3
Gas turbines	
Gas plant	200 lb/10^6 ft^3
Pipeline	200 lb/10^6 ft^3
Refinery	200 lb/10^6 ft^3
Waste disposal	
Open burning	11 lb/ton
Conical incinerator	0.65 lb/ton
Municipal incinerator	2 lb/ton
On-site incinerator	2.5 lb/ton
Other combustion	
Coal refuse banks	8 lb/ton
Forest burning	11 lb/ton
Agricultural burning	2 lb/ton
Structural fires	11 lb/ton
Chemical industries	
Nitric acid manufacture	57 lb/ton HNO_3 product
Adipic acid	12 lb/ton product
Terephthalic acid	13 lb/ton product
Nitrations	
large operations	0.2–14 lb/ton HNO_3 used
small batches	2–260 lb/ton HNO_3 used

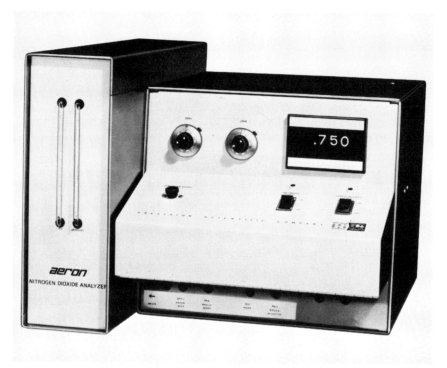

FIGURE 5.4.

Continuous analyzer for nitric oxide and nitrogen dioxide. (*Courtesy Precision Scientific Co.*)

The various methods available for controlling nitrogen oxide emissions from stationary sources have been published by the National Air Pollution Control Administration [58]. The most suitable means for reducing emissions from large oil- and gas-fired steam-electric plants is injection of a major portion (90–95%) of the theoretical combustion air directly into the flame leaving the burner. Addition at this point not only cools the flame but prevents a local excess of oxygen from contacting the combustible material. The remaining air is added as secondary air. By this combination of low excess air and 2-stage combustion, oxides of nitrogen emissions may be reduced by at least 50%, in some cases considerably more. A combination of low-excess air and flue gas recirculation is said to give comparable results. Whether or not either system is applicable to pulverized-coal firing is unknown at present.

A catalytic reduction method has been applied successfully to the brown NO_2 plumes emitted from nitric acid manufacturing plants and other chemical processes emitting similar plumes. A reducing gas (methane, carbon monoxide, or hydrogen) is first added in the stack and burned to remove excess oxygen. The deoxidized gases are then passed over a catalyst where NO_2 is reduced to NO or N_2O. A colorless plume results. Prior removal of oxygen is necessary to maintain catalyst activity. In many cases the cost of fuel to remove the excess oxygen is exorbitant, and high stacks are the best means of reducing ground-level emissions.

Methods proposed for reducing oxides of nitrogen from motor-vehicle emissions are discussed in Chapter 9.

Air Quality Criteria and Emission Regulations

Intensive studies of air quality criteria were made by the California Department of Health in 1965 [10] and in 1969 [11] and more recently by the Environmental Protection Agency [19a].

As a result of the 1965 study, the California State Board of Health adopted an ambient air quality standard for nitrogen dioxide of 0.25 ppm for 1 hour. Basis for the standard was a calculated discoloration of the atmosphere if 0.25 ppm NO_2 were present over a 10-mile path length on a day of 20-mile visibility, i.e., when atmospheric aerosol concentration was too low to affect light absorption by nitrogen dioxide. In 1969, the same standard was re-adopted by the California Air Resources Board on the stated basis of "prevention of undesirable discoloration (of the atmosphere) and the provision of some margin of safety for serious health effects." Various criteria were listed in the report, e.g., 2.5 ppm for 7 hours damages certain vegetation [51]; 25 ppm for 2 hours (repeated) causes emphysematous changes in rabbits and guinea pigs [41]; 0.5 ppm continually increases the susceptibility of mice to experimental infection [19]; 0.25 ppm to 1 ppm for 1 to 4 hours causes reversible lysis of rat lung mast cells [82].

No standards were set for total oxides of nitrogen (NO_x). However, by early 1970, two states (North Dakota and Colorado) had set standards of 0.1 ppm for 1 hour for NO_x. Basis of the standard was not given.

Inasmuch as nitric oxide has no effect on man or his property, certainly below 5 ppm, the low values of NO_x must have been chosen on the basis of photochemical smog formation (see Chapter 8).

The standard recently adopted by the Environmental Protection Agency is shown in Table 1.3.

CARBON MONOXIDE

Carbon monoxide is a colorless, essentially odorless, highly poisonous gas that is the product of incomplete combustion of carbon and its compounds. It is generally classed as an asphyxiant because of its strong combination with hemoglobin in the blood. Innumerable deaths have resulted from carbon monoxide in coal mines and in enclosed spaces where it has been released from improperly adjusted heating equipment or internal-combustion engines. The MAC in air is 100 ppm.

On the other hand, an increase in the body of CO as carboxyhemoglobin (COHb) can occur with ambient levels greater than a few ppm. For instance, 30 ppm CO for 8 to 12 hours will produce an equilibrium value of 5% COHb in the blood [73]. Similar data led the state of California to adopt 30 ppm for 8 hours (and its equivalent, 120 ppm for 1 hour) as an ambient air quality standard in 1959. Basis for the standard was that although 10% COHb would significantly affect oxygen transport by the blood, the likelihood of exposures other than ambient air made it reasonable to select 5% as the maximum contribution from community air pollution [12]. In 1969 the standard was changed to 20 ppm for 8 hours on the basis that community air pollution should not contribute more than 2% COHb [18]. Others call for even lower values on the basis that cardiac cripples should be protected. The whole matter is further confused by the presence of 400–475 ppm CO in cigarette smoke.

The concentration of carbon monoxide in city streets and vehicular tunnels has ranged up to 90 ppm and more [44, 53, 64]. Cholak [14] has reported average values in several American communities of 4 to 10 ppm.

Since 1962, the U.S. Public Health Service has measured CO continuously at one station in each of eight cities. Typical data for the period 1962–1967 are shown in Table 5.5. Concentrations reported by the Los

TABLE 5.5

Carbon Monoxide Concentrations in Various U.S. Cities (1962–1967)*

City	Highest max. hourly in any year (ppm)	Lowest max. hourly in any year (ppm)	Hourly geometric mean for entire period (ppm)
Chicago	59	28	13.2
Cincinnati	34	20	4.8
Denver	55	40	6.7
Los Angeles	47	35	9.7
Philadelphia	54	37	6.9
St. Louis	29	25	5.5
San Francisco	38	22	4.8
Washington	41	25	3.5

* From Ref. 53.

Angeles County APCD at 28 stations for the years 1956 to 1967 show hourly mean concentrations ranging from 3.7 to 14.9 ppm and hourly maxima ranging from 16 to 68 ppm [53].

The U.S. Public Health Service has suggested that a prudent air quality standard for carbon monoxide would be 10 ppm or less for an eight-hour period [53]. The standard adopted by EPA is shown in Table 1.3.

Sources

As mentioned previously, the chief source of carbon monoxide in the atmosphere is combustion. However, except for motor vehicles and other internal-combustion engines, very little carbon monoxide is found in the effluents from properly adjusted, properly operated installations. Although certain industrial operations, such as electric and blast furnaces, some petroleum refining operations, gas manufacturing plants, and coal mines, are potential contributors of carbon monoxide to the atmosphere, automobile exhaust is by far the most important source.

In 1968, the Los Angeles County Air Pollution Control District estimated that 98% of the carbon monoxide emissions in Los Angeles County came from gasoline-powered motor vehicles.

Analytical Methods

The standard method for analyzing stack gases and auto exhaust systems for their carbon monoxide content is by Orsat analysis, which is precise to about 0.1% and hence not suitable for atmospheric analyses. The method makes use of a saturated gas sample and determines carbon dioxide, oxygen, carbon monoxide, and sometimes hydrocarbons.

The most common method used for atmospheric analysis is the non-dispersive infrared gas analyzer [53]. The method most used for spot analyses in the atmosphere is a colorimetric method, based on the fact that when highly purified silica gel impregnated with ammonium molybdate and a solution of palladium sulfate is exposed to carbon monoxide, molybdenum blue is formed [84]. Special tubes containing this reagent are available commercially (Fig. 5.5). At atmospheric concentrations the method is accurate only to 15 to 25% of the true concentration [53].

Abatement Methods

The most commonly used method of abating carbon monoxide is to burn it, i.e., oxidize it, to carbon dioxide. Where concentrations and quantities are large enough, the heat evolved may be recovered economically. In the petroleum refining industry, waste heat boilers utilizing carbon monoxide are often a part of catalytic cracking units; in the steel industry, blast furnace gas, essentially carbon monoxide, is cleaned and returned to the furnace as a heating gas.

Where the concentration and temperature of the carbon monoxide are too low to sustain a flame but the quantity is still large, the gas is often burned catalytically (see next section on "Hydrocarbons"). In the case

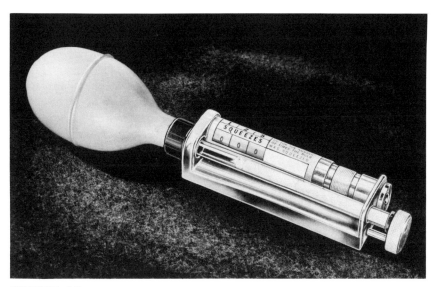

FIGURE 5.5.
Colorimetric carbon monoxide tester. (*Courtesy Mine Safety Appliances Company.*)

where both the concentration and the total quantity are small, the carbon monoxide may still be oxidized catalytically and the effluent carbon dioxide gas exhausted to the atmosphere. A very effective catalyst at normal atmospheric temperature is hopcalite, a special formulation of manganese dioxide and copper oxide. Hopcalite decomposes rapidly, however, at high temperatures and in highly humid atmospheres. If carbon monoxide must be removed selectively from gases containing other combustible constituents, as in some chemical operations, it may be removed by scrubbing with copper ammonium formate or similar solution [20].

The removal of carbon monoxide from auto exhaust poses special problems. Several commercial concerns sell catalytic mufflers for lift trucks used in mines and warehouses. The mufflers contain platinum or palladium catalysts and are highly effective with units using white (nonleaded) gasoline or liquefied petroleum gases for fuel. Methods used for removing carbon monoxide from motor vehicle exhaust gases are discussed in Chapter 9.

HYDROCARBONS AND OTHER ORGANIC GASES

The term *hydrocarbons* covers a wide variety of organic compounds containing only carbon and hydrogen. The simplest hydrocarbon is methane (CH_4), the principal constituent of natural gas. Other common gaseous hydrocarbons are acetylene (C_2H_2), ethylene (C_2H_4), propane (C_3H_8), and butane (C_4H_{10}). Propane and butane are the principal constituents of liquefied petroleum gas (LPG).

As the number of carbon atoms in the hydrocarbon molecule increases, its boiling point becomes higher. Hence, hydrocarbons containing 1 to 3 carbon atoms are normally gaseous; those with 4 carbon atoms are on the borderline between a gas and a liquid at ordinary temperatures; those with 5 to about 16 carbon atoms are liquids; and hydrocarbons with more than 16 to 20 carbon atoms are solids. There is, of course, some overlapping depending on the chemical structure of the molecule.

Chemically, the common hydrocarbons may be classified by type as: paraffins or saturated straight-chain (and branched-chain) hydrocarbons, e.g., methane (CH_4), propane (C_3H_8), isopentane (C_5H_{12}), hexane (C_6H_{12}); olefins or unsaturated straight-chain (and branched-chain) hydrocarbons, e.g., ethylene (C_2H_4), propylene (C_3H_6), pentene (C_5H_{10}), hexene (C_6H_{12}), isooctene (C_8H_{16}); aromatics or unsaturated ring compounds, e.g., benzene (C_6H_6), toluene (C_7H_8), naphthalene ($C_{10}H_8$); and naphthenes or saturated ring compounds, e.g., cyclopentane (C_5H_{10}), cyclohexane (C_6H_{12}), methyl cyclohexane (C_7H_{14}).

Methane is the only hydrocarbon that may be considered a normal constituent of the atmosphere, usually in concentrations of 2 ppm. Other hydrocarbons result from human activity. Monthly average atmospheric concentrations in urban communities usually vary from 2.5 to 4.5 ppm (calculated as methane) with maxima in the 6- to 10-ppm range.

The importance of hydrocarbons in air pollution lies almost entirely in their reactivity with nitrogen dioxide in the presence of sunlight to form photochemical smog (see Chapter 8). In this connection, it appears that the more reactive types of hydrocarbons (i.e., the olefins and aromatics) are by far the most important.

In addition to hydrocarbons, various oxygenated, chlorinated, and other organic compounds may also be found in the air. The oxygenated organics include simple alcohols, aldehydes, acids, ketones, esters, and ethers. The other compounds may be more complex. The amounts of these nonhydrocarbon organic compounds in the atmosphere are usually much less than the amounts of hydrocarbons. Emissions in Los Angeles County are estimated to be about 20% of total organics.

As mentioned previously, the importance of atmospheric organics is a function of their reactivity in photochemical oxidation reactions (see Chapter 8). For control purposes, the Los Angeles County APCD and the Bay Area APCD have classified organic solvents according to reactivity, but the two lists do not agree. Actually too few data are available to make a meaningful list. Work at Stanford Research Institute [90] has indicated that the reactivity of a compound varies depending on whether it is in a single-component or multi-component system, and on its relative atmospheric concentration.

Sources

The ultimate sources of atmospheric hydrocarbons are natural gas and crude petroleum. Specific sources involve all places where these materials or their simple products are handled. Some hydrocarbons are lost from natural gas and crude oil in the oil fields; more are lost in treating and refining operations. Some are lost in transporting and marketing petroleum products, e.g., gasoline, kerosene, and Stoddard solvent, and a relatively large quantity of hydrocarbons is emitted to the atmosphere from automobile exhaust and paint and lacquer solvents.

Hydrocarbon losses from oil fields have not been adequately catalogued, but they probably include breathing and filling losses from tanks during field storage, casinghead gas, and miscellaneous losses that occur during the separation of bottom sediment and water.

The greatest amount of hydrocarbons lost by evaporation and handling of petroleum probably occurs in the refinery. Over-all losses in refineries have been estimated at 0.1 to 0.2% of the total crude oil processed [79]. The largest losses, brought about by diurnal temperature change, are breathing losses from gasoline and intermediate product storage tanks. The magnitude of the loss depends on the volatility of the stored liquid and the magnitude of the temperature change. Similar losses occur during filling of the tank. These losses may be evaluated by calculation [35].

Other sources of hydrocarbon losses are oil-water separators [88]; flares; vacuum jets; relief valves; leakage from pump glands, valves, and fittings; air blowing of light oils; blind changing; and accidental spills.

In the marketing of gasoline, vapor losses result from filling tanks at the bulk loading station, filling tank cars or tank trucks, filling service station tanks, and filling automobile tanks.

The largest single loss of hydrocarbons is from the motor vehicle (see Chapter 9). Smaller amounts are lost from aircraft during ground taxiing, takeoff, and landing (also see Chapter 9).

Other substantial emissions of hydrocarbons to the atmosphere arise from the use of petroleum solvents in painting (more properly, surface coating), dry cleaning, and rotogravure printing. The Los Angeles County APCD estimates daily emissions of solvents from these sources in Los Angeles County amount to 500 tons.

Analytical Methods

The most commonly used methods for determining the hydrocarbon content of the atmosphere and of process streams are spectrographic, chromatographic, and flame ionization methods. All three methods are used for the comparatively high concentrations in process streams. Both dispersive and nondispersive infrared instruments are available, but the nondispersive instrument is normally used in process streams. It is sensitized for the particular hydrocarbon of interest so that quantitative analyses can be made. Figure 5.6 shows a schematic diagram of the nondispersive gas analyzer.

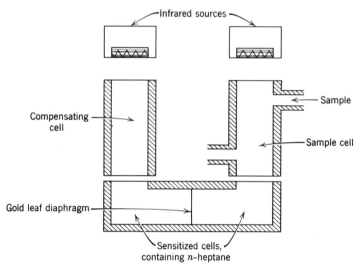

FIGURE 5.6.
Diagram of nondispersive infrared analyzer.

The analysis is made by passing the exhaust gas through the right-hand cell. The compensating cell, in parallel with the sample cell, contains air and serves as a blank. Two sensitized cells, shown in the bottom of Fig. 5.6, are placed in series with the parallel sample and blank cells. These sensitized cells contain the compound which is being determined, in the present case, a representative hydrocarbon.

Infrared radiation passes down through both cells. As there is no hydrocarbon in the blank cell, this infrared radiation is absorbed by the sensitized cell at the characteristic wavelength of hydrocarbons (chiefly 3.42 μ). This absorption of energy heats the gas in the left-hand portion of the sensitizing cell. Similar infrared radiation passes through the sample cell, but is partially absorbed by the gasoline-type hydrocarbons in the exhaust. The percentage of absorption depends on the concentration of hydrocarbons in the exhaust gas. The infrared radiation which is not absorbed by the hydrocarbons in the exhaust sample is subsequently absorbed by the hydrocarbon in the sensitizing cell. Since the amount of radiation remaining to be absorbed is less, the amount of heat produced in the sensitizing cell on the right is less. A diaphragm between the sensitizing cells forms one side of a condenser. Differential heating of the sensitizing gas produces a deflection of the diaphragm, changing the condenser capacity. After suitable amplification, the change is indicated on a meter or a strip chart recorder [86].

The nondispersive analyzer may also be used to measure the approximate total hydrocarbon content of a mixture of hydrocarbons. It may be sensitized with the mixture itself or, if the components of the mixture vary widely, with a single hydrocarbon or a mixture of selected hydrocarbons. In these latter cases precision may be poor because of the different response of the various molecular species.

This has been quite a problem in the analysis of auto exhaust. Some investigators have sensitized their instruments with hexane, in which case certain hydrocarbons, e.g., methane, acetylene, and ethylene, and the aromatics show little, if any, absorption. Another group sensitized the instrument with a special mixture of benzene, acetylene, and ethylene, in which case most hydrocarbons absorb infrared radiation more or less proportionally to the number of carbon atoms in the molecule. Unfortunately, however, under these conditions, certain hydrocarbons, e.g., ethane, absorb twice the calculated amount of radiation predicted.

The mass spectrometer shows several specific mass peaks for each hydrocarbon and, therefore, has a high order of precision for simple mixtures. However, when the mixture is complex or the organic molecule has a high molecular weight (more than 20 carbon atoms) the resulting spectrogram is extremely complex and of doubtful utility.

In the past 10 years, gas-liquid chromatography has come to the fore as a means of separating complex mixtures into individual components for subsequent determination, usually by physical methods. The separation process is based on the differential migration of components of a mixture through a porous medium. The discrete separated substances appear in a carrier gas as a function of time as the carrier gas passes through the adsorption column. Detection of the separated components takes place as the carrier gas emerges from the column. Either concentrated air sampling (e.g., freeze-out traps) [77] or ambient sampling (e.g., plastic bags) may be used to collect material for analysis, depending on the concentration of the pollutant mixture in the air.

Flame ionization hydrocarbon analyzers [5] were introduced in the early 1960's and are now used widely both in atmospheric and auto exhaust analyses. An air or gas stream containing hydrocarbons (or other organics) is mixed with a hydrogen fuel and passed through a small jet. Air is supplied in the annular space around the jet to support combustion. Any hydrocarbon carried into the flame results in the formation of ions. An ion current is generated which is proportional to the rate at which hydrocarbon carbon atoms are introduced into the flame. Values are therefore reported as total hydrocarbons in "ppm carbon atoms."

The flame ionization analyzer also responds to oxygen but this response can be reduced markedly by appropriate operating conditions [31]. The analyzer has rapid response, high sensitivity, and a wide linear range.

Hydrocarbon Abatement

Methods for abating hydrocarbon emissions fall into two categories: (1) those where emission is prevented or where the hydrocarbon is recovered and (2) those in which the hydrocarbon in the emission stream is destroyed, e.g., by burning to carbon dioxide and water.

Prevention of emission, which includes recovery methods, may be provided by proper process design. In the case of petroleum refineries, this includes floating roofs for large gasoline storage tanks; installation of vapor recovery systems on smaller tanks and oil-water separators; venting of vacuum jets, relief valves, and pressure control valves through a manifold system to smokeless flares; and proper maintenance of pumps. Refinery abatement procedures have been described in detail in a federal "control techniques" document [57].

The vapor-recovery methods used in refineries are useful when the concentration of hydrocarbons in the effluent stream is high and relatively uncontaminated. Thus, these methods are adaptable to filling tank trucks and tank cars and even to tanks in service stations. In the last-mentioned case, however, cost has so far been prohibitive.

When, for economic or other reasons, hydrocarbon recovery methods are not feasible, burning or complete oxidation of the hydrocarbons is a satisfactory method of abatement. If the temperature of the gas stream is sufficiently high and a flame can be sustained, direct burning may be practiced. If the hydrocarbon concentration is below the flammable range, the only resort is catalytic oxidation. Combustion can be initiated and sustained below the flammable range, even down to traces, in several commercially available catalytic units that utilize platinum as the active constituent.

Among the many industrial operations utilizing catalytic fume combustion are paint and enamel bake ovens, varnish kettles, catalytic cracking installations, air-blowing of asphalt, phthalic anhydride manufacture (tail gases), and ethylene plants. When the oxidation reaction evolves sufficient heat, heat recovery units are combined with the catalytic burner. A diagram of a catalytic installation on a cooking kettle line in a paint manu-

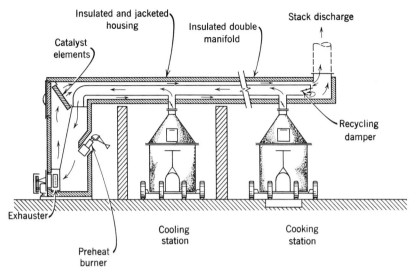

FIGURE 5.7.

Catalytic combustion unit for waste fume control from paint manufacturing.
(*Courtesy Catalytic Combustion Corporation.*)

facturing plant is shown in Fig. 5.7. A view of catalyst units stacked in an oxidation installation is shown in Fig. 5.8.

An entirely different and more complex problem is that of controlling hydrocarbons emitted in automobile exhaust, which will be discussed in Chapter 9.

Emission Regulations

Inasmuch as photochemical smog arises primarily from the reactive hydrocarbons emitted along with oxides of nitrogen from motor vehicles, there is some doubt as to need to control hydrocarbons from stationary sources. The first hydrocarbon-control regulations directed toward stationary sources were those adopted in Los Angeles in the early 1950's before it was realized that motor vehicle exhaust was the primary source of photochemical smog. The regulations generally specified equipment that must be used to reduce evaporative emissions, i.e., floating roofs or vapor recovery units on gasoline storage tanks, covered oil-water separators, vapor collection and disposal systems on gasoline loading and unloading facilities, submerged fill-pipes in filling-station gasoline tanks, and venting of process equipment to smokeless flares.

Leaking valves and pump seals were corrected by continuous inspection made mandatory through the permit system. In the San Francisco Bay Area

FIGURE 5.8.
Catalyst units stacked in oxidation installation. (*Courtesy Oxy-Catalyst, Inc.*)

this maintenance requirement was made mandatory by a specific regulation requiring use of "modern maintenance and operating practices used by superior operators of like equipment and which may reasonably be applied under the circumstances."

Both the Los Angeles and Bay Area APCD's have also adopted regulations to minimize emissions of organic compounds from process equipment and the drying of surface coatings. These regulations are described in Ref. 57. The California regulations have not been widely adopted.

Radioactive Gases

As mentioned previously (see "Radioactive Particles," Chapter 4) the most important potential source of radioactive emissions to the atmosphere is the nuclear power reactor and related fuel-handling facilities.

Two kinds of radioactive gases are produced in nuclear power reactors, activation gases and fission gases. Activation gases (nitrogen, oxygen, argon, hydrogen, etc.) are those which become radioactive in passage through the reactor core where they are exposed to the neutron flux. The fission gases, usually krypton, xenon, and some tritium, come from fuel leaks. All of these gases, except tritium, have short half-lives and can be partially abated by holding them for a sufficient time, then releasing them to a tall stack. The flow diagram of the off-gas system of a boiling water reactor is shown in Fig. 5.9.

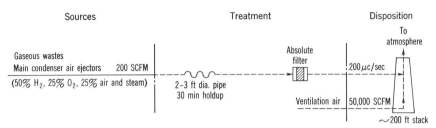

FIGURE 5.9.

Off-gas system of typical boiling water reactor. Note: Activity due half to noble-gas fission and half to activation products.

All modern nuclear power reactors are also designed in such manner that the plants can be shut down safely in case of accident. Among other additional precautions, a containment structure is required which will prevent or hold up the escape to the atmosphere of any fission products accidentally released from the fuel. A further aid is the installation of a heat-removal system and an air-cleaning system in the containment shell. The fission products in the containment atmosphere following an accident can be divided into three groups with respect to air cleaning: particulates, iodine and other chemically reactive gases, and noble gases. Particulate filtration has already been discussed (see Chapter 4).

The most important reactive gases are molecular iodine, hydrogen iodide, and methyl iodide, of which about 90% is molecular iodine. Two methods, or a combination of the two, are currently used to remove the iodine from the containment atmosphere, by adsorption on activated charcoal and by reaction with a spray of caustic-soda or sodium thiosulfate solution.

The only suitable means for minimizing the normal emissions of noble gas radionuclides is by holdup for radioactive decay, followed by controlled discharge through a high stack. For details of reactor air-cleaning systems, see Ref. 40.

Another source of radioactive particulates and gases that is increasing in importance is the nuclear fuel reprocessing plant. Spent fuels from power reactors must be reprocessed to isolate and separate the uranium and plutonium from fission products and from each other. Processing involves removal of the fuel element cladding, followed by a variety of chemical

200 Gases

and physical operations. Highly reactive particulates and gases are generated at almost every point in the process and must be captured to prevent release to the atmosphere. The methods used are essentially the same as those already described, i.e., glass fiber, HEPA and charcoal filters, chemical treatment, and (in the case of the noble gases) holding to allow decomposition and subsequent release from high stacks.

REFERENCES

1. Adams, D. F., "An Automatic Hydrogen Fluoride Recorder Proposed for Industrial Hygiene and Stack Monitoring," *Anal. Chem.,* **32,** 1312–1316 (1960).
2. Adams, E. M., and E. J. Schneider, "Eye Irritants Formed by the Interaction of Styrene and Halogen in the Atmosphere," *Proc. Air Pollution Smoke Prevention Assoc. Am.,* 61–64 (1952).
3. Agate, J. N., et al, "Industrial Fluorosis: A Study of the Hazard to Man and Animals Near Fort William, Scotland," *Medical Research Council (British) Mem.,* 22 (1949).
4. American Society of Mechanical Engineers, "Recommended Guide for the Control of Emission of Oxides of Sulfur—Combustion for Indirect Heat Exchangers," ASME Standard APS-2, American Society of Mechanical Engineers, New York (1970).
5. Andreatch, A. J., and R. Feinland, "Continuous Trace Hydrocarbon Analysis by Flame Ionization," *Anal. Chem.,* **32,** 1021–1024 (1960).
6. Anon., "Profile of Air Pollution Control in Los Angeles County," Los Angeles County APCD (1969).
7. Bamesberger, W. L., and D. F. Adams, "Improvements in the Collection of Hydrogen Sulfide in Cadmium Hydroxide Suspension," *Environ. Sci. and Technology,* **3,** 258–261 (1969).
8. Battigelli, M. C., "Sulfur Dioxide and Acute Effects of Air Pollution," *J. Occupational Medicine,* **10,** No. 9, 68–79 (1968).
9. Blume, J. H., D. R. Miller, and L. A. Nicolai, "Remove Sulfur from Fuel Oil at Lowest Cost," *Hydrocarbon Processing,* 131–136 (Sept. 1969).
10. California Air Resources Board, "Recommended Ambient Air Quality Standards," A Report to the Air Resources Board by the Technical Advisory Committee (of the Board) in Conjunction with the State Department of Health (May 21, 1969).
11. California Dept. of Public Health, "The Oxides of Nitrogen in Air Pollution" (July 15, 1965).
12. California Dept. of Public Health, "Technical Report of California Standards for Ambient Air Quality and Motor Vehicle Exhaust," State of California Dept. of Public Health, Berkeley, California, 69–73 (1960).

13. Cholak, Jacob, "Analytical Methods," in *Air Pollution Handbook,* Paul L. Magill, et al, eds., McGraw-Hill, New York (1956).

14. Cholak, Jacob, "The Nature of Atmospheric Pollution in a Number of Industrial Communities," *Proc. Natl. Air Pollution Symposium,* 2nd Symposium, Pasadena, Calif., 1952.

15. Cholak, Jacob, L. J. Shafer, and R. F. Hoffer, "Results of a Five-Year Investigation of Air Pollution in Cincinnati," *Arch. Ind. Hyg. Occupational Med.,* **6,** 314–325 (1952).

16. Code of Federal Regulations, Chapter IV, Title 42, Part 410, National Primary and Secondary Ambient Air Quality Standards, Appendix A, Method for Determination of Sulfur Dioxide (Pararosaniline Method) (1971).

17. Dickinson, Janet E., "The Operation and Use of the Titrilog and the Autometer," *J. Air Poll. Control Assoc.,* **6,** No. 4, 224–227 (1957).

18. Dinman, B. D., "Pathophysiologic Determinants of Community Air Quality Standards for Carbon Monoxide," *J. Occupational Medicine,* **10,** 14–24 (1968).

19. Ehrlich, R., and M. C. Henry, "Chronic Toxicity of Nitrogen Dioxide: I. Effect on Resistance to Bacterial Pneumonia," *Archives of Environmental Health,* **17,** 860–865 (1968).

19a. Environmental Protection Agency, "Air Quality Criteria for Nitrogen Oxides," Publication No. AP-84, U.S. Govt. Printing Office, Washington, D.C. (1971).

20. Faith, W. L., Donald B. Keyes, and Ronald L. Clark, *Industrial Chemicals,* 3rd Ed., Wiley, New York (1965).

21. Faith, W. L., Donald B. Keyes, and Ronald L. Clark, *Industrial Chemicals,* 3rd Ed., Wiley, New York, 741 (1965).

22. Farrah, G. H., "Manual Procedures for the Estimation of Atmospheric Fluorides," *J. Air Poll. Control Assoc.,* **17,** No. 11, 738–741 (1967).

23. Foster, E. G., and F. Daniels, "Recovery of Nitrogen Oxides by Silica Gel," *Ind. Eng. Chem.,* **43,** 986–992 (1951).

24. Goldman, F. H., and M. B. Jacobs, *Chemical Methods in Industrial Hygiene,* Interscience, New York, 205–212 (1953).

25. Haggard, H. W., "Action of Irritant Gases Upon the Respiratory Tract," *J. Ind. Hyg.,* **5,** 390–398 (1924).

26. Harkins, J., and S. W. Nicksic, "Studies on the Role of Sulfur Dioxide in Visibility Reduction," *J. Air Poll. Control Assoc.,* **15,** 218–221 (1965).

27. Hemeon, W. C. L., "The Estimation of Health Hazards from Air Pollution," *Arch. Ind. Health,* **11,** 397–402 (1955).

28. Hendrickson, E. R., "Dispersion and Effects of Airborne Fluorides in Central Florida," *J. Air Poll. Control Assoc.,* **11,** 220–225, 232 (1961).

29. Hirschler, D. A., et al, "Particulate Lead Compounds in Automobile Exhaust Gas," *Ind. Eng. Chem.,* **49,** No. 7, 1131–1142 (1957).

30. Hodge, H. C., and F. A. Smith, "Air Quality Criteria for the Effects of Fluorides on Man," *J. Air Poll. Control Assoc.,* **20,** 226–232 (1970).

31. Jackson, M. W., "Analysis for Exhaust Gas Hydrocarbons—Non-dispersive Infrared Versus Flame Ionization," *J. Air Poll. Control Assoc.,* **11,** 697–702 (1966).

32. Jacobs, M. B., *Chemical Analysis Series, Vol. I,* Interscience, New York, 354 (1949).

33. Jimeson, R. M., "Utilizing Solvent Refined Coal in Power Plants," *Chem. Engr. Progress,* **62,** 53–60 (1966).

34. Johnstone, H. F., and D. R. Coughanowr, "Absorption of Sulfur Dioxide from Air and Oxidation in Drops Containing Dissolved Catalysts," *Ind. Eng. Chem.,* **50,** 1169–1172 (1958).

35. Kanter, C. V., et al, "Interim Progress Report, July 1, 1956," Joint District, Federal and State Project for the Evaluation of Refinery Emissions, Los Angeles, Calif.

36. Kanter, C. V., R. G. Lunche, and A. P. Fudurich, "Techniques of Testing for Air Contaminants from Combustion Sources," *J. Air Poll. Control Assoc.,* **6,** No. 4, 191–199 (1957).

37. Katz, M., "Atmospheric Pollution Studies with Particular Reference to the Detroit-Windsor Area," *Air Repair,* **4,** No. 4, 176–183 (1955).

38. Katz, M., "Sources of Pollution," *Proc. Natl. Air Pollution Symposium,* 2nd Symposium, Pasadena, Calif., 1952.

39. Katz, B., and R. D. Oldenkamp, "Integration of Molten Carbonate Process for Control of Sulfur Oxide Emissions from a Power Plant," ASME Preprint 69-WA/APC-6, American Society of Mechanical Engineers, New York (1969).

40. Kielholtz, G. W., C. E. Guthrie, and G. C. Battle, Jr., "Air Cleaning as an Engineered Safety Feature in Light-Water-Cooled Power Reactors," Oak Ridge National Laboratory, Oak Ridge, Tenn. (Sept. 1968).

41. Kleinerman, J., and G. W. Wright, "The Reparative Capacity of Animal Lungs after Exposure to Various Single and Multiple Doses of Nitrite," *Amer. Review of Respiratory Diseases,* **83,** 423–424 (1961).

42. Kohl, A. L., "Selective Hydrogen Sulfide Absorption—A Review of Available Processes," *Petrol. Processing,* **6,** 26–31 (1951).

43. Little, A. D., Inc., "Research on Chemical Odors, Part 1—Odor Thresholds for 53 Commercial Chemicals," Manufacturing Chemists' Association, Washington, D.C. (1968).

44. Los Angeles County Air Pollution Control District, "Survey of Carbon Monoxide Concentrations Along Major Traffic Arteries," Los Angeles, Calif. (1956).

45. McCabe, L. C., and G. D. Clayton, "Air Pollution by Hydrogen Sulfide in Poza Rica, Mexico," *Arch. Ind. Hyg. Occupational Med.,* **6,** 199–213 (1952).

46. McCrea, D. H., J. H. Field, and E. R. Bauer, Jr., "The Alkalized Alumina System for SO₂ Removal: Design and Operation of a Continuous Pilot Plant," ASME Preprint 68-WA/FU3, American Society of Mechanical Engineers, New York (1968).

47. McCune, D. C., "Fluoride Criteria for Vegetation Reflect the Diversity of Plant Kingdom," *Environ. Sci. and Technology,* **8,** 720, 727–28, 731–32, 735 (1969).

48. Malette, F. S., ed., *Problems and Control of Air Pollution,* Reinhold, New York (1955).

49. Maurin, P. G., and J. Jonakin, "Removing Sulfur Oxides from Stacks," *Chemical Engineering,* Deskbook Issue (April 27, 1970).

50. Meetham, A. R., *Atmospheric Pollution,* 2nd Ed., Pergamon Press, London, 215 (1956).

51. Middleton, J. T., E. F. Darley, and R. F. Brewer, "Damage to Vegetation from Polluted Atmospheres," *J. Air Poll. Control Assoc.,* **8,** 9–15 (1958).

52. Mills, J. L., K. D. Luedtke, P. F. Woolrich, and L. B. Perry, "A Summary of Data on Air Pollution by Oxides of Nitrogen Vented from Stationary Sources." Final Report. Report No. 4. Emissions of Oxides of Nitrogen from Stationary Sources in Los Angeles County, L.A. County Air Pollution Control District, Los Angeles (1961).

53. National Air Pollution Control Administration, "Air Quality Criteria for Carbon Monoxide," NAPCA Publication AP-62, U.S. Govt. Printing Office, Washington, D.C. (1970).

54. National Air Pollution Control Administration, "Air Quality Criteria for Sulfur Oxides," NAPCA Publication No. AP-50, U.S. Govt. Printing Office, Washington, D.C. (1969).

55. National Air Pollution Control Administration, "Air Quality Data—1966," NAPCA Publication No. APTD 68-9, National Air Pollution Control Administration, Durham, N.C. (1968).

56. National Air Pollution Control Administration, "Control Techniques for Carbon Monoxide, Nitrogen Oxide, and Hydrocarbon Emissions from Mobile Sources," NAPCA Publication No. AP-66, Chapter 8, U.S. Govt. Printing Office, Washington, D.C. (1970).

57. National Air Pollution Control Administration, "Control Techniques for Hydrocarbon and Organic Solvent Emissions from Stationary Sources," NAPCA Publication No. AP-68, U.S. Govt. Printing Office, Washington, D.C. (1970).

58. National Air Pollution Control Administration, "Control Techniques for Nitrogen Oxide Emissions from Stationary Sources," NAPCA Publ. No. AP-67, U.S. Govt. Printing Office, Washington, D.C. (1970).

59. Natonal Air Pollution Control Administration, "Control Techniques for Sulfur Oxide Air Pollutants," NAPCA Publ. No. AP-52, U.S. Govt. Printing Office, Washington, D.C. (1969).

60. Palmer, H. F., C. E. Rodes, and C. J. Nelson, "Performance Characteristics

of Instrumental Methods for Monitoring Sulfur Dioxide. 2. Field Evaluation," *J. Air Poll. Control Assoc.*, **19**, 778–786 (1969).

61. Peterson, J. E., H. R. Hoyle, and E. J. Schneider, "The Analysis of Air for Halogenated Hydrocarbon Contaminants by Means of Absorption on Silica Gel," *Am. Ind. Hyg. Assoc. Quart.*, **17**, 429–433 (1956).

62. Pratt, D. C. F., and A. Rutherford, "Removal of Hydrogen Sulfide from Exhaust Air from a Viscose Staple Fibre Factory," *Chem. & Ind. (London)*, **41**, 1281–1286 (1955).

63. Reed, R. M., and N. C. Updegraff, "Removal of Hydrogen Sulfide from Industrial Gases," *Ind. Eng. Chem.*, **42**, 2269–2277 (1950).

64. Renzetti, N. A., "Analysis of Air Near Heavy Traffic Arteries," Air Pollution Foundation (San Marino, Calif.) Rept. 16 (December 1956).

65. Rodes, C. E., H. F. Palmer, L. A. Elfers, and C. H. Norris, "Performance Characteristics of Instrumental Methods for Monitoring Sulfur Dioxide. 1. Laboratory Evaluation," *J. Air Poll. Control Assoc.*, **19**, 575–584 (1969).

66. Saltzman, Bernard E., "Colorimetric Microdetermination of Nitrogen Dioxide in the Atmosphere," *Anal. Chem.*, **26**, 1949–1955 (1954).

67. Sands, A. E., et al, "The Determination of Low Concentrations of Hydrogen Sulfide in Gas by the Methylene Blue Method," United States Bureau of Mines Rept. Invest. 4547 (1949).

68. Schrenk, H. H., "Air Pollution in Donora, Pa.," Public Health Bull. 306, United States Public Health Service, Washington, D.C. (1949).

69. Semrau, K. T., "Feasibility Study of New Sulfur Oxide Control Processes for Application to Smelters and Power Plants; Part IV: The Wellman-Lord SO_2 Recovery Process for Application to Power Plant Flue Gases (Final Report)," Report PB 197208, National Technical Information Service, Springfield, Va. (1969).

70. Sensenbaugh, J. D., and W. C. L. Hemeon, "A Low-Cost Sampler for Measurement of Low Concentrations of Hydrogen Sulfide," *Air Repair*, **4**, 5–7, 25 (1954).

71. Shupe, J. L., "Levels of Toxicity to Animals Provide Sound Basis for Fluoride Standards," *Environ. Sci. and Technology*, **8**, 721–726 (1969).

72. Slack, A. V., and H. L. Falkenberry, "Sulfur Dioxide Removal from Power Plant Stack Gas by Limestone Injection," ASME Preprint No. 69-Pwr-2, American Society of Mechanical Engineers, New York (1969).

73. Smith, R. G., "A Study of Carboxyhemoglobin Levels Resulting from Exposure to a Low Concentration of Carbon Monoxide," *J. Occupational Medicine*, **10**, 456–63 (1968).

74. Smith, M. E., "Reduction of Ambient Air Concentrations of Pollutants by Dispersion from High Stacks," Proceedings: The Third National Conference on Air Pollution, Washington, D.C., PHS Publication 1649, 151–160 (1967).

75. Southwest Research Institute, "Air Pollution Survey of the Houston Area," Tech. Rept. No. 4 for Houston Chamber of Commerce, Texas (1957).

76. Specht, R. C., and R. R. Calaceto, "Gaseous Fluoride Emissions from Stationary Sources," *Chem. Engr. Progress,* **63,** 78–84 (1967).

77. Stephens, E. R., and F. R. Burleson, "Analysis of the Atmosphere for Light Hydrocarbons," *J. Air Poll. Control Assoc.,* **17,** 147–153 (1967).

78. Stites, J. G., Jr., W. R. Horlacher, Jr., J. L. C. Bachofer, Jr., and J. S. Bartman, "The Catalytic-Oxidation System for Removing SO_2 from Flue Gas," ASME Preprint 68-WA/APC-2, American Society of Mechanical Engineers, New York (1968).

79. Swearingen, J. S., and H. Levin, "Hydrocarbon Losses from the Petroleum Industry in Los Angeles County," Air Pollution Foundation (San Marino, Calif.) Rept. 5 (November 1954).

80. Tabor, E. C., and C. C. Golden, "Results of Five Years' Operation of the National Gas Sampling Network," *J. Air Poll. Control Assoc.,* **15,** 7–11 (1965).

81. Teller, A. J., "Control of Gaseous Fluoride Emissions," *Chem. Engr. Progress,* **63,** 75–79 (1967).

82. Thomas, H. V., P. K. Mueller, and R. Wright, "Response of Rat Lung Mast Cells to NO_2 Inhalation," *J. Air Poll. Control Assoc.,* **17,** 33–35 (1967).

83. Thomas, M. D., "Effects of Air Pollution on Plants" in *Air Pollution,* Columbia Univ. Press, pp. 244–260 (1961).

84. Thomas, M. D., and J. O. Ivie, *Ind. Eng. Chem., Anal. Ed.,* 383–387 (June 1946).

85. Thomas, M. D., J. A. MacLeod, R. C. Robbins, R. C. Goettelman, R. W. Eldridge, and L. H. Rogers, "Automatic Apparatus for the Determination of Nitric Oxide and Nitrogen Dioxide in the Atmosphere," *Anal. Chem.,* **28,** 1810–1816 (1956).

86. Twiss, S. B., et al, "Application of Infrared Spectroscopy to Exhaust Gas Analysis," *J. Air Poll. Control Assoc.,* **5,** No. 2, 75–83 (1955).

87. Urone, Paul, and W. H. Schroeder, "SO_2 in the Atmosphere," *Environ. Sci. and Technology,* **3,** 436–445 (1969).

88. Viles, P. S., and L. T. Jones, "Estimating Oil Losses by Atmospheric Evaporation from Refinery Separator Surfaces," *Petrol. Refiner,* **31,** 117 (1952).

89. West, P. W., and G. C. Gaeke, "Fixation of Sulfur Dioxide as Disulfitomercurate (II) and Subsequent Colorimetric Estimation," *Anal. Chem.,* **28,** 1816–1819 (1956).

90. Wilson, K. W., "Photoreactivity of Trichloroethylene," Stanford Research Institute, Menlo Park, Calif., 2–3 (Sept. 1968).

91. Zimmerman, P. W., "Impurities in the Air and Their Influence on Plant Life," *Proc. Natl. Air Pollution Symposium,* 1st Symposium, Pasadena, Calif., 1949.

92. Zimmerman, R. E., "Economics of Coal Desulfurization," *Chem. Engr. Progress,* **62,** 61–66 (1966).

6

ODORS

Odor is undoubtedly the most complex of all air pollution problems. Not only is odor caused by very minute quantities of substances, but also the only good measuring device is the human nose, which is notoriously undependable. Further, people have mixed reactions with respect to the offensiveness of odors. One needs only to recall the perfume worn by some women to realize that a vast difference of opinion must exist as to what is fragrant and what is, to be charitable, less fragrant.

Besides disagreement as to the offensiveness of certain odors, two other facts stand out. First, an unfamiliar odor is more likely to cause complaint than a familiar one. Second, given sufficient time, one can become accustomed to almost any odor and only notice it when it varies in intensity.

Time and place are also important psychological factors in one's reaction to odors. Nearly everyone likes the odor of chocolate, but if one lives near a chocolate factory and smells it 24 hours a day, it is less pleasing. Similarly, the farmer who moves to the city would be the first to complain if his neighbor piled fresh manure under his dining-room window.

Obviously, the effect of disagreeable odors on people is primarily a nuisance effect, and it is usually treated as such. There are cases, however, when secondary effects may be quite important. Certain intense odors may lead to nausea. Moreover, persistent odors that regularly interfere with sleep cannot help but interfere with one's well-being. On the economic front, the loss of property values near poorly operated dumps and slaughterhouses is partly a consequence of offensive odors.

SOURCES OF ODORS

The most common chemical sources of odors are organic compounds that contain nitrogen or sulfur in their molecular configuration. Amines,

206

mercaptans, and the complex decomposition products of proteins (containing both nitrogenous and sulfurous compounds) are typical representatives of this general class.

Among the odor-producing compounds that contain no nitrogen or sulfur are phenols, cresols, and some fatty acids (butyric, valeric, and caproic acids). A list of odorous industrial operations is given in Table 6.1.

TABLE 6.1

Odorous Industrial Operations (Ref. 9)

Industry	Odorous material
Chemical manufacture	Hydrogen sulfide, ammonia, amines, alcohols, aldehydes, phenols, mercaptans, esters, chlorine and chlorinated organics, etc.
Coke ovens	Sulfurous, ammoniacal, and phenolic compounds
Fertilizer	Bone meal, organic nitrogen compounds, ammonia
Food and kindred products	Dairy wastes, cannery wastes, fish, baking bread, chocolate, flavors, packinghouse wastes, meat products for rendering, coffee roaster effluents, cooking odors, etc.
Foundries	Core-oven odors, quenching oils
General industrial	Burning rubber, forming and molding plastics, incinerator smoke, solvents and lacquers, asphalt
Petroleum	Sulfur compounds from crude oil, cresols, asphalt
Pharmaceuticals	Biological extracts and wastes, spent fermentation liquors
Pulp and paper	Sulfurous compounds
Soap and toiletries	Perfumes, animal fats
Tanneries	Hair, flesh, hides

Odorous compounds may also arise from a variety of human activities. Among the more important nonindustrial sources of odor are garbage dumps, sewage works, agricultural operations, motor vehicles, and, occasionally, gases of natural origin (sulphur springs, decaying vegetation, etc.).

MEASUREMENT OF ODOR

The measurement of odor falls into two categories: determination of the threshold concentration of odoriferous gases, and determination of the type and intensity of atmospheric odors.

Considerable research has been carried out on ways of measuring the odor threshold of various substances and characterizing the intensity of various odors. Byrd and Phelps [2] have described various odor-measuring devices that have been used in such work. Most of these osmometers are based on a vapor-dilution method, in which the amount of odor-producing substance added to purified air is gradually increased until an observer or "sniffer" can just barely detect the odor.

Several investigators have prepared lists of the odor thresholds of various chemicals, but from the different values reported by different investigators it is apparent that the precision of the methods used was not high. Table 6.2 shows a listing published by the Manufacturing Chemists Association in 1968 [11].

The identification and determination of odors in the open air is even less precise than the determination of threshold values in the laboratory. The best method is the appraisal of odor by a jury or panel. A group of 5 to 10 persons in normal health sniff the air at a given location at the same time and report individually the nature and intensity of the odor. Some experience may be required to identify an odor, but intensity is usually stated according to a predetermined rating system. Odor intensity, like most physiological responses varies as the logarithm of the concentration. Thus a doubling of intensity requires a tenfold increase in concentration. A widely used scale for odor intensity is the following: 0, no odor; 1, threshold level (barely perceptible); 2, definite odor; 3, strong odor; and 4, overpowering odor. Half scores may be used when the observer is undecided. By averaging the values recorded by members of the panel, a single value can be assigned to the odor intensity at a given location. If measurements are made by the same panel at various locations, the source of the particular odor may sometimes be found. If the members of the panel are sufficiently discerning, the determination may be limited to a particular type of odor, but care must be used because mixtures of two or more different odors can be very misleading.

Several descriptive words have been used to characterize the type or *quality* of an odor. Some of these are fruity, flowery, burnt, fragrant, sweet, foul, nauseating, etc. A system devised by Crocker and Henderson [3] uses four reference qualities (fragrant, acid, burnt, and caprylic), each on a nine-point intensity scale.

The expert panel method is fairly expensive and limited to short periods of use. One method used by control officers is to have residents in certain areas note the time of day a particular odor is noticed and the number of times over a specific period. These data may be collected and reported graphically [10].

Intermediate between determination of the odor threshold level of individual compounds and the estimation of atmospheric odors is the determination of the concentration of odoriferous compounds in ducts and stacks. A syringe technique for this measurement has been developed by

TABLE 6.2

Odor Thresholds of Common Chemicals (ppm by volume in air)

Chemical	Response	
	50%	100%
Acetaldehyde	0.21	0.21
Acetic acid	0.21	1.0
Acetone	46.8	100.0
Acrolein	0.1	0.21
Acrylonitrile	21.4	21.4
Allyl chloride	0.21	0.47
Amine, dimethyl	0.021	0.047
Amine, monomethyl	0.021	0.021
Amine, trimethyl	0.00021	0.00021
Ammonia	21.4	46.8
Aniline	1.0	1.0
Benzene	2.14	4.68
Benzyl chloride	0.01	0.047
Benzyl sulfide	0.0021	0.0021
Bromine	0.047	0.047
Butyric acid	0.00047	0.001
Carbon disulfide	0.1	0.21
Carbon tetrachloride (chlorination of CS_2)	10.0	21.4
Carbon tetrachloride (chlorination of CH_4)	46.8	100.0
Chloral	0.047	0.047
Chlorine	0.314	0.314
p-Cresol	0.00047	0.001
Dimethylacetamide	21.4	46.8
Dimethylformamide	21.4	100.0
Dimethyl sulfide	0.001	0.001
Diphenyl ether (perfume grade)	0.1	0.1
Diphenyl sulfide	0.0021	0.0047
Ethanol (synthetic)	4.68	10.0
Ethyl acrylate	0.0001	0.00047
Ethyl mercaptan	0.00047	0.001
Formaldehyde	1.0	1.0
Hydrochloric acid gas	10.0	10.0
Hydrogen sulfide (from Na_2S)	0.001	0.0047
Hydrogen sulfide gas	0.00021	0.00047
Methanol	100.0	100.0
Methyl chloride	[Above 10 ppm]	

TABLE 6.2 (*continued*)

Chemical	Response	
	50%	100%
Methylene chloride	214.0	214.0
Methyl ethyl ketone	4.68	10.0
Methyl isobutyl ketone	0.47	0.47
Methyl mercaptan	0.001	0.0021
Methyl methacrylate	0.21	0.21
Monochlorobenzene	0.21	0.21
Nitrobenzene	0.0047	0.0047
Perchloroethylene	4.68	4.68
Phenol	0.021	0.047
Phosgene	0.47	1.0
Phosphine	0.021	0.021
Pyridine	0.01	0.021
Styrene (inhibited)	0.047	0.1
Styrene (uninhibited)	0.047	0.047
Sulfur dichloride	0.001	0.001
Sulfur dioxide	0.47	0.47
Toluene (from coke)	2.14	4.68
Toluene (from petroleum)	2.14	2.14
Tolylene diisocyanate	0.21	2.14
Trichloroethylene	21.4	21.4
p-Xylene	0.47	0.47

Byrd of the Procter and Gamble Company [1]. Measured samples of odor-laden gas are taken in hypodermic syringes and brought to an odor-free room. A small quantity of the gas is then added to a syringe into which odor-free air is then drawn. The observer holds the syringe to his nostrils and drives the mixture into his nasal passages by means of a plunger. He can then determine the amount of dilution necessary to bring the odor to a barely perceptible level. Concentration is expressed in terms of *odor unit*, which is defined as the amount of odor necessary to contaminate 1 cu ft of air to the threshold level.

Odorants in gas streams or in the atmosphere may also be collected by passing known volumes of gas or air through a column of activated carbon [13] or by condensation techniques [12]. The collected odorants may then be taken to a laboratory and released by vacuum desorption and distillation, respectively.

ODOR ABATEMENT

Source control is the most effective means of abating odor, as well as any other form of air pollution. In many cases this requires only good sanitation practices, as the most persistent and offensive odors arise from putrefaction. In many manufacturing operations in the chemical and process industries, a raw material, the main product, a by-product, or an unavoidable intermediate product may be odoriferous. Inasmuch as most odorants are gases, they may be emitted from vents or stacks. Accordingly, they may be controlled by the same means described for the control of various gases (see Chapter 5). These are absorption or scrubbing, adsorption (on porous solids), and incineration (combustion).

Absorption is applicable when the odorous vapors are soluble or emulsifiable in a liquid or react chemically in solution. Addition of an oxidizing agent such as potassium permanganate [8] or chlorine gas [7] to a circulating water stream has proved effective in some cases. Because of the cost of the oxidizing agent, the system is usually adapted only to gases containing low concentrations of oxidizable substances.

Adsorption, particularly on activated carbon, has been widely used in odor control because activated carbon has a preferential attraction and high retentivity for organic vapors. Furthermore, the retained material may be desorbed comparatively easily, and the carbon reactivated and used again. On the other hand, economic considerations often limit adsorption methods to cases when the odorous material in the gas stream is less than 5 ppm.

When considerably higher concentrations of organic odorants are present, direct-flame or catalytic combustion is widely used. A discussion of commercial units may be found in Chapter 5. Current installations of catalytic units for odor control handle effluent gases from coffee roasters, refuse incinerators, wire-enamelling ovens and dryers, foundry core-baking ovens, paint- and enamel-baking ovens, chemical plants, varnish manufacture, rendering plants, etc.

Direct-flame incineration is the most commonly used method of odor abatement if the odor-containing air stream can be contained. If so, most odors can be destroyed by subjecting them to 1200° to 1400°F for not less than 0.3 second. The greatest drawback to flame incineration is the cost of fuel to produce the required temperature. Accordingly, it is imperative that the odorous stream be kept to as low a volume as possible. In some cases, heat from the hot incinerated gases may be recovered for other uses. An incinerator unit which incorporates heat recovery is shown diagrammatically in Fig. 6.1. In a few cases it has been possible to recirculate odorous air streams into boiler furnaces as secondary air and thus destroy the odor without the use of additional fuel [5].

FIGURE 6.1.

Gas incinerator with heat recovery equipment. (*Courtesy Hirt Combustion Engineers.*)

When the cost of heating the gas stream to 1200°F is prohibitive, catalytic combustion may be considered. In a catalytic unit, oxidation may be initiated at a much lower temperature than necessary for flame combustion. On the other hand, if the temperature of the gases entering the catalyst unit is too high or the concentration of the organic constituent above a certain value, heat may be liberated so rapidly that the catalyst will sinter and become ineffective.

Gases entering the unit should also be free from dusts that might cause catalyst loss by abrasion and from other compounds that might impair catalyst activity by coating the catalyst or poisoning its "active centers."

Another important consideration in odor control by oxidation, particularly catalytic oxidation, is the degree of oxidation attained. As an example, butanol vapors have only a mild odor but upon oxidation first form butyraldehyde which has a fairly intense odor, and then butyric acid which has a very offensive odor. The next step in the oxidation process is to burn the butyric acid to odorless carbon dioxide and water vapor. It is the *complete* oxidation that solves the odor problem in this case; anything less would intensify the problem. In a similar way incomplete oxidation might produce corrosive or lachrymatory compounds.

A means sometimes used to abate an odor problem is to release the odorous stream from a *tall stack* so as to allow the normal dispersion characteristics of the atmosphere to take over and decrease ground-level concentrations below the threshold value. Care must be taken that all likely meteorological situations are considered in determining stack height, lest under some conditions the odor may only be transferred to a downwind location further away. Unless the stack effluent contains considerable heat to add buoyancy to the stack plume, atmospheric dispersion will probably be insufficient.

Detailed methodology of industrial odor abatement in Los Angeles has been published [6].

NEUTRALIZATION OF ODORS

Odor masking and a related technique, odor counteraction, are becoming popular in odor control, because of their effectiveness in abating odors from area sources.

Odor masking is based on the principle that when two odors are mixed, the stronger one will predominate. Thus, when a sufficient amount of a pleasant odor is mixed with an unpleasant one, the latter will become unnoticeable. However, the solution to an odor problem is more complicated than the simple addition of any pleasant odor to an odorous stream, because of the need for stability under a variety of conditions and the unpredictable reactions of people to unfamiliar or unusual odors. Further, one could not morally add a masking agent to an odorous toxic gas. Care must also be taken that the additive is not allergenic, flammable, or corrosive. Because of the complexities of odor masking, the selection of the amount and type of added material should be left to specialists in the field.

There are several methods of using masking compounds to abate odor. The compound is often vaporized or atomized directly into a stack, and it may also be added in a water solution. Under some conditions the masking agent may be added directly to material being processed, such as to a rendering kettle in which bones and meat scraps are being cooked. In such cases, of course, the masking agent must not affect the product deleteriously or be destroyed in the cooking process.

Masking agents are also used to control odors originating outdoors in such places as dumps or waste lagoons. Here, the masking compound must vaporize rapidly enough to overcome the unpleasant odor and slowly enough to last for a reasonable length of time.

Odor counteraction, on the other hand, is based on the principle that certain pairs of odors, in appropriate relative concentrations, are antagonistic. Thus, when the two odors are mixed the noticeability of each is greatly diminished. This principle was first applied in air conditioning, and is familiar to everyone who has used household odor counteractants in aerosol form. Selection of the proper counteractant for a given odor is even more difficult than the selection of masking agents, so again the choice should be left to experts. Several reliable commercial organizations specialize in odor masking and counteraction.

The methods of adding counteractants are similar to those for adding masking agents. The comparatively simple equipment required is shown in Fig. 6.2.

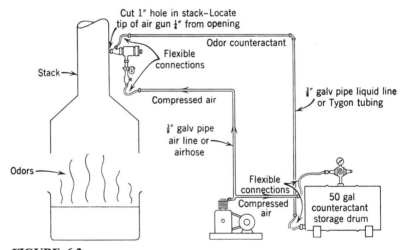

FIGURE 6.2.

Typical layout for adding odor counteractant to stack.

AIR QUALITY AND EMISSION STANDARDS

Although odors are the most common cause of air pollution complaints, few regulations have been directed toward abatement of specific sources. The main reason has been the lack of generally accepted methods for quantifying the obnoxiousness of a given odor. Intensity may be measured at least semiquantitatively, as mentioned previously, but repugnance is highly subjective. Accordingly, most control authorities have resorted to nuisance laws to abate odors, but nuisance regulations are notoriously ineffective except in obviously intolerable situations.

Recently there have been attempts to state both emission and air quality standards in terms of allowable dilutions, e.g., city of St. Louis, state of New Mexico. Only experience will show the effectiveness of this approach.

The most effective rules are those which can be directed toward specific emissions, e.g., the Los Angeles regulation against odors from animal reduction processes (incineration is required, see Ref. 4), and the Oregon Sanitary Authority's regulation of pulp mill odors which limits emissions of sulfur compounds. Similar regulations directed toward other specific industries may well be the most effective means of regulating odorous emissions.

REFERENCES

1. Byrd, J. F., "Demonstration—Syringe Odor Measurement Technique," *J. Air Poll. Control Assoc.,* **7,** No. 1, 58–59 (1957).

2. Byrd, J. F., and A. H. Phelps, Jr., "Odor and Its Measurment," Chapter 23, Vol. II, Second Edition of *Air Pollution,* A. C. Stern, ed., Academic Press, New York (1968).

3. Crocker, E. C., and L. F. Henderson, "Analysis and Classification of Odors," *Am. Perfumer Essent. Oil Rev.,* **22,** 325 (1927).

4. Danielson, John A., ed., "Air Pollution Engineering Manual," p. 6, U.S. Public Health Service Publication No. 999-AP-40, Government Printing Office, Washington, D.C. (1967).

5. Ibid., pp. 187–192.

6. Ibid., pp. 746–784.

7. Ibid., p. 770.

8. Emanuel, A. G., "Potassium Permanganate Offers New Solutions to Air Pollution Control," *Air Engineering* (Sept. 1965).

9. Faith, W. L., "Air Pollution from Industrial Operations and Its Control," Proceedings National Conference on Air Pollution, Public Health Service Publication No. 1022, pp. 88–92, U.S. Government Printing Office, Washington, D.C. (1963).

10. Gruber, Charles W., "Odor Pollution from the Official's Viewpoint," *Symposium on Odor, Am. Soc. Testing Materials Spec. Tech. Publ.,* **164,** 56–68, 1954.

11. Little, A. D., Inc., "Research on Chemical Odors, Part 1—Odor Thresholds for 53 Commercial Chemicals," Manufacturing Chemists Assoc., Washington, D.C. (1968).

12. Pantaleoni, R., "Odors and Fragrances," *Chem. Eng. News,* **31,** 1730–1734 (1953).

13. Turk, Amos, "Obnoxious Odors," *Ind. Wastes,* **3,** No. 1, 9–14 (1958).

7

STATIONARY SOURCES

Stationary sources of air pollution may be divided conveniently into three general categories: space heating (residential, institutional, and commercial); combustible refuse disposal; and industrial processes, including power generation.

Air pollution problems from *residential, institutional,* and *commercial* heating are almost entirely related to the burning of fossil fuels: coal, oil, and natural gas. Institutional and commercial establishments are commonly considered together, and include public and private institutions, schools and hospitals, large office buildings, stores, banks, etc. Basically the air pollution contributions from heating plants are particulate matter and sulfur dioxide. There are some nitrogen oxides formed, but the amounts are small compared with motor vehicles and large stationary power sources. On the other hand particulate matter and sulfur dioxide emissions from heating plants are often the chief source of these contaminants in urban communities. Major contributing factors are low-level emission points and inefficient operation of the heating plant (see Chapter 3). Because of this situation, one school of thought which is increasing in size, believes that all heating and power plants less than an arbitrary size should use only desulfurized and de-ashed fuels.

Combustible refuse disposal has already been discussed in Chapter 3. Current thinking is that open burning should be abolished and that incineration should be restricted to well-designed multiple-chamber incinerators. If these restrictions come about, marked improvements in visibility in urban communities will result. The greatest obstacle to bans on open and inadequate burning of trash is the cost of municipal collection and consequent increases in taxes.

Of all the stationary sources of air pollution, *industrial processes* and *power generation* bear the brunt of present air pollution control regulations and environmental concern. The greatest deterrents to effective control are economic and technological in nature, with considerable variation between old and new plants, large and small plants, from one locality to another, and from one industry to another. The problem is compounded by a desire for "across-the-board" control regulations.

The rest of this chapter deals with the major industries with air pollution potential. The process is described, and major air pollution problems and possible solutions are identified.

Electric Power Generation

One of the hallmarks of modern civilization is extensive use of electric power. Demand is increasing by nearly 100% every 10 years. Only 30 years ago total electrical power consumption in the United States was 180 billion kilowatt-hours. In 1970 it was 1500 billion kwh. By the year 2000 consumption will be nearly 10,000 billion kwh.

The chief sources of this energy at present are combustion of fossil fuels, nuclear power plants, and water power. Although utilization of nuclear energy is increasing rapidly, there will be considerable demand for fossil-fuel-fired plants for many years (see Fig. 7.1). Coal is by far the most common fossil fuel used; smaller quantities of residual fuel oil and natural gas are used. In a few instances, e.g., pulp mills, wood waste is still used.

The basic process of converting coal to electric energy is simple. Coal is burned in a furnace to produce hot combustion gases. The hot gases in turn are used to convert water to high-pressure steam; the steam drives a turbine wheel connected to an electric generator. Details vary considerably. In the combustion process, the carbon and hydrogen content of the fuel are oxidized to carbon dioxide and water vapor which are discharged to the atmosphere after removal of heat. The gases are commonly discharged at 260 to 300°F. Although both carbon dioxide and water vapor are normal constituents of the atmosphere and both are removed eventually by atmospheric processes, there is some concern that increased carbon dioxide emissions may result in an atmospheric buildup that would upset the global heat balance [22]. This so-called "greenhouse effect" could lead to increased heating of the earth over a long period of time and subsequent effects on global ecology.

The main air pollution problem of coal-fired power plants is related to impurities in coal, principally sulfur compounds and ash. Most coals used

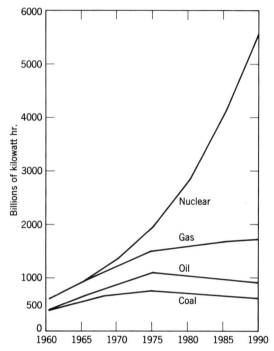

FIGURE 7.1.

Estimated fuel demand for U.S. electric generating plants.

in steam-electric plants vary from 1 to 3% sulfur (some higher) and up to 20% ash. In a modern highly efficient power plant most of the sulfur in the coal is burned to sulfur dioxide (plus a small amount of sulfur trioxide), and 50 to 90% of the ash enters the stack. Inasmuch as the chief air pollutants of public concern in most large urban communities are sulfur oxides and suspended particulate matter, the large coal-burning power plant has become almost synonymous with air pollution.

It is interesting to note that unlike many industries, the air pollution problem of the steam-electric plant is not inefficient operation. On the contrary, the more efficient a unit is the more sulfur and particulate matter will be released. It is true that grossly inefficient burning will lead to black smoke evolution, but this type of operation is usually restricted to comparatively small operations (see Chapter 3).

Another factor that contributes to the air pollution problems of the thermal-electric power plant is the large size of the plant. Thus a 500-megawatt plant (500,000 kw) using 12,000 Btu-per-pound coal containing 3% sulfur and 8% ash (80% up the stack) would have potential emissions of 12.5 tons SO_2 and 13.3 tons of particulate matter per hour. Even if a 99% efficient electrical precipitator were used, particulate emission would be 266 lb per hour.

Another pollutant that is of increasing interest is nitrogen oxide (see Chapter 5). Nitrogen oxide emissions are not a function of the nitrogen content of the fuel but arise rather from oxidation of some of the nitrogen in the air used for combustion.

Residual fuel oil rather than coal is used as a power-plant fuel in areas where economic factors are favorable. Inasmuch as most residual fuel oils also contain sulfur compounds and ash, problems similar to those in coal-burning units are encountered. Natural gas is also used occasionally in areas where it is available. It is free from sulfur compounds and ash and therefore is generally considered a nonpolluting fuel. Nitrogen oxide is formed, however, just as it is with coal and oil.

The central station electric industry has been a leader in the control of emissions of particulate matter. It is reported that the power industry has used 90% of the output of dust collector manufacturers, particularly mechanical collectors, electrostatic precipitators, or the two in combination. Efficiencies are usually greater than 90%; in some metropolitan areas efficiencies of 99.5% have been achieved with some coals. A recent innovation has been the adaptation of Venturi scrubbers to remove fly ash from large power plant stacks.

Attempts have been made to apply both electrostatic precipitators and baghouses to units using residual fuel oil, but neither have been successful. Work is also underway trying to adapt bag filters to coal burners.

So far, methods for reducing sulfur dioxide in flue gases have not been successful (see Chapter 5). Accordingly, stacks have had to be built sufficiently high to allow adequate dispersion in the upper atmosphere. The high stack is also useful in dispersing finely divided fly ash that escapes the electrostatic precipitator.

In a few metropolitan areas, regulations have required the use of natural gas in place of other fuels in order to reduce sulfur dioxide emissions. This alternative is expensive and in the long run is limited by gas supplies. Some jurisdictions have forced utilities to build coal-fired plants in uninhabited areas near the source of the coal (mine-mouth plants) and then transmit high-voltage electricity to the urban market. Cost prohibits wider use of this alternative at the present state of the art.

Many believe that the mine-mouth plant of the future will convert coal to a clean, gaseous or liquid fuel that may be transported by pipeline to the urban power plant. The most appealing long-run solution to air pollution from power plants is the nuclear power plant, particularly the fast-breeder reactor which is a net producer rather than user of nuclear fuel. If safety problems can be resolved and there appears to be every indication that they can be, nuclear power will ultimately replace fossil fuels in central power plants.

In the distant future, solar energy, controlled thermo-nuclear fusion, and magnetohydrodynamic conversion may change our present world.

Performance standards suggested by EPA for new fossil-fuel fired steam generators and some other new stationary sources are shown in Table 7.1.

TABLE 7.1
Standards of Performance for New Stationary Sources*

Source	Emission standard (max. 2-hr. average, except opacity)
Fossil-Fuel Fired Steam Generators	
Particulate matter	0.1 lb/MM Btu heat input
	20% opacity
Sulfur dioxide	
Liquid fuel	0.8 lb/MM Btu heat input
Solid fuel	1.2 lb/MM Btu heat input
Nitrogen oxides	
Gaseous fuel	0.2 lb/MM Btu heat input
Liquid fuel	0.3 lb/MM Btu heat input
Solid fuel	0.7 lb/MM Btu heat input
Incinerators	
Particulate matter	0.08 grain (gr)/scf
	10% opacity
Portland Cement Plants	
Particulate matter	
Kiln	0.30 lb/ton kiln feed
	10% opacity
Clinker cooler	0.10 lb/ton kiln feed
	no visible emission
Nitric Acid Plants	
Nitrogen oxides	3 lb/ton acid produced (as NO_2)
	10% opacity
Sulfuric Acid Plants	
Sulfur dioxide	4.0 lb/ton acid produced
	10% opacity
Acid mist	0.15 lb/ton acid produced (as H_2SO_4)
	10% opacity

* Title 40, Code of Federal Regulations, Part 60 (Dec. 23, 1971).

Rock Products Processing Industry

From an air pollution standpoint, the rock products processing industry has four major components: Portland cement, lime, gypsum, and glass and clay products. In each of these industries nonmetallic minerals are processed at high temperatures to produce the desired products. The chief air pollution problem is usually the emission of dusts from various points in the process but particularly from high-temperature operations.

The manufacture of *Portland cement* in the United States is carried out in 180 plants most of which have in excess of 100 employees. Portland cement production in 1968 was about 400 million barrels (one barrel is the equivalent of 376 pounds).

The raw materials required for the manufacture of Portland cement are a source of calcium and a source of silica, iron, and aluminum. The calcium sources (calcareous substances) are limestone, marble, marl, oyster shells, or waste sludge from lime plants. The sources of silica, aluminum, and iron (argillaceous materials) are clay, shale, bauxite, silica sand, iron ore, waste from aluminum plants, and slag from blast furnaces. The particular raw materials available to a plant are quarried, crushed, ground, mixed intimately, and fed to the cement kiln. During the grinding, blending, and handling operations, dust is emitted at various points. Emissions may be controlled with fabric filters or other suitable means.

In a typical process, a ground mixture of limestone and clay is fed into the upper end of a rotary kiln either as a slurry (wet process) or as a dry mix (dry process). A typical kiln is 12 ft or more in diameter and may be more than 500 ft long. It is inclined very slightly with a slope of ⅜- to ½-inch per foot of length. As the kiln revolves, the raw materials move slowly toward the lower end of the kiln countercurrent to hot combustion gases from burning gas, oil, or coal. As the raw materials are heated, they first lose moisture, then the carbonates decompose to produce carbon dioxide gas, and the mass finally reaches incandescence at 2300 to 2700°F. At this temperature, the remaining solids combine chemically to produce small marble-size balls of cement "clinker" which are discharged from the lower end of the kiln into a cooler, and finally to mills and the finishing process.

The spent combustion gases leaving the kiln pass through a dust collector to remove entrained particles of unreacted raw material and are then discharged from a stack. Dust concentration in the effluent gases varies considerably depending on specific operating variables. In the wet process, dust concentration in the gas stream normally ranges from 3 to 5 grains per standard cubic foot (gr/SCF); in the dry process dust loadings are 5 to 9 gr/SCF. Thus with dust collection of 99% efficiency, outlet grain loading will vary from 0.03 to 0.09 gr/SCF.

The problems of the cement industry center around plume visibility, old plants, and raw material variability. Even with highly efficient electrostatic precipitators (preferred by wet-process plants), plumes from high stacks are visible, mainly because of their great diameter (up to 20 ft). To a large segment of the public, a visible plume means no control.

The problem of old plants is not unique to the cement industry, but it is often more troublesome. The large ground area needed for precipitators and baghouses is often not available near cement kilns in old plants. "Shoe-horning" them in place may be an expensive undertaking. The problem can, of course, be dealt with readily in designing new plants.

Raw material variability is of a dual nature. Some raw materials are naturally dusty, and larger quantities (up to 20 gr/SCF) are released to the spent combustion gases. In such a case, 99% collection will yield 0.2 gr/SCF to the effluent stack gases. The other problem is an uncontrollable change in raw material characteristics which affects the efficiency of the collector, sometimes markedly.

Performance standards proposed by EPA for new Portland cement plants are shown in Table 7.1.

Lime manufacture entails calcining limestone rock in various forms to produce quicklime and carbon dioxide. About 16 million tons of lime per year is produced in 100 plants. Although 80% of production comes from large rotary kiln operations (up to 500 tons daily capacity), there are still a number of vertical shaft kilns in operation.

The main problem of the industry is emission of dust with the effluent flue gases from the kiln. Dust collection methods have not been standardized, so high-efficiency scrubbers, baghouses, and electrostatic precipitators are all used in the industry. In isolated areas, high-efficiency mechanical collectors have been found acceptable.

Typical dust loadings in effluent gases from vertical kilns are 0.3 to 1.0 gr/SCF [19]. In rotary kilns, loadings may be as high as 20 gr/SCF.

Another source of dust is the hydrator where quicklime is converted to hydrated lime by addition of water. Dust emission may be controlled by the use of water scrubbers. Dust loadings in the effluent vary from 0.01 to 1 grain per cubic foot depending on collector efficiency. Even with highly efficient scrubbers, the discharge is usually opaque because of condensing water vapor. A survey of control problems in lime plants has been published [21].

In the *gypsum and plaster industry,* pulverized gypsum rock is calcined in vertical "kettles" to remove three-fourths of the water of crystallization and produce powdered gypsum plaster. In a few cases, rotary kilns are used instead of vertical calciners. Approximately 70 plants are distributed widely throughout the United States and produce nearly 8 million tons of calcined gypsum annually.

Cyclones and electrostatic precipitators are used to control emissions of dust from the kettle stacks. In a few cases baghouses have been used successfully.

Glass and ceramics manufacturing involves a broad range of products from sand and clays. These products include brick, structural clay products, refractories, pottery, porcelain enamel, and various glass products. The processes used entail high-temperature operations in kilns or furnaces. Potential emissions are dust and fumes. Occasionally fluoride problems are encountered. If coal is used as a fuel for the kilns, there may also be periods of black smoke. Generally, however, aside from the glass industry, the air pollution problems of the ceramics industry are minor.

The glass industry presents a different situation. Three phases of the industry are of interest with respect to air pollution. They are flat glass manufacture, glass container manufacture, and glass fiber manufacture. The air pollution problems of the two first-mentioned categories are related to plume visibility requirements and may be minimized by operational and raw materials control. However, the fineness of emitted particles [31] makes elimination of a visible plume difficult, perhaps impossible.

The manufacture of glass fiber is the one division of the glass industry that has received considerable attention because of air pollution potential. The chief problem is a highly opaque white plume which so far has been difficult to abate. Considerable effort is being applied to the problem, and industry scientists believe the problem may be solved by careful attention to operating procedures and raw material selection.

Steel and Related Industries

The steel industry, as considered here, covers a very wide range of plant sizes from small gray-iron foundries to large, integrated steel mills.

An integrated mill usually includes a coke plant, a sinter plant, blast furnaces, several open-hearth or basic-oxygen furnaces, and related facilities for converting steel ingots to finished shapes, e.g., sheets, bars, rods, pipe, wire, etc.

Emissions from coke plants consist of coal gases which escape from charging holes and leaks in ovens, plus particulate matter expelled with steam during the quenching of red-hot coke. The particulate matter is no great problem in that it usually consists of large particles which fall out on the mill property. The odorous gases from the coke ovens are another matter. Close attention to operating and maintenance practices usually yields the greatest degree of control. Shrouding and collection of gases evolved during coal charging and coke pushing, and closed pipe-line charging show promise of alleviating the air pollution problem of the coking operation. Entirely new processes are also being studied [2].

The sintering plant, where iron ore fines and blast furnace flue dust are burned with coke breeze to yield a product more suitable for charging the blast furnace, emits particulate matter which may be controlled by electrostatic precipitators, baghouses, cyclones, or scrubbers [27].

The blast furnace is a relatively clean operating unit, except when slippage of the furnace charge causes puffs of dust to be expelled. Scrubbers and other types of control equipment are used to abate these emissions. It is normal practice to clean the blast furnace gas thoroughly so that it may be used as fuel in the furnace. Good air pollution control results.

The open-hearth furnace, whether operated in a normal manner or with an oxygen lance emits varying amounts of dust during different phases of the operation. The emitted particulates or fume are chiefly red iron oxide. Since variable gas flow and dust loading during the various operations in a "heat" make control difficult, it has not been standardized. Electrostatic precipitators and Venturi scrubbers appear to be preferred, although baghouses also show promise.

Basic oxygen processes, in which the metal is refined by blowing oxygen against the surface of the metal in the convertor, are rapidly replacing the open-hearth process all over the world. Particulate emissions are considerably greater than those from the open hearth and thus require even more extensive control equipment. Dust concentration at the mouth of the convertor may be greater than 20 grains per cubic foot. Electrostatic precipitators and scrubbers again appear to be preferred.

Electric-arc furnaces are used to manufacture steel, usually alloy steel, from steel scrap. Compared with the making of steel from ore, electric-steel manufacture is less bothersome from an air pollution standpoint. Baghouses are commonly used for controlling emissions of particulate matter, although electrostatic precipitators and Venturi scrubbers are also used. In any case, it is necessary to hood the furnace well so that the particulate-laden gases may be directed into the control equipment.

The gray-iron foundry industry [34] is engaged in the production of cast iron from scrap and pig iron, usually in a coke-fired cupola furnace. During the melting of the charge, particulate matter ranging from sub-micron to large readily settleable particles is evolved. Of the 4200 foundries in the United States, 90% operate cupolas (the remainder melt the iron electrically), but only 15% of the cupolas are equipped with suitable air pollution controls (baghouses, scrubbers, or electrostatic precipitators).

Foundries vary greatly in size, and there are a great many small operations in the industry. However, sixteen companies with very large or multiplant foundries produce about 75% of all gray-iron castings. Total industry production is estimated at 18 million tons annually. Operations of the larger companies are well controlled. The bulk of the small companies face a difficult situation. They will have to convert to electric furnaces or install effective control equipment on their cupolas. In many cases this added capital expenditure may well make operation of the foundry uneconomical.

Nonferrous Metal Industry

The principal divisions of the nonferrous metal industry from an air pollution standpoint are the copper industry, lead and zinc smelters, aluminum reduction plants, and nonferrous foundries [23].

The copper industry entails mining copper ore, milling and concentrating the ore, smelting the concentrate to produce metallic copper, refining the crude copper, and mechanically transferring the refined copper to various articles of commerce. Mining, milling, and concentrating the ore are usually conducted in remote locations where air pollution is not a problem. If there is a problem, the usual methods of particulate control are applicable.

The chief potential source of air pollution is the copper smelter where sulfur dioxide and particulate matter are the pollutants of interest. Both materials are emitted from the roasters, reverberatory furnaces, and convertors which comprise the smelting process. Nearly all smelters are equipped with electrostatic precipitators to recover valuable metals, and many have acid plants in which sulfur dioxide may be recovered. During periods when the sulfur dioxide concentration in the effluent gases is too low to be converted to acid, the gases are directed to high stacks designed to allow the sulfur dioxide to disperse into the atmosphere without causing undesirable ground concentrations.

The refining of crude (blister) copper is usually done electrolytically and poses no air pollution control problem. Gases from furnace treatment of slimes (for recovery of valuable metals) are controlled by scrubbers followed by electrostatic precipitators.

Primary production of lead and zinc is somewhat similar to primary production of copper. Sulfide ores or concentrates are smelted at elevated temperatures in a variety of processes which produce effluent gases containing both particulate matter and sulfur dioxide. Particulates are usually collected in baghouses. Zinc sulfide ores are sufficiently high in sulfur content to permit the sulfur dioxide in the smelter gases to be used as raw material for the manufacture of sulfuric acid. The gaseous effluent from lead smelters is so low in sulfur dioxide that acid manufacture is uneconomical. Because of the low concentrations the gases may be dispersed adequately from stacks of moderate height. There is considerable federal pressure for further reduction of SO_2 from all nonferrous smelters [28].

Secondary smelters, and some primary ones, use oxides or metals as raw materials and thus have no sulfur dioxide problem. Evolved particulates may be collected in baghouses.

Nonferrous foundries are engaged in producing various alloys, e.g., brass, bronze, and zinc die-casting metals. The principal emissions are metallic fumes and (if the charge is oily or greasy) smoke. Both baghouses and electrostatic precipitators may be used to abate emissions provided the furnaces, ladles, and molds are properly hooded. As in the gray-iron foundry industry (q.v.) there are a great many small operators to whom the cost of control equipment is a major item. Consequently, the 2000 plants which comprise the nonferrous foundry industry will probably become fewer.

The *aluminum industry* is made up of several segments; winning of alumina from the ore, bauxite; reduction of alumina to metallic aluminum; and conversion of the metal ingots to a variety of shapes and articles.

Grinding and treating the bauxite involves several dusty operations. Most bauxite is imported and brought to the United States in ships. The material is light in weight and difficult to keep in place during ship unloading, conveying, grinding, and subsequent mixing with caustic soda solution. Consequently, bag filters are widely used at all transfer points to prevent loss of the powder to the atmosphere.

After purified aluminum hydrate has been recovered from the autoclaved caustic solution of bauxite, it is calcined in a rotary kiln to produce pure alumina, the raw material for the reduction plant. Particulates evolved during the calcining operation are collected in electrostatic precipitators or other appropriate equipment.

In the reduction plant, a fused mixture of alumina and certain fluoride compounds is electrolyzed in carbon-lined cells containing heavy carbon anodes. The chief air pollution problem in reduction plants is the emission of fluorides in both particulate and gaseous forms. Control requires not only the handling of large amounts of cell gases admixed with air and efficient treatment of the air-gas mixture to reduce fluorides to a minimum, but also extensive monitoring of the air, vegetation, and animals outside the plant.

One company has developed a fluidized-bed reactor to contact pot gases from pre-bake cells with incoming feed alumina. Bag filters are used to separate entrained solid materials from pot gases. Ninety-five per cent of the pot gases go to the reactor with 99% fluorine recovery [5].

At present there are twenty-four aluminum reduction plants in the United States and at least two more are under construction. In 1968, production of aluminum was approximately 3,500,000 tons.

The Petroleum Industry

The petroleum industry may be divided into three major divisions: production, refining, and marketing. The production division includes operations related to locating and drilling oil wells, removing oil from the ground, treating the oil at the well site, and transporting the oil to the refinery. Refining consists of converting the crude oil to salable products, e.g., gasoline, furnace oils, light and heavy fuel oils, asphalt, paraffin wax (see Fig. 7.2). Modern refineries also include the manufacture of various chemicals from petroleum fractions (petrochemical industry). Marketing consists of the distribution and sale of petroleum products. As of January 1, 1969 there were 262 operating refineries in the United States with an annual throughput capacity of 4.2 billion barrels of crude oil.

The major air contaminants evolved from petroleum-related operations are hydrocarbons, sulfur compounds, carbon monoxide, and particulate matter. Crude oil is a heavy, viscous liquid or semiliquid consisting of a variety of hydrocarbons ranging from light, normally gaseous materials to normally-solid, waxlike materials. The crude is contaminated with small amounts of sulfur compounds, water, and ash. In the handling, separation, purification, and distribution of the crude oil, refinery intermediates, and refinery products, hydrocarbons are lost to the atmosphere principally by evaporation from storage tanks, processing units, pumps and valves (leakage), compressors, transfer operations, loading and unloading, and waste-effluent handling. For methods of estimating hydrocarbon losses from various refinery operations, see Ref. 32.

Sulfur compound losses from the petroleum industry are usually in the form of hydrogen sulfide, sulfur dioxide, and, in minor quantities, mercaptans. Crude oils themselves vary widely in their sulfur content, and the emissions of sulfur compounds both in the oil field and the refinery reflect the quantity in the crude. Most of these emissions are in the form of hydrogen sulfide and mercaptans, which (along with some nitrogen compounds) are largely responsible for the malodors in and around oil fields and refineries. In refining operations, particularly stills, separators, and cracking units, the hydrogen sulfide evolved is accompanied by considerable quantities of hydrocarbon gas which makes the gas highly suitable as a fuel. Inasmuch as large amounts of fuel are needed in refining operations, the gas is burned in the heaters and boilers throughout the refinery. In burning, any unremoved hydrogen sulfide is converted to sulfur dioxide and emitted in the flue gases. Estimates of sulfur emissions from refineries are so much a function of the sulfur content of the crude oil processed that general or average emission factors are meaningless. Only a detailed survey of the specific refinery in question can yield good estimates.

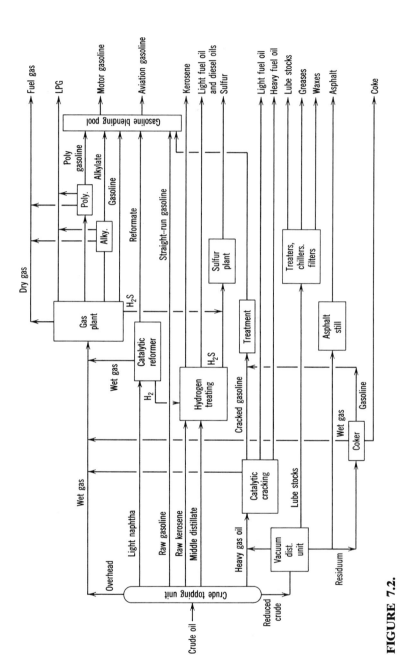

FIGURE 7.2.

Processing plan for modern refinery. (Ref. 32)

The principal source of carbon monoxide and particulate matter in refinery operations is the catalyst recovery unit operated in conjunction with the catalytic cracking unit. The quantities emitted are usually minor in comparison to other sources in the community. Black smoke may be emitted occasionally from overloaded flares or burning of waste residues.

The nature of abatement equipment used in the petroleum industry is fairly well standardized. Emissions can be controlled to a degree that will permit refinery operations in any kind of a community where a buffer zone of a few blocks can be allowed for dissipation of residual odors.

There is some doubt as to the degree to which hydrocarbon emissions from petroleum operations need to be controlled since most of the gases emitted are comparatively unreactive from a photochemical standpoint. Of course, in areas where photochemical smog is not a problem (see Chapter 8), the need for hydrocarbon control is lacking. In those communities where photochemical smog is important, hydrocarbon control may involve the following: use of floating roofs, pressure tanks, or vapor-recovery systems [6] for storage facilities; vapor return systems or splash-proof nozzles for loading and unloading gasoline from tank trucks, tank cars, marine vessels, and stationary tanks [8]; frequent inspection and modern maintenance practices for valves, flanges, pipeline blinds, and pump seals (to minimize leakage); use of properly designed flares for gases from pumps, compressors, safety valves, rupture discs, blow down systems, vacuum jets, and condensers [7]; and covering of oil-water separators [11].

In large refineries where sufficient hydrogen sulfide is available, the vapor may be concentrated by absorption (see Chapter 5) and used as a raw material for the manufacture of elemental sulfur or sulfuric acid. Smaller amounts of hydrogen sulfide or gases with low concentrations may be flared. The best control for sulfur dioxide emissions is minimization of the amount of hydrogen sulfide burned.

Dust from catalyst regeneration units may be recovered by cyclones or electrostatic precipitators [9]. When sufficient carbon monoxide is produced it may be used to fuel a waste-heat boiler. A payout of six years after taxes is said to be attractive [10].

Petroleum residues may be burned with a minimum of smoke evolution in trench incinerators (see Chapter 3). Another incinerator designed especially for liquid chemical wastes has been described [4].

An operation more closely related to the rock products processing industry but classified by the U.S. Census of Manufacturers as part of the petroleum industry is the manufacture of asphaltic concrete in so-called "hot-mix" plants. Here hot-mix asphalt paving material is made by mixing various aggregates with hot asphalt. Prior to mixing, the aggregate is heated to 250 to 350°F in a rotary dryer. It is the dryer that produces the dust plume so noticeably emitted from the hot-mix plants. Control is not difficult since the plume can be almost entirely eliminated by the use of scrubbers or baghouses [13, 20]. The chief problem of the industry is the large number of small, uncontrolled portable units that follow highway construction. Large plants are usually well controlled.

The Chemical Industry

The chemical industry is a highly complex one whose products reach every facet of modern life. Its basic raw materials are minerals, coal, oil, natural gas, air, and water. By various means, these raw materials are converted to plastics, fibers, pigments, paints, pharmaceuticals, solvents, plasticizers, flavoring materials, herbicides, fungicides, insecticides, explosives, soaps and detergents, fertilizers, dyestuffs, synthetic elastomers, and myriads of chemicals for industrial and home use.

As a rule, the air pollution problems of the chemical industry are odors and dusts. A few operations must deal primarily with specific pollutants, e.g., sulfur dioxide, hydrogen sulfide, fluorides, certain acid mists, etc. Where fossil fuels are used for heat and power, there are emissions of combustion products.

A 1967 survey made by the Manufacturing Chemists Association covered some 9000 different processes. One-third had no routine emissions to the atmosphere. Of the other two-thirds, 80% were controlled to varying degrees. Some of the specific problems of the industry are outlined in the following paragraphs.

Alkali and chlorine manufacture is an important segment of the chemical industry. The problem in chlorine plants is largely vent gas odors which may be contained by absorption in milk of lime. Mercury in vent gases is also a problem (see Chapter 4). In caustic plants, the chief effluent is caustic mist which may be abated by mist eliminators. Soda ash kilns usually discharge dust which is susceptible to all types of particulate control equipment.

Inorganic pigment manufacture involves calcining, milling, and drying operations which have a potential for dust and fume emissions. However, because of the high value of the product, bag filters have been used for product recovery for years.

Other inorganic chemicals manufacturing processes include the manufacture of salts of sodium, potassium, aluminum, copper, chromium, magnesium, etc., and various other compounds, e.g., calcium carbide, hydrogen peroxide, phosphates, sodium silicate, ammonium compounds, various acids, phosphorus, metallic sodium, etc. The scope of air pollution problems in this segment of the industry may be shown by consideration of three different products. Calcium carbide is made in a furnace by reacting quicklime and coke. Both particulate matter and odorous gases that are difficult to control are evolved.

Sulfuric acid plants will evolve some sulfur dioxide and some acid mist. The latter can be controlled by mist eliminators, the SO_2 is difficult to control except by use of tall stacks [33]. Process changes to allow greater conversion efficiency may also decrease SO_2 emissions [3]. Performance standards for new sulfuric acid plants have been published by EPA (see Table 7.1).

The manufacture of sodium phosphates, used chiefly in synthetic detergents, mainly involves handling solutions and slurries and has an air pollution potential only at the final drying step.

Organic chemicals manufacture includes a broad segment of the synthetic organic chemical industry, e.g., alcohols, aldehydes, acids, ketones, ethers, esters, and metallic organic compounds, used either as such or as intermediates for the manufacture of plastics, fibers, dyes, elastomers, and other chemical end-products. More than 300 plants, ranging from one to several thousand employees, operate in the United States.

Most of the simple organic chemicals are liquids, so most of the atmospheric emissions are gaseous. In order to reduce both product loss and air pollution, scrubbers are commonly used in product streams. The major problem here is odors which may at times be caused by very low concentrations of gases. In such cases, afterburners may be required to abate emissions.

When the products of organic chemicals manufacture are solids, e.g., phthalic anhydride (PA) and urea, scrubbers or other particulate control equipment may be used. However, PA plants commonly use afterburners because of the emission of an odorous by-product gas.

Paints and allied products cover some 1200 plants in the United States. The only operations with marked air pollution potential are the manufacture of alkyd resins and varnish cooking. Odorous plumes from the cooking kettles, if not abated, can be definite nuisances. Control may be effected by oxidizing scrubbers, flame incinerators, or catalytic combustion equipment [12].

Fertilizer manufacture is a varied industry from an air pollution control standpoint, ranging from ammonia plants with little or no manpower allocated to air pollution control to phosphatic fertilizer plants with their highly publicized dust and fluoride problems. Scrubbers, bag filters, and electrostatic precipitators are used throughout the industry [30].

There are also a number of *miscellaneous* and other classifications of interest within the chemical industry (see U.S. Census of Manufacturers). Some have no problems of consequence. Others have problems of varying magnitudes. Rayon plants, for instance, must curtail carbon disulfide emissions; the soap and detergent industry must cope with odors and particulate emissions [25]. Carbon black plants have highly visible black plumes for which control is often effected only by atmospheric dispersion because of remote location of the plant. Where more control is necessary, baghouses are effective [15]. Other plants with varying air pollution problems include leather tanning, glue and gelatin, explosives, printing ink, and fatty acids.

The Pulp and Paper Industry

The pulp and paper industry embraces those operations involved in the processing of wood to pulp and eventually to paper and paper products. To some extent the general lumber and wood industry is related and will be discussed at the end of this section.

Wood consists of countless individual fibers bound together by lignin compounds. In the pulp and paper industry, logs are delivered to the pulp mill where they are de-barked and converted to chips which are then processed to remove the lignin compounds and separate the individual fibers in the form of fibrous wood pulp or cellulose pulp. The pulp is converted to paper by mechanical means. The bark is burned. Three pulp processes are commonly used, kraft pulping, sulfite pulping, and mechanical. In the United States the most important process is the sulfate (kraft) process used in 94 mills which produced 22,700,000 tons of pulp in 1967. The sulfite process produced 4,116,000 tons in 56 plants, and semichemical processes were used in 39 mills to produce 3,273,000 tons.

In the kraft process, wood chips are treated in digesters, under pressure, with a solution of caustic soda and sodium sulfite to solubilize the lignin. Although the sulfite is present primarily as a buffer, small amounts react with some of the lignin to produce malodorous sulfur compounds (hydrogen sulfide, mercaptans, or organic sulfides). When the "cook" is completed the pulp is discharged into a blow tank by opening the digester or cooker and literally blowing the pulp from the digester by the pressure of the steam in the digester. At this point, some of the odorous vapors are discharged into the atmosphere. The rest of the process is essentially multiple washing of the pulp to purify it and to remove the spent pulping chemicals. During the washing, more volatile gases and vapors are evolved. The solution of spent chemicals ("black liquor") must be treated to recover its chemical content in order for the process to be economically feasible.

The dilute black liquor is recovered by evaporation to a thick "strong liquor" which may be burned in a recovery furnace. The resulting molten ash called "smelt," is redissolved, treated with lime, clarified, and re-used.

The evaporator-furnace operation is a complex one in which hot flue gases from the furnace are used in direct contact with the black liquor to concentrate it (see Fig. 7.3).

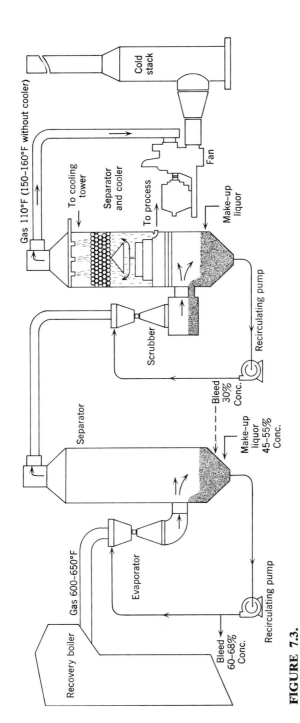

FIGURE 7.3.

Two stage evaporator-scrubber system for black liquor recovery.

The chief air pollution problems of the kraft pulp industry are related to emissions of malodorous gases from the digesters, evaporators, and the recovery furnace and emissions of particulate matter from the recovery furnace, the lime kiln, and the bark-burning boilers. The U.S. Public Health Service has published rough factors for estimating the emissions of odorous gases and particulates from various operations in the kraft pulping process [18]. More recent estimates are given in Ref. 29.

The abatement of kraft odors is a particularly difficult air pollution control operation for three reasons: (1) the low concentration of odorous gas that is noticeable, (2) evolution of odorous gases at so many points in the process, and (3) a wide range of gas flows from various operations. One method that has been widely adopted to reduce odor levels is the *black liquor oxidation process* in which the sulfur compounds in the weak black liquor are oxidized with air prior to admission to the multiple-effect evaporators. The resulting oxidized sulfur compounds are less volatile and less odorous than those in the unoxidized liquor. For further odor reduction the evaporator, digester relief, and blow gases may be treated by any of the methods discussed in Chapter 6, "Odors" (q.v.). Just which method or combination of methods is best to use is largely an economic problem. Chlorine is often used to oxidize off-gases because chlorine is readily available. Many kraft mills have their own chlorine manufacturing facilities, primarily for bleaching the pulp. To care for the wide variations in digester gases, spherical collection tanks are used as surge tanks. A concerted industry program is under way to further alleviate odor problems [14].

Emissions of particulate matter may present problems at the bark burner, the recovery furnace, and the lime kiln. Scrubbers, mechanical collectors, and electrostatic precipitators have all been used successfully in specific instances. One major obstacle in some mills is overloading the recovery furnace.

The sulfite process does not present nearly the air pollution problem that the kraft process does. The cooking liquor is usually magnesium bisulfite and sulfurous acid. To be competitive economically, sulfur dioxide from the blow gases and the chemical recovery furnace is recovered in scrubbers for re-use. The magnesium oxide particulate in the flue gas is recovered simultaneously. Bark-burning emits considerable particulate matter, especially if burned in tepee burners or similar poorly controlled units.

Wood pulp from either process described is tan to dark brown in color. If white pulp is desired the pulp must be bleached. Chlorine is used in one form or another in most bleaching processes, so there is always the possibility of release of chlorine off-gases. These may be readily controlled by alkaline scrubbing.

All in all, the major air pollution problem of the pulp and paper industry is the abatement of the kraft mill odor. Considerable progress has been made; more is needed. The solution is undoubtedly complex and will take time.

In the lumber and wood products industry, the only real air pollution problem arises from the disposal of waste bark, chips, shavings, and sawdust. Traditionally these waste materials have been burned in the open or in tepee burners. Neither method will be tolerated any longer in urban communities. If suitable by-product outlets, e.g., pressed wood, cannot be found the small operator must resort to multiple-effect incineration (which may be expensive) or sanitary landfill. Large operators can use bark and sawdust in specially designed boilers to produce steam, unless visible smoke is a problem. Economic means of burning waste-wood as fuel are not available.

The control of sawdust and wood flour from woodworking operations presents no great problem.

Food and Feed Industries

Food and feed are primarily products of agriculture. Hence the industry really includes operations on the farm or ranch, processing of the agricultural product, and its distribution to the consumer. All the way along the line, from preparation of the soil for seeding to cooking a hot meal, pollutants are emitted into the atmosphere. Two types of pollutants are most common, particulate matter and odorous gases.

The agricultural practices that contribute most to air pollution are: soil preparation, crop spraying, weed burning, orchard smudging, fruit and vegetable harvesting, and animal production. The best way of controlling pollution from these necessary activities is strict adherence to good practices [16]. Among these good practices are restriction of plowing and harrowing operations to periods of low wind speed. Adley, et al, have shown very clearly the great increase in wind-blown dust at wind speeds above 15 mph [1]. Fig. 7.4 is based on their data. Similarly, weed burning and insecticide spraying should be limited to periods of favorable meteorological conditions. Factors for estimation of pollutant emissions from open burning of agricultural wastes are shown in Table 7.2. Smudging of fruit crops to prevent freezing is no longer necessary and has been replaced by wind machines and smokeless heaters.

One of the most noticeable and objectionable air pollution problems is the odor from improperly kept cattle feed yards and animal pens. The problem can be alleviated by good housekeeping practices, adequate drainage, and frequent removal of offal [17].

Dust problems from food and feed processing are most commonly found in the handling and milling of grains, drying operations, mixing of dry

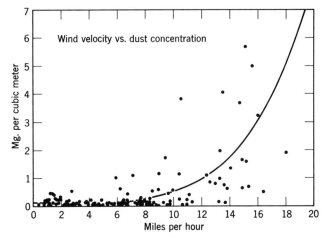

FIGURE 7.4.

Effect of wind speed on atmospheric dust loading. (adapted from Ref. 1)

TABLE 7.2

Factors for Estimation of Pollutant Emissions from Open Burning of Agricultural Wastes

Material burned	Particulate matter	CO	Hydrocarbons
		(pounds per ton of fuel)	
Grass and straw	16	100	12
Fruit prunings	—	60	14
Native brush	17	65 (dry)	7
		130 (green)	

feeds and foods, and a few specialty operations, e.g., coffee roasting. The dustiest of all these is the loading and unloading of grains from transport vehicles and grain cleaning. The basic problem is improperly designed and undersized pickup systems to move dusty air to the dust collector.

Odors from food and feed processing are generally related to the bacterial decomposition of proteins and carbohydrates, occasionally by overheating and burning. The greatest problems arise from the meat and meat processing industry, fish processing, and the disposal of fruit and vegetable wastes. The most difficult of all the odor problems arises from inedible fat processing. Meat scrap and other animal residues, including dead animals, are ground to a hamburgerlike consistency and charged to kettles where the material is cooked under pressure in the presence of steam. The fat cells are broken down to water, grease, and proteinaceous solids. The solids are then separated from the grease which may subsequently be purified to produce tallow. The solids are dried and ground to produce a proteinaceous meal used as animal feed (see Fig. 7.5a).

The major odor problem is caused by the high volume of odorous gases emitted during the cooking process. These gases need not create a problem inasmuch as they can be passed through a condenser and thence to odor control equipment as discussed in Chapter 6. Direct-flame incineration is highly effective but sometimes ruled out on the basis of fuel cost. Odors are also emitted when the cooker is dumped, during the pressing of the cooked meat to remove the last bit of tallow, and during drying of the pressed "cracklings." These odors may be picked up at the point of origin and delivered to odor abatement equipment. When air pickup is inadequate, all room air must be treated. In both cases, chemical scrubbing and odor counteraction are said to be effective. Localized odor problems develop in the transport and storage of the raw material caused by bacterial decomposition which begins at the death of the animal. Temperature control and good ventilation will reduce odors during prolonged storage. The

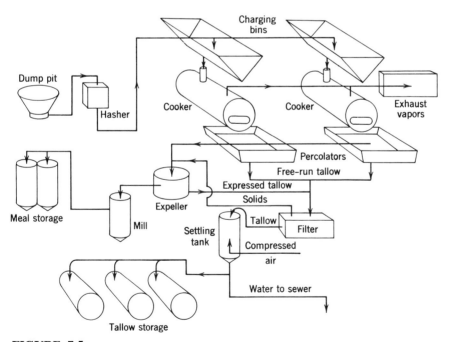

FIGURE 7.5a.
Typical flow sheet—batch rendering plant.

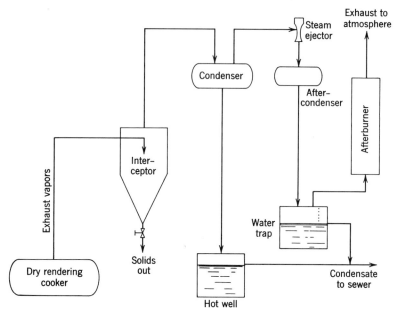

FIGURE 7.5b.
Odor control system for cooker vapors.

main problem in many plants is merely sloppy operation or poor house-keeping. Good sanitation practices are remarkably effective. Many of the odor problems of the inedible fat rendering industry are said to be miti-gated by use of modern continuous rendering processes. *Fish processing* presents many of the same problems as those just described, and similar control methods are indicated.

In the *processing of fruits and vegetables,* the disposal of hulls, leaves, rinds, pods, and cuttings may create an odor problem if disposal is delayed too long. Even some disposal methods, e.g., lagooning, may produce odors during temporary upsets.

The *coffee roasting* industry is representative of a common odor situa-tion where "one man's food is another man's poison." Many people delight in the odor of roasting coffee. Others abhor it. Similar problems exist near bakeries and chocolate factories. However, control of coffee roasters is directed to both odors and smoke. In a typical coffee roasting operation (see Fig. 7.6), green coffee beans are cleaned and then roasted by contact with hot combustion gases. Moisture, odorous gases, and particulate mat-ter are emitted from the roasters. Both smoke and odor may be controlled by passing the roaster effluent gases through cyclone collectors and thence

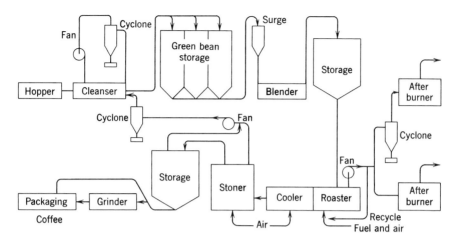

FIGURE 7.6.
Flow diagram of coffee roasting plant.

to either an afterburner or to the combustion zone of the roaster. Particulate emission factors have been published [24].

Mechanical and Related Industries

There is a large group of industries whose operations are essentially mechanical in nature, but which produce some atmospheric contaminants. These include metal working and finishing, machinery construction, assembly plants for various mechanical consumer products, rubber compounding, wood products and woodworking, the textile industry, and a number of service industries. Generally, emissions are minor, but there are a few special cases that warrant mention.

The wood products and woodworking industry is comprised of saw mills, planing mills, and producers of millwork, plywood, prefabricated wood products, cooperage, and allied wood products. The most common emissions are wood flour and sawdust from cutting operations. Cyclones of various types are used to collect the dust and prevent excessive emissions to the atmosphere. The collected sawdust, along with other remnants of logs and boards, must be disposed of either by processing or by burning. If burning is practiced, smoke and charred particles will be emitted unless highly efficient burners are used.

A special situation exists in the drying operations of some plywood veneer plants where visible plumes are emitted. Either fume incineration or process changes may be required to alleviate the problem.

The rubber industry involves two general types of operations, (1) rubber compounding, in which raw rubber is mixed with various chemicals to improve the properties of the rubber or to make it more attractive, and (2) forming it into a desired shape (tires and tubes, footwear, toys, novelties, etc.) and curing the product under controlled conditions of time and temperature.

During compounding, dusts, fumes, oil mists, and odors may be evolved. In many cases, the emissions of particulate matter are minor and not controlled. When control is necessary, baghouses and, occasionally, scrubbers are used. In other operations, few pollutants are emitted to the atmosphere, except in some forming processes where solvents are used and released into the air. At times, these solvents may be the occasion of odor complaints. Direct-flame incineration has proved to be effective [26].

Assembly plants and metal finishing operations generally do not produce air pollutants except in a minor way from degreasing operations and applications of surface coatings. Emissions from these operations are organic in nature and may cause local nuisances. In two localities, Los Angeles and the San Francisco Bay Area, these emissions are controlled because of possible contribution to photochemical smog. To allay the situation one may either use so-called "unreactive" or complying solvents or incinerate any effluent vapors before they reach the air.

REFERENCES

1. Adley, F. E., and W. E. Gill, "Atmospheric Particulate Background in a Rural Environs," *Am. Ind. Hygiene Assoc. J.,* **19,** 271–275 (1958).

2. Anon., "A Cleaner Way to Make Coke," *Business Week,* 41 (July 31, 1971).

3. Browder, T. J., "Modern Sulfuric Acid Technology," *Chem. Eng. Prog.,* **67,** No. 5, 45–50 (May, 1971).

4. Coleman, L. W., and L. F. Cheek, "Liquid Waste Incineration," *Chem. Eng. Prog.,* **64,** 83–87 (1968).

5. Cook, C. C., G. R. Swany, and J. W. Colpitts, "Operating Experience with the Alcoa 398 Process for Fluoride Recovery," *J. Air Poll. Control Assoc.,* **21,** 479–487 (1971).

6. Danielson, J. A., ed., "Air Pollution Engineering Manual," PHS Publication No. 999-AP-40, U.S. Govt. Printing Office, Washington, D.C., 565–606 (1967).

7. Ibid., pp. 606–629.

8. Ibid., pp. 629–640.

9. Ibid., pp. 642–649.

10. Ibid., p. 651.

11. Ibid., pp. 652–655.

12. Ibid., pp. 681–693.

13. Ibid., pp. 325–334.

14. Davis, John C., "Pulpers Apply Odor Control," *Chem. Eng.*, **78**, 52–54 (June 14, 1971).

15. Drogin, I., "Carbon Black," *J. Air Poll. Control Assoc.*, **18**, 216–228 (1968).

16. Faith, W. L., "Food and Feed Industries," Chapter 40, Vol. 3, 2nd edition, *Air Pollution*, A. C. Stern, ed., Academic Press, New York (1968).

17. Faith, W. L., "Odor Control in Cattle Feed Yards," *J. Air Poll. Control Assoc.*, **14**, 459–460 (1964).

18. Kenline, P. A., and J. M. Hales, "Air Pollution and the Kraft Pulping Industry," Public Health Service Publication No. 999-AP-4, Robert A. Taft Sanitary Engineering Center, Cincinnati, Ohio (1963).

19. Lewis, C. J., and B. B. Crocker, "The Lime Industry's Problem of Airborne Dust," *J. Air Poll. Control Assoc.*, **19**, 31–39 (1969).

20. Loquercio, P. A., and C. F. Skinner, "Abating Pollution from Asphalt Mixing Plants Especially by Means of a Baghouse," Paper No. 70-144, 63rd Annual Meeting Air Poll. Control Assoc., St. Louis, Mo. (1970).

21. Minnick, L. J., "Control of Particulate Emissions from Lime Plants—A Survey," *J. Air Poll. Control Assoc.*, **21**, 195–200 (1971).

22. Moller, F., "On the Influence of Changes in the CO_2 Concentration in Air on the Radiation Balance of the Earth's Surface and on the Climate," *J. Geophys. Res.*, **68**, 3877–3886 (1963).

23. Nelson, K. W., "Nonferrous Metallurgical Operations," Chapter 37, Vol. 3, 2nd ed., *Air Pollution*, A. C. Stern, ed., Academic Press, New York (1968).

24. Partee, F., "Air Pollution in the Coffee Roasting Industry," Public Health Service Publ. No. 999-AP-9 (1964).

25. Phelps, A. H., Jr., "Air Pollution Aspects of Soap and Detergent Manufacture," *J. Air Poll. Control Assoc.*, **17**, 505–507 (1967).

26. Sandomirsky, A. G., D. M. Benforado, L. D. Grames, and C. E. Pauletta, "Fume Control in Rubber Processing by Direct Flame Incinerators," *J. Air Poll. Control Assoc.*, **16**, 673–676 (1966).

27. Sebesta, W., "Ferrous Metallurgical Processes," Chapter 36, Vol. 3, 2nd ed., *Air Pollution*, A. C. Stern, ed., Academic Press, New York (1968).

28. Semrau, K. T., "Control of SO_2 Emissions from Primary Copper, Lead, and Zinc Smelters—A Critical Review," *J. Air Poll. Control Assoc.*, **21**, 185–194 (1971).

29. Shah, I. S., "Pulp Plant Pollution Control," *Chem. Eng. Prog.*, **64**, No. 9, 66–77 (1968).

30. Stern, A. C., ed., *Air Pollution*, Vol. 3, 2nd ed., 221–227 (1968).

31. Stockham, John D., "The Composition of Glass Furnace Emissions," Paper

No. 70–72, 63rd Annual Meeting Air Pollution Control Association, St. Louis, Mo. (1970).

32. *U.S. Public Health Service, "Atmospheric Emissions from Petroleum Refineries," PHS Publication No. 763, U.S. Govt. Printing Office, Washington, D.C. (1960).

33. U.S. Public Health Service, "Atmospheric Emissions from Sulfuric Acid Manufacturing Processes," PHS Publication No. 999-AP-13, Robert A. Taft Sanitary Engineering Center, Cincinnati, Ohio (1965).

34. Weber, H. J., "Air Pollution Problems of the Foundry Industry, Informative Reports 1–7 of APCA TI-7 Committee," *in* Technical Manual No. 1, Air Pollution Control Assoc., Pittsburgh, Pa. (1963).

8

PHOTOCHEMICAL AIR POLLUTION

Before World War II, air pollution around the world was the result of smoke, dust, fumes, and odors from factories and homes. In the late 1940's, however, a new type of air pollution became noticeable in the Los Angeles area. It was characterized mainly by an unusual odor, intense eye irritation, and highly restricted atmospheric visibility even when relative humidity was low. Concurrently, a new type of damage to certain leafy truck crops appeared shortly after intense sieges of "Los Angeles smog."

In the early 1950's, Prof. A. J. Haagen-Smit, a Caltech biochemist, announced his then revolutionary theory [10] that the smog was the result of a reaction between hydrocarbons and nitrogen oxides triggered by the bright southern California sunlight. The two pollutants, hydrocarbons and nitrogen oxides, first became concentrated in an atmosphere made stagnant by low wind speeds and an intense atmospheric inversion (see Chapter 2); the sunlight-induced reaction then followed. Later it was shown that the major source of both hydrocarbons and oxides of nitrogen was the exhaust gases from the many motor vehicles operated in the Los Angeles Basin (See Chapter 9).

Since 1950, photochemical air pollution has been identified in various degrees of intensity throughout the world. Generally it is most intense in highly motorized subtropical regions at times when stagnant air masses prevail. One or more symptoms of photochemical smog occurs in nearly all urban areas. Went [27] reported in 1955 that he had found vegetation showing typical smog damage in many European and South American cities. Middleton and others have documented photochemical smog symptoms throughout the United States [16].

243

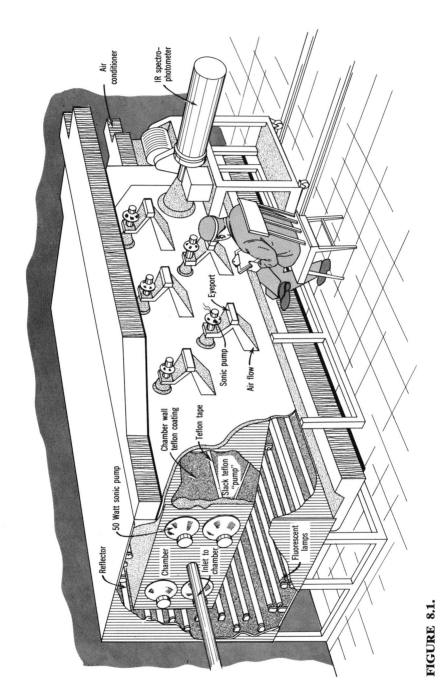

FIGURE 8.1.

Sketch of photochemical smog chamber at Stanford Research Institute, Menlo Park, California.

The Photochemical Smog Reaction

Basically, photochemical smog arises from the photochemical oxidation of gaseous organic compounds, chiefly hydrocarbons. Since Haagen-Smit's early experiments, a great deal of scientific research has been carried out seeking to unravel the great complexity of atmospheric photochemical reactions [2, 13, 18]. Figures 8.1 and 8.2 show typical environmental chambers in which this type of research is conducted.

The initial reaction in atmospheric photooxidation processes is the absorption of ultraviolet light energy by NO_2 (see Equation 1, below). The highly energized molecule (NO_2^*) then decomposes (photolyzes) to nitric oxide and atomic oxygen (Eq. 2). The atomic oxygen quickly reacts with molecular oxygen to form ozone (O_3). However, unless some other energy-absorbing molecule (M) is present the ozone will rapidly decompose. But with a third body (M) present, a stable ozone molecule is formed (Eq. 3). If nitric oxide is present it reacts with ozone to form NO_2 and a molecule of

FIGURE 8.2.

Photochemical smog chamber at Battelle Memorial Institute, Columbus, Ohio.

oxygen (Eq. 4). These reactions are shown by the following chemical equations:

$$NO_2 + h_v \rightarrow NO_2* \tag{1}$$

$$NO_2* \rightarrow NO + O \tag{2}$$

$$O + O_2 + M \rightarrow O_3 + M \tag{3}$$

$$O_3 + NO \rightarrow NO_2 + O_2 \tag{4}$$

If these were the only reactions taking place, the net result would be nothing because the two reactants, NO_2 and O_2, are regenerated. However, in the presence of certain hydrocarbons, other reactions take place. Some of the atomic oxygen, some ozone, and some nitric oxide react with the hydrocarbons to form a variety of products and intermediates with which

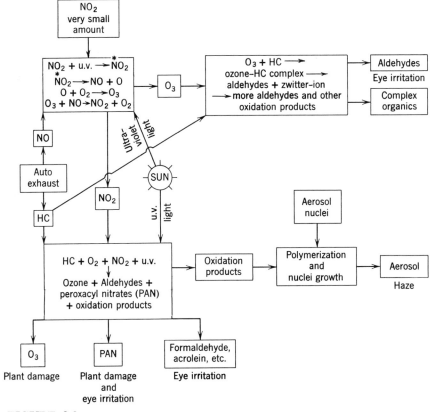

FIGURE 8.3.

Diagram showing photochemical smog formation.

even further reaction can take place. Among the products that have been identified are: aldehydes (e.g., formaldehyde and acrolein), peroxides, hydroperoxides, alkyl nitrates, carbon monoxide, peroxyacetylnitrate (PAN), and peroxybenzoyl nitrate (PB_2N). A schematic diagram of these complex reactions is shown in Fig. 8.3.

Another important facet of the photochemical reaction is the rate and sequence of the reaction of various compounds and formation of the various products. Figure 8.4 is a graphic representation of the reactions taking place

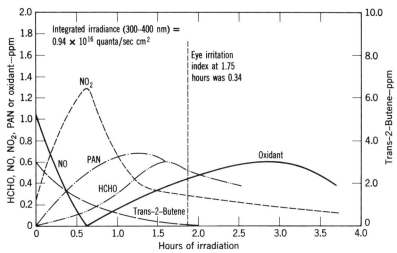

FIGURE 8.4.

Photochemical reactions in smog chamber. (*Courtesy Stanford Research Institute.*)

in an environmental chamber in which the atmospheric smog reaction is partially simulated. The chamber is charged with 3 ppm of a reactive hydrocarbon and 1.3 ppm of nitrogen oxides, chiefly nitric oxide. When lights of suitable intensity and proper wave length (3000–4000 Å) are turned on at time zero, reaction begins, hydrocarbon and nitric oxide begin to disappear (react), and NO_2 begins to form. Similarly aldehydes, PAN, and other organic products begin to form. When the NO has disappeared and NO_2 concentration reaches a peak, ozone forms, and its concentration begins to increase. Eventually the NO_2 and ozone disappear and there remains a mixture of organic products and excess hydrocarbon.

The reaction in the atmosphere is generally of the same nature but infinitely more complex. Reactants are continually being added to the reacting mixture, the air mass is moving with the light breeze, and the products which are being formed continually are also being diluted by atmospheric diffusion processes.

The specific nature of the photochemical reactions taking place also depends on a variety of factors, e.g., light intensity, its spectral distribution, hydrocarbon reactivity, meteorological variables, the ratio of hydrocarbon to nitric oxide, the presence of light absorbers other than NO_2 [19], and others.

Some insight into the effect of these parameters has been gained by studies of the manner in which smog develops in the Los Angeles Basin. On a typical smog day, the concentration of ozone (as measured by oxidant value (see Fig. 8.5) varies with time. In the early morning, oxidant concentrations are low and start increasing about 7 a.m., when heavy traffic starts moving. The oxidant concentration reaches a peak about noon, then de-

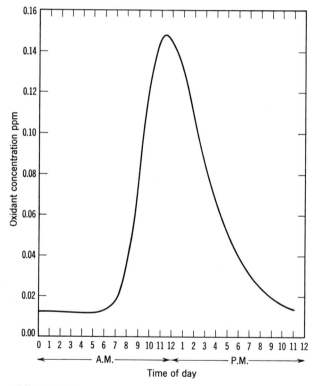

FIGURE 8.5.

Diurnal variation of hourly oxidant concentration in Los Angeles, August 1964 and 1965. (From Air Pollution Foundation data.)

creases to the 8 a.m. value by 8 p.m. and to the early morning value by 11 p.m. Thus as Angelenos say "the smog goes away at night." The noonday peak is largely a function of sunlight intensity, but is modified somewhat by the sea breeze (west to east in California) which arrives in downtown Los Angeles at about 10 a.m. and pushes the smog cloud east. At Pasadena, 15 miles east, the peak oxidant values occur around 1 p.m. The city of River-side, 50 miles east, experiences 2 peaks, one at noon (from local traffic sources) and another higher peak at 4 p.m. when the cloud from Los Angeles moves in on the sea breeze (see Fig. 8.6). The curves in Figs. 8.5 and 8.6 depict then the influence of sunlight intensity and air movement on

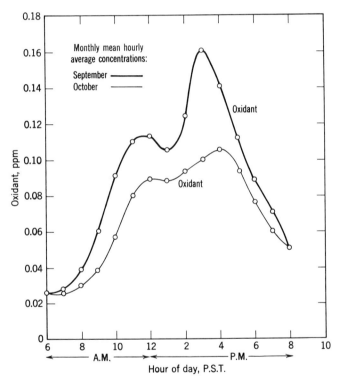

FIGURE 8.6.

Oxidant concentrations in Riverside, California, September and October, 1966.

smog formation. Another important factor is the height and intensity of the atmospheric inversion. Everything else being equal, the lower the inversion base, the more intense the smog (as measured by oxidant value).

Although smog may occur on any day of the year in the Los Angeles

Basin, it is most intense from August 15 to October 15, the time of year when wind speeds are lowest, the inversion is the strongest, and the sunlight is still intense. The low incidence of smog in the winter months is occasioned by less intense sunlight, more clouds, an unstable atmosphere, and absence of a persistent low-level inversion.

These same factors explain the difference in the number of smog occurrences in eastern cities, say Philadelphia, as compared with Los Angeles. In a typical year (1967), Philadelphia experienced three days when oxidant exceeded 0.15 ppm; Los Angeles had 209. Yet traffic is nearly as dense in

TABLE 8.1

Summary of Total Oxidant Concentrations Recorded at Camp Sites, 1964–1967 (Ref. 18)

City	Year	Days of valid data	Number of days with at least 1 hourly average equal to or exceeding			Maximum hourly average, ppm
			0.05 ppm	0.10 ppm	0.15 ppm	
Chicago	1964	254	149	15	0	0.13
	1965	275	120	9	0	0.13
	1966	235	52	6	3	0.19
	1967	255	113	16	1	0.16
Cincinnati	1964	303	137	36	5	0.26
	1965	310	182	19	5	0.17
	1966	208	54	1	0	0.10
	1967	228	122	24	1	0.20
Denver	1965	285	226	51	14	0.25
	1966	298	187	46	9	0.19
	1967	166	76	12	4	0.21
Philadelphia	1964	269	124	37	9	0.20
	1965	266	109	23	4	0.33
	1966	315	145	52	19	0.52
	1967	282	124	28	3	0.17
St. Louis	1964	253	156	26	6	0.26
	1965	329	206	33	8	0.35
	1966	292	174	33	5	0.22
	1967	289	185	38	4	0.20
Washington, D.C.	1964	293	163	40	4	0.20
	1965	284	150	25	3	0.21
	1966	325	134	27	2	0.16
	1967	322	137	27	5	0.26

Philadelphia as in Los Angeles. Table 8.1 shows the number of smog days in other cities.

Hydrocarbon Reactivity. Of the other variables that affect the photochemical reaction, hydrocarbon reactivity is one of the most important inasmuch as all hydrocarbons do not take part in the photochemical oxidation reaction. Generally olefins react most rapidly, followed by aromatics with side chains. Paraffins are much less reactive, and acetylene is relatively inert. One aromatic hydrocarbon (benzene) is also inert. Oxygenated hydrocarbons also vary in reactivity. Some approach aromatic hydrocarbons in reactivity, others are more like paraffins, and many are inert.

It is possible to rate the reactivity of various hydrocarbons by rate of disappearance, by rate of oxidation of NO to NO_2 when irradiated in the presence of a specific hydrocarbon, by yield of various products, or by intensity of various effects. Unfortunately, the relative ranking of hydrocarbons is not the same by all methods, but differences are not as great as occasional differences between scientists would indicate. It is generally agreed that the olefins and a few aromatics are the real culprits in photochemical smog formation. Several schemes for rating the relative reactivity of individual hydrocarbons have been proposed [1, 8, 11].

The variations in relative reactivity proposed by different investigators probably stem from the great complexity of the photochemical smog reaction. Two factors of importance in addition to those mentioned previously are the NO_x : hydrocarbon ratio and the action of light absorbers other than NO_2.

Most studies in environmental chambers indicate that molar ratios of NO_2 to hydrocarbon between 1:1 and 1:3 give the most rapid reactions and lead to most intense effects. As a consequence the suggestion has been made that NO_x should be allowed to remain at a comparatively high level in the atmosphere in order to quench the photochemical reaction [9].

Another complexity which enters the photochemical smog picture is the role of light absorbers other than NO_2 as initiators of photochemical oxidation. Aldehydes are known to absorb light albeit at much lower efficiencies than NO_2. It has also been postulated that in the absence of NO_2, aldehydes may be more important than they are at present.

Another possible mechanism of photochemical oxidation of hydrocarbons is by means of singlet oxygen produced by photosensitization of atmospheric organic compounds in the near-ultraviolet region of the spectrum [21].

Measurement of Photochemical Smog

Inasmuch as photochemical smog is a complex mixture of both undesirable and neutral compounds, methods of assessment of the presence and intensity of smog must be arbitrary. Of its various constituents the most nearly unique are ozone and PAN, both present in smoggy atmospheres at concentrations less than 1 ppm. Although ozone is probably the major oxidizing agent in smog there are small amounts of others which vary in amount and interfere with the ozone determination. Consequently, all oxidizing compounds in smog with oxidizing power greater than molecular oxygen are combined and called "oxidant."

Most commonly, "oxidant" is measured by passing an oxidant-containing airstream into a solution of neutral potassium iodide [26], where the oxidant reacts to release iodine and the tri-iodide ion, which may be measured colorimetrically. Another instrument uses a coulometric method [15]. The methods are said to be comparable when both instruments are calibrated by the user. Principal interfering gases are NO_2 and SO_2. The degree of interference in the colorimetric procedure is a function of KI concentration; the coulometric cell analyzers register about 10% of the NO_2 concentration as ozone.

A more serious interference in the oxidant determination by the potassium iodide method is the negative response to sulfur dioxide. In solution, SO_2 will react with ozone and reduce the oxidant reading by an amount equivalent to its concentration. The interference may be eliminated by inserting chromium trioxide scrubbers in the air tube leading to the oxidant analyzer. Details of the method have been published by the Environmental Protection Agency as a part of the federal air quality standard for photochemical oxidants [4]. PAN may be measured either by infrared spectroscopy or by gas chromatography [5, 22].

Effects of Photochemical Smog

The importance of any atmospheric pollutant is its effect on man, his property, and his activities, at concentrations and dosages that may be expected to occur in the open atmosphere. A great deal is known about the effects of photochemical smog in those areas where it is most prevalent (e.g., Los Angeles). Other information of value has been obtained by simulating photochemical smog in environmental chambers [6, 7, 25].

The principal effects of photochemical air pollution are eye irritation, damage to sensitive vegetation, a reduction in atmospheric visibility, deterioration of materials, and possible effects on the respiratory systems of men and animals.

Eye irritation is perhaps the most annoying manifestation of photochemical smog, particularly in California where its intensity is greatest. Apparently it is seldom noticed elsewhere. Inasmuch as eye irritation is a subjective effect, it is very difficult to quantify. In laboratory experiments elaborate statistical methods have been devised to eliminate psychological factors from panels of subjects who were used to measure the appearance and intensity of eye irritation. Even in Los Angeles, it is estimated that at least 25% of the population has never experienced eye irritation. Others report eye irritation when photochemical smog as measured by oxidant index is almost nonexistent. Among the substances responsible for the eye-irritating aspects of photochemical smog are formaldehyde, acrolein, PAN, and PB_zN, but they do not account completely for the effect. In 1959, the California Department of Public Health estimated that eye irritation was generally noticeable in Los Angeles when the oxidant index of smog exceeded 0.15 ppm. On the other hand, under some meteorological conditions eye irritation has been noticed when the oxidant index was around 0.1 ppm, so there has been a tendency to lower the previous value of 0.15 ppm.

Vegetation damage resulting from smog was first noticed in Los Angeles in 1944 [17] on leafy vegetables (romaine lettuce, spinach, etc.) and ornamentals (particularly petunias). First noticeable effects are silvering and bronzing of the underside of leaves. Cell collapse and necrosis follows. Where the value of a crop depends on either the appearance or nutritive value of the leaves, economic damage results.

Since 1944, considerable research has been conducted to determine the factors responsible for smog damage to various types of vegetation. The three principal phytotoxicants are ozone, NO_2, and PAN. Ozone usually affects the top side of leaves and this is largely responsible for damage to some varieties of tobacco, white pine, tomato, pinto beans, and spinach [12]. All are said to take place at concentrations of 0.1 ppm or less for periods ranging from ½ to 8 hours. Growth retardation of citrus has also been attributed to ozone.

The principal effects of PAN are the silvering, bronzing, and glazing of the lower leaf surface. Petunias are affected at very low concentrations (0.005 ppm for 8 hours). Pinto beans are affected at 1.0 ppm for 30 minutes.

Not as much is known about the effects of NO_2, but long-time exposure to 0.5 ppm is said to damage tomato plants [24].

Estimation of the dollar damage to crops from photochemical smog is largely an educated guess, but estimates of nationwide damage have been placed at $100 million by NAPCA.

Visibility reduction resulting from the photochemical aerosol is perhaps the most widely noticeable effect of photochemical smog. But even in Los Angeles, it is sometimes difficult for the public to distinguish the photochemical aerosol from fog, haze, and dust except on bad days when other manifestations of photochemical smog (eye irritation and oxidant odor) are apparent. In other areas smoke obviously interferes. The nature of the aerosol is not completely known. Particle size centers around 0.3 microns, an optimum size for scattering light. Chemically the aerosol contains compounds of carbon, hydrogen, oxygen, sulfur, nitrogen, and halides [14]. The liquid phase is largely organic matter and the water content appears to be a function of relative humidity.

Attempts to devise a quantitative relationship between visibility and oxidant index have been unsuccessful, although there is no doubt that the two are often associated.

Deterioration of materials by photochemical smog is thought to be primarily the result of the ozone content of smog. Thus the most specific effect of smog on materials is the cracking of rubber. In fact an early, specific analytical method for the determination of atmospheric ozone was the depth of cracks on stretched rubber bands. An important economic effect of smog in Los Angeles in the 1940's was the deterioration of the sidewalls of automobile tires. Addition of an antiozonant has alleviated the problem. A variety of other rubber products (hoses, gaskets, wire insulation) are still affected. Another important effect believed to be caused by ozone is the fading of dyes [20].

Effects on the respiratory system of men and animals is an area of controversy in the photochemical smog picture that does not appear likely to be resolved in the near future. An NAPCA document lists references which indicate a variety of effects on mice at concentrations near 0.1 ppm oxidant [18]. It also reports impairment of pulmonary function in emphysema patients, impaired performance of high school athletes, and aggravation of asthma at similar concentrations (0.1 ppm oxidant). On the other hand, Tabershaw in 1968 [23] pointed out "[For] 20 and more years, millions of Los Angeles residents have been exposed to levels injurious to smaller animals. The present and proposed measures will not substantially lower atmospheric oxidants before 1975. By that time the people of Los Angeles will be exposed for 25 to 30 years or longer to levels considered above the threshold. Presumably the chronic effects, if the experimental evidence is at all correct, should be visible now or should soon become evident."

The only other constituent of photochemical smog that might be expected to have an effect on health is NO_2. It has been discussed in Chapter 5 (q.v.).

Air Quality Standards for Photochemical Smog. The first air quality standards for photochemical smog were adopted by the state of California in 1959. Many different methods of stating the standard were proposed and studied, but the State Department of Health finally selected "oxidant index" as the preferred measure of smog and adopted a value of 0.15 ppm for 1 hour as an ambient air quality standard. Based on data available at the time it was felt that by keeping the oxidant index below 0.15 ppm, averaged for 1 hour, the adverse effects of eye irritation, plant damage, and reduced visibility would be prevented.

The standard turned out to be academic because as late as 1968, the standard was violated in Los Angeles County on 188 days.

California adopted new standards in 1969, this time 0.1 ppm for 1 hour, although no confirmed new data were available. The standard is slightly higher than that adopted by the Environmental Protection Agency (see Table 1.3).

REFERENCES

1. Altshuller, A. P., "An Evaluation of Techniques for the Determination of the Photochemical Reactivity of Organic Emissions," *J. Air Poll. Control Assoc.,* **16,** 257–260 (1966).

2. Altshuller, A. P., and J. J. Bufalini, "Photochemical Aspects of Air Pollution: A Review," *Photochem. Photobiol.,* **4,** 97–146 (1965).

3. Barrett, L. B., and T. E. Waddell, "The Cost of Air Pollution Damage," National Air Pollution Control Administration (1970).

4. Code of Federal Regulations, Chapter IV, Title 42, Part 410, Appendix D (1971).

5. Darley, E. F., K. A. Keltner, and E. R. Stephens, "Analysis of Peroxyacyl Nitrates by Gas Chromatography with Electron Capture Detection," *Anal. Chem.,* **35,** 589–591 (1963).

6. Dimitriades, B., "Methodology in Air Pollution Studies Using Irradiation Chambers," *J. Air Poll. Control Assoc.,* **17,** 460–466 (1967).

7. Doyle, G. J., "Design of a Facility (Smog Chamber) for Studying Photochemical Reactions under Simulated Tropospheric Conditions," *Environ. Sci. and Tech.,* **4,** 907–916 (1970).

8. Glasson, W. A., and C. S. Tuesday, "Hydrocarbon Reactivities in the Atmospheric Photooxidation of Nitric Oxide." Paper presented at 150th Meeting of American Chemical Society, Atlantic City, N.J. (Sept. 12–17 1965).

9. Glasson, W. A., and C. S. Tuesday, "Inhibition of Atmospheric Photooxidation of Hydrocarbons by Nitric Oxide," *Environ. Sci. and Tech.,* **4,** 37–43 (1970).

10. Haagen-Smit, A. J., "Chemistry and Physiology of Los Angeles Smog," *Ind. Eng. Chem.,* **44,** 1342–1346 (1952).

11. Heuss, J. M., and W. A. Glasson, "Hydrocarbon Reactivity and Eye Irritation," *J. Environ. Sci. and Tech.,* **2,** 1109–1116 (1968).

12. Jacobson, J. S., and A. C. Hill, eds., "Recognition of Air Pollution Injury to Vegetation: A Pictorial Atlas," Air Pollution Control Assoc., Pittsburgh, Pa. (1970).

13. Leighton, P. A., "Photochemistry of Air Pollution," Academic Press, New York (1961).

14. Mader, P. P., R. D. Macphee, R. T. Lofberg, and G. P. Larson, "Composition of Organic Portion of Atmospheric Aerosols in the Los Angeles Area," *Ind. Eng. Chem.,* **44,** 1352–1355 (1952).

15. Mast, G. M., and H. E. Saunders, "Research and Development of Instrumentation of Ozone Testing," *Instr. Soc. Amer. Trans.,* **1,** 325–328 (1962).

16. Middleton, J. T., and A. J. Haagen-Smit, "The Occurrence, Distribution, and Significance of Photochemical Air Pollution in the United States, Canada and Mexico," *J. Air Poll. Control Assoc.,* **11,** 129–134 (1961).

17. Middleton, J. T., J. B. Kendrick, Jr., and H. W. Schwalm, "Injury to Herbaceous Plants by Smog or Air Pollution," *Plant Disease Reptr.,* **34,** 245–252 (1950).

18. National Air Pollution Control Administration, "Air Quality Criteria for Photochemical Oxidants," NAPCA Publication No. AP-63, U.S. Govt. Printing Office, Washington, D.C. (1970).

19. Pitts, J. N., "The Role of Singlet Molecular Oxygen in the Chemistry of Urban Atmospheres," *Chemical Reactions in Urban Atmospheres* (edited by C. S. Tuesday), American Elsevier Publishing Co., New York, 3–34 (1971).

20. Salvin, V. S., "Relation of Atmospheric Contaminants and Ozone to Light Fastness," *Am. Dyestuff Reporter,* **53,** 33–41 (1964).

21. Steer, R. P., J. L. Sprung and J. N. Pitts, Jr., "Singlet Oxygen in the Environmental Sciences: Evidence for the Production of O_2 ($^1\Delta$g) by Energy Transfers in the Gas Phase," *Environ. Sci. and Tech.,* **3,** 946–947 (1969).

22. Stephens, E. R., "Absorptivities for Infrared Determination of Peroxyacyl Nitrates," *Anal. Chem.,* **36,** 928–929 (1964).

23. Tabershaw, I. R., F. Ottoboni, and W. C. Cooper, "Oxidants: Air Quality Criteria Based on Health Effects," *J. Occupational Med.,* **10,** No. 9, 1–135 (1968).

24. Taylor, O. C., and Eaton, F. M., "Suppression of Plant Growth by NO_2," *J. Plant Physiol.,* **41,** 132–135 (1966).

25. Tuesday, C. S., B. A. D'Alleva, J. M. Heuss, and G. J. Nebel, "The General Motors Smog Chamber," General Motors Research Pub. GMR-490 (1965).

26. U.S. Public Health Service, "Selected Methods for the Measurement of Air Pollutants," Division of Air Pollution, Cincinnati, Ohio. PHS Publication No. 999-AP-11 (1965).

27. Went, F. W., "Global Aspects of Air Pollution as Checked by Damage to Vegetation," *Proc. Natl. Air Pollution Symposium,* 3rd Symposium, Pasadena, Calif., 1955, pp. 8–11.

9

TRANSPORTATION SOURCES

The means of transportation that produce atmospheric pollutants comprise automobiles, buses and trucks with gasoline engines, diesel-powered trucks and buses, propeller and jet-powered aircraft, railroad locomotives, and marine vessels. By far the most important from an air pollution standpoint is the spark-ignited internal-combustion engine using gasoline as fuel.

THE INTERNAL-COMBUSTION ENGINE

In the modern spark-ignited internal-combustion engine, a sequence of five events takes place:

1. Air and fuel are mixed in predetermined proportions in a carburetor or by fuel injection.

2. The mixture of fuel, in vapor and fine-droplet form, and air is drawn into a combustion chamber (cylinder) equipped with a movable piston.

3. The mixture is compressed and then ignited by means of a controlled spark.

4. Controlled combustion takes place, producing power that is translated mechanically to the drive mechanism (crankshaft).

5. The spent gases are exhausted to the atmosphere.

The heart of the automotive power plant is the combustion chamber, shown diagrammatically in Fig. 9.1. Modern automobiles usually have six or eight cylinders, arranged to operate in proper sequence to deliver power smoothly to the crankshaft. Each cylinder head is fitted with a spark plug

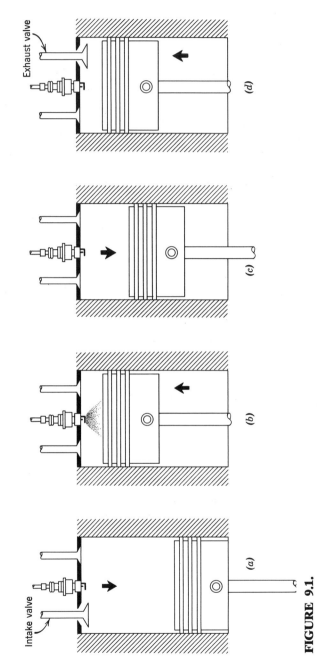

FIGURE 9.1.

Four-stroke cycle of internal-combustion engine. (a) Intake stroke (b) compression stroke (c) power stroke and (d) exhaust stroke.

to initiate combustion, an inlet valve for air and fuel, and an exhaust valve for discharge of burned gases. The bottom of the combustion chamber is a movable piston, which is attached to the crankshaft by a connecting rod.

The operating cycle of the combustion chamber may be understood by reference to Fig. 9.1. In Fig. 9.1*d* the piston is at top dead center (TDC), leaving only a small free volume in the cylinder. As the piston is pulled down by action of the crankshaft, the inlet valve opens (the exhaust valve is closed) and a proportioned mixture of gasoline and air from the carburetor is pulled into the cylinders via the intake manifold. When the piston reaches bottom dead center (BDC) shown in Fig. 9.1*a*, the inlet valve closes. Action of the crankshaft then moves the piston upward, compressing the gaseous mixture of fuel and air. The actual pressure at this point depends on the compression ratio of the engine, which is the ratio of the volume of the cylinder at BDC to the volume at TDC. Just before the piston reaches TDC, an electric spark is passed between the points of the spark plug (Fig. 9.1*b*). The compressed gasoline vapor then burns rapidly (explodes), causing an increase in volume of gases due to both the formation of carbon dioxide and water vapor and the increase in temperature. The pressure from combustion drives the piston downward (Fig. 9.1*c*), and power is transmitted to the crankshaft. Just before the piston reaches BDC, the exhaust valve opens, and the piston completes its downward travel and starts upward again. The hot exhaust gases are forced into the exhaust manifold, through the muffler, and out the tailpipe of the automobile. The exhaust valve closes, the inlet valve opens, and the cylinder is ready to receive a fresh charge of gasoline and air.

This is the basic operation of the four-stroke-cycle engine: intake, compression, power, and exhaust. With each series of four strokes the crankshaft undergoes two complete revolutions. Inasmuch as engine speeds vary from 400 rpm at idle to about 4000 rpm at high speed, each cylinder is fired 200 to 2000 times each minute.

Besides its basic requirement of delivering power, the automotive engine must produce it in *varying* amounts at the whim of the operator. This has led to the establishment of four phases of the operating cycle of the engine: idling, acceleration, cruising, and deceleration. The characteristics of each of these phases, commonly called "operating modes" are of extreme importance in determining the combustion efficiency of the internal-combustion engine, so they will be described in more detail.

Operating Modes

At *idle*, only enough power is needed to keep the engine running smoothly. The usual air-fuel mixture (11:1 to 12.5:1) is on the "rich side," compared with the stoichiometric 15.2:1. The engine idles at about 400 to 500 rpm, or slightly higher in cars with air conditioning and other power-absorbing equipment. The airflow is low (6 to 8 cfm), and vacuum in the cylinder is 16 to 20 in. Hg. At the low airflow used, distribution of fuel to the 6 or 8 cylinders is not uniform, so combustion in a given cylinder may vary considerably from the average. Of course, a misfiring spark plug will affect combustion efficiency to a much greater extent than maldistribution. Because the cylinders are operated on the rich side during idle, the exhaust gases will contain carbon monoxide and some unburned hydrocarbons. The magnitude and variations in the composition of the exhaust gases and temperature attained will be discussed later.

Under *accelerating* conditions, the flow of both fuel and air is increased markedly, engine speed increases, spark timing is advanced, and combustion conditions continually change. Exhaust composition depends on many factors, such as rate of acceleration, over-all accelerating time, and change in gear ratio. In a typical situation, airflow is 30 to 35 cfm, air-fuel ratio is 11:1 to 13:1, engine speed goes from 400 rpm to 3000 or even 4000 rpm. Cylinder vacuum will vary from 0 to 7 in. Hg. The range of conditions of the gases in the cylinders is largely a function of the carburetor-transmission combination. In cars with standard transmission and manual shift, fuel flow is interrupted at the shift point, with subsequent leaning of the mixture and a corresponding temperature increase; with automatic transmissions, the shift is less noticeable. The use of the four-barrel carburetor permits very high fuel flow and extremely rapid acceleration. One 1957-model automobile consumed nearly a half pint of fuel in 15 sec under rapid acceleration. More moderate accelerations, such as 0 to 25 mph in 13 sec and 15 to 30 mph in 11 sec, will consume considerably less fuel.

Under *cruise* conditions, a steady state is reached. Airflow normally varies from 15 to 35 cfm, depending on speed. In many modern cars the air-fuel ratio is even on the "lean side," or greater than 15.2:1. The dead weight of the car need no longer be overcome, so the power developed will normally range from 10 to 50 hp, depending on cruising speed.

Deceleration is a continually varying condition. The deceleration may be for only a short period of time with a partially closed throttle, e.g., going from 70 down to 60 mph in the open country; it may be a rapid braking action from a high speed to a complete stop in as short a time as possible.

We all know from experience that different traffic situations require different types of deceleration. In the typical Los Angeles traffic pattern, two types of deceleration are recognized—a moderate deceleration (30 to 15 mph in 10 sec) and a heavy deceleration (30 to 0 mph in 14 sec). In the latter case the only fuel reaching the cylinder is that from the idling jet, although drying of the wet intake manifold may flash a high momentary vapor concentration into the cylinder. Air flow during deceleration is about the same as during idle. Cylinder vacuum rises above 20 in., sometimes as high as 25 in. Combustion efficiency is very poor—anywhere from 20 to 60% of the supplied fuel may pass through the cylinders unburned (*Note:* corrected since 1966).

MOTOR FUEL

Physical Properties

Motor gasoline, the universal fuel for spark-ignited internal-combustion engines, is a complex mixture of relatively volatile hydrocarbons that vary widely in their chemical and physical properties. Because of the wide range of conditions under which motor gasoline is used, its properties must be properly balanced to give satisfactory performance.

Drivers demand the following gasoline qualities: easy starting, quick warmup, fast acceleration, good mileage, and antiknock properties.

Undesirable properties of a gasoline are high vapor pressure, which causes vapor lock; low octane number, causing detonation and preignition power loss; high gum content, in which case the fuel is not stable in storage and gums up the carburetor; incorrect distillation range, causing engine roughness and flat spots; and high sulfur content, causing corrosion. Accordingly, the properties of gasoline which must be closely controlled are volatility (through Reid vapor pressure and distillation range), octane number, gum content, and sulfur content.

Volatility is an extremely important property of motor fuel in that the fuel must be sufficiently volatile to allow the engine to start easily but not so volatile that vapor lock results at high temperatures. The fuel must also be vaporized sufficiently to permit equal distribution to all cylinders. It must also be sufficiently volatile not to dilute the lubricating oil in the crankcase. Because gasoline volatility varies with temperature and atmospheric pressure, allowances must be made in blending stocks for different geographical areas, different seasons of the year, and different altitudes. A common method of increasing volatility during the winter season is to blend butane with normal fuel.

Probably the most important property of gasoline is its antiknock quality. If a fuel burns unevenly or too rapidly in the combustion chamber, an audible ping or knock is heard and there is loss of power. The knocking characteristics of hydrocarbons vary widely with their molecular structure, so the tendency of gasoline to knock will vary as the composition varies. The antiknock quality of motor gasoline is measured by octane number, which is defined as the percentage of isooctane (2,2,4-trimethylpentane) in a blend with *n*-heptane which will give the same detonation characteristics as the gasoline. Pure *n*-heptane would have an octane number of zero; pure isooctane would be 100.

There are two recognized methods for obtaining the octane number of a fuel in the laboratory—the Research Method and the Motor Method. No straight-line method of correlating the two values exists, because they are obtained under different conditions. The Research Method usually provides a better guide as to performance on the road.

Until 1970, the trend in engine design was toward higher octane requirements. In 1970 the major American automobile manufacturers announced that beginning with 1971 models, engine octane requirements would be lowered. Purpose of the change was said to be necessary to allow the use of catalytic exhaust-gas convertors. The entire matter is highly controversial.

At present, the preferred means of producing motor fuel of adequate antiknock quality is the addition of up to 3 milliliters per gallon of special antiknock fluids which are primarily a mixture of tetraethyl or tetramethyl lead, ethylene dichloride, and ethylene dibromide. The specific amount added depends on the quality of the base gasoline and the octane number desired.

Thus, in December 1968, the octane numbers (Research Method) of premium gasolines varied from 96.9 in Denver to 100.6 in Boston and Cleveland; regular gasolines varied from 90.8 in Denver to 94.9 in Boston, Detroit, Houston, and Philadelphia. Although the average antiknock quality of regular gasolines was identical in Boston and Houston, the average antiknock fluid content (in terms of metallic lead) was 1.87 and 2.55 grams per gallon [15]. This difference reflects differences in the hydrocarbon composition of the base gasoline.

Supposedly, lowering the octane requirement of new engines will allow lead compounds to be eliminated from gasoline and thus permit long-lived catalysts to be developed for use in exhaust control devices. If lead addition to gasoline is prohibited by law, some other means of obtaining adequate antiknock quality must be devised. The problem is highly complex and the lead issue has unfortunately become a political football.

Gum content is a measure of the sticky, hard nonvolatile residue left when gasoline is evaporated. It is held to a minimum commercially to reduce manifold and carburetor deposits and sticking of intake valves.

The sulfur content of gasoline is usually less than 0.1%. Generally, West Coast gasolines are higher in sulfur than those found elsewhere, because of the higher concentration in the crude oil.

Chemical Composition

The specific types of hydrocarbons in gasoline depend on the crude oil from which it is derived and the particular refining process used, with the latter predominating.

The three major types of hydrocarbons in gasoline are aromatic, olefinic, and saturated hydrocarbons. The aromatics include benzene and various branched-chain derivatives of benzene, e.g., toluene, xylenes, ethyl benzene, etc.; the olefins are both normal and branched-chain types; the saturates include normal and branched-chain paraffins as well as naphthenes (saturated cyclics).

A general classification of gasoline stocks by refining process and typical hydrocarbon content is shown in Table 9.1. It must be realized that none of

TABLE 9.1

Types of Hydrocarbons in Gasoline Stocks*

Process	Major hydrocarbon type
Straight-run	Paraffins (straight, branched, and cyclic), C_5 and C_6
Catalytically-cracked	Paraffins and olefins, C_4 to C_9
Hydro-cracked	Paraffins, C_5 to C_6
Alkylate	Paraffins, C_4 to C_9
Reformed	Aromatics, C_6 to C_9; cyclic paraffins
Isomerized	Paraffins (iso and cyclic), C_5 to C_6
Extracted (from reformate)	Aromatics, C_6 to C_9

* From *Chem. Eng. News*, 59 (Nov. 9, 1970).

these represents a complete gasoline, because the physical characteristics required can be obtained only by blending. In November 1957 the Ethyl Corporation analyzed 190 samples of gasoline purchased at filling stations throughout the United States. In these gasolines aromatics ranged from 11 to 40%; olefins, 1 to 48%; and saturates, 31 to 80%. Of course, these broad hydrocarbon types cover a wide range of specific hydrocarbons. For instance, the specific aromatic hydrocarbons in reformates are quite different from those in catalytically cracked stock; similarly, the saturates in alkylates are chiefly isoparaffins (branched compounds), whereas those in straight-run gasolines are naphthenic.

The importance of specific hydrocarbon types from an air pollution standpoint is a function of the relative reactivities of the various types of hydrocarbons in photochemical smog formation (see Chapter 8). This is particularly true in cases where the gasoline evaporates into the atmosphere or is carried relatively unchanged through the engine into the exhaust. On the other hand, most of the hydrocarbons showing up in exhaust gas are different from those in the gasoline used as fuel because of partial oxidation and thermal decomposition (cracking) resulting from the high temperature in the cylinders [23, 24].

A more detailed analysis of a typical gasoline is shown in Table 9.2.

TABLE 9.2

Specific Hydrocarbons in a Specific Full-Range
Motor Gasoline*

Component	Composition % by wt
Propane	0.01
Isobutane	0.37
Isobutylene and butene-1	0.04
n-Butane	4.29
trans-2-butene	0.20
cis-2-butene	0.17
Pentanes	16.59
Pentenes	3.50
Hexanes	9.67
Hexenes	2.70
Benzene	0.81
Heptanes	8.80
Heptenes	7.94
Toluene	12.20
Octanes	9.67
Octenes	—
Nonanes	4.88
Nonenes	—
Ethyl benzene	1.70
Xylenes	4.60
Subs. benzenes (9 carbons)	4.00
Decanes	1.64
Subs. benzenes (10 carbons)	1.85
Undecanes	0.09
Subs. benzenes (11 carbons)	0.11
Naphthalene	0.10
Dodecanes	0.05
Subs. benzenes (12 carbons)	

* Adapted from Ref. 30.

Additives

Besides tetraethyllead, several other additives are present in many commercial gasolines. Among these are the preignition preventers, tricresyl phosphate, certain boron-containing esters, and complex organic compounds added to reduce spark-plug fouling and engine deposits; antioxidants, e.g., aminophenols and alkylated phenols, to inhibit gum formation; organic metal deactivators to inhibit the oxidizing properties of metal impurities; antirust additives, usually surface-active agents; anti-icing agents, e.g., isopropyl alcohol, to prevent freezing of water in tanks and strainer bowls and to prevent ice formation in the carburetor; upper cylinder lubricants to aid valve stem and upper cylinder lubrication; and dyes, used for identification of various grades of gasoline.

LUBRICATING OIL

The role of lubricating oil in air pollution is particularly important in engines with worn piston rings where excessive amounts of oil are sucked upward past the rings into the cylinder, particularly during deceleration. The heavy oil is only partially burned and is then emitted as smoke with the exhaust gases. During acceleration rich exhaust gases also force their way past weak piston rings and carry oil vapors into the air. Most oils contain various additives, the most important of which are complex barium and calcium sulfonates, used for their detergent properties. Small amounts of these metals will enter the exhaust stream when oil consumption is excessive.

COMPOSITION OF AUTO EXHAUST

The major components of auto exhaust are, of course, the complete oxidation products of the fuel, carbon dioxide and water, and the nitrogen that accompanied the air fed to the combustion chamber. Because oxidation is incomplete, carbon monoxide is always present. Minor constituents, but important ones from an air pollution standpoint, are hydrogen, oxygen, unburned hydrocarbons, partially oxidized hydrocarbons, nitric oxide, and sulfur dioxide. The normal range of some of these minor constituents is shown in Table 9.3. Despite the fact that automobiles are mass-produced items, the emissions of no two cars, even of the same model, are identical.

At the present time (1972), tremendous variations exist in the composition of exhaust gases from motor vehicles, depending largely on the age of the car. It was only in 1959 that California passed the first regulations requiring reduction of hydrocarbon and carbon monoxide emissions. At that time little attention was given to other automotive sources of hydrocarbon emissions, i.e., the crankcase, carburetor, and fuel tank. However, in 1960, domestic automobile manufacturers installed crankcase blowby devices on most 1961 model automobiles sold in California. In 1963 the device (slightly modified) became standard nationwide. The nature of the device is discussed later in this chapter (see p. 276).

Thus only pre-1963 model cars still on the road are completely uncontrolled cars. On the average, exhaust emissions from these cars, in 1962, were: hydrocarbons—600 ppm (calculated as hexane); carbon monoxide —2.3%; oxides of nitrogen—1000 ppm (calculated as NO_2). There may have been some deterioration (increased emissions) since that time, but

TABLE 9.3

Effect of Engine Operating Conditions on Composition of Auto Exhaust (uncontrolled cars)*

	Idle	Acceleration	Cruising	Deceleration
Air-fuel ratio	11:1–12.5:1	11:1–13:1	13:1–15:1	11:1–12.5:1
Engine speed, rpm	400–500	400–3000	1000–3000	3000–400
Airflow, cfm	6–8	30–35	15–35	6–8
Cylinder vacuum, in. Hg	16–20	0–7	7–19	20–25
Exhaust analysis:				
CO, %	4–6	0–6	1–4	2–4
NO, ppm	10–50	1000–4000	1000–3000	10–50
Hydrocarbons, ppm	500–1000	50–500	200–300	4000–12,000
Unburned fuel, % of supplied fuel	4–6	2–4	2–4	20–60

* rpm—revolutions per minute   in. Hg—vacuum, inches of mercury
  ppm—parts per million, by volume   CO—carbon monoxide
  cfm—cubic feet per minute   NO—nitric oxide

no data are available. Exhaust emissions of 1963–1967 models will also vary but little from the above composition.

Inasmuch as the values given are averages, there could be (and are) marked variations from one model to another, from one make to another, and even from one make and model to another of the same make and model. What one might call the "normal range" of composition of pre-1963 models is shown in Table 9.3. By inspection of the table, one may see that the manner in which a vehicle is operated can greatly affect total emissions. Thus a great deal of stop-and-go driving means longer periods in high-emission transient modes (acceleration and deceleration) than in sustained steady-speed driving. Similarly the "jack-rabbit" driver who engages in quick starts and stops increases emissions still more. Cars operated on hilly terrain (e.g., San Francisco) will emit more pollution than those operated at steady speed on a level freeway (e.g., Los Angeles).

A great many engine variables also affect emissions. Details are beyond the scope of this chapter, but a simple listing would include mechanical factors related to carburetion, the ignition system, and related equipment [22].

Carburetion variables which increase hydrocarbon emissions are: rich mixtures (low air-fuel ratio), improper idle speed, a stuck choke, external carburetor leaks, and plugged air cleaners. Ignition system factors which lead to increased emissions are: improper timing, malfunctioning distributor, defective ignition points and condensers, distributor cap and rotor malfunctions, worn spark plugs, and defective ignition wires or terminals. Related equipment failures include dirty PCV system (q.v.), stuck heat-riser valves, defective thermostat, worn rings, leaking valves, and intake manifold leaks. All of these defects and malfunctions affect emissions. As will be seen later, some may be corrected by good maintenance practices or by design changes.

1968 and 1969 Model Cars

Beginning with 1968 models, new motor vehicles sold in the United States were required to meet federal standards as follows:

Vehicle engine displacement, in.^{3}	Exhaust composition	
	Hydrocarbons, ppm	Carbon monoxide, %
50–100	410	2.3
101–140	350	2.0
141 and up	275	1.5

Dynamometer test procedures, number of cars to be tested, type of fuel to be used, analytical methods, and required calculations, etc., were specified in a federal regulation [10]. A portion of the cars, less than those required for obtaining emission data, also had to be tested for emission deterioration in a 50,000-mile durability test. The data from the emission-data cars were then corrected for the deterioration factor found. Emissions from all cars of a given make and class were averaged, and, if the average met the regulation, the particular class of car was certified and could be offered for sale. This "averaging" concept caused quite a political hassle in California where an identical regulation had been in force beginning with 1966 model cars. Some segments of the community thought that every car leaving the assembly line should be tested and should meet the standard. The California standards for 1973 models require such testing.

1970 Model Cars

Beginning with 1970 models, stricter federal regulations became effective [11]. The regulation further restricted emissions of hydrocarbons and carbon monoxide from the exhausts of light-duty motor vehicles. It limited evaporative fuel losses (from fuel tank and evaporator) from light-duty motor vehicles; and it extended 1968 exhaust-emission limitations to heavy-duty vehicles (trucks and buses).

In setting the 1970 exhaust standards, the federal government for the first time recognized the difference between mass emissions and concentration of pollutants. In the 1968 regulation, a large engine, with, say, 173-cubic-inch displacement, was allowed to emit a greater amount of pollutants than a smaller engine (e.g., 141 cubic inches), as long as the concentration in the exhaust did not exceed 275 ppm hydrocarbons and 2.3% CO. The

1970 regulation was therefore stated in terms of allowed mass emissions, i.e., hydrocarbons—2.2 grams per vehicle mile; carbon monoxide—23 grams per vehicle mile. The effect of the mass standards on allowable concentrations is shown in Table 9.4.

TABLE 9.4

Allowable Concentrations of Hydrocarbons and Carbon Monoxide in Exhaust from 1970 Model Vehicles

Loaded vehicle weight* (lb)	Carbon monoxide concentration (%)		Hydrocarbon concentration (ppm)	
	Automatic	Manual	Automatic	Manual
Up to 1625	2.29	2.55	411	457
1626–1875	1.95	2.17	350	389
1876–2125	1.71	1.90	307	341
2126–2375	1.53	1.70	274	305
2376–2625	1.39	1.55	250	278
2626–2875	1.28	1.42	230	256
2876–3250	1.19	1.33	214	238
3251–3750	1.06	1.18	190	212
3751–4250	.97	1.07	173	193
4251–4750	.90	1.00	162	180
4751–5250	.85	.95	153	170
5251–6000	.82	.91	147	164

* Loaded weight is curb weight (weight of a vehicle ready for operation) plus 300 lbs.

A comparison of the standards for 1968 and 1970 cars is shown in Table 9.5. Thus the concentration limits imposed by the 1970 standards were closer to equitable regulations of new car exhaust than the 1968 standards. However, because vehicle weight and exhaust volume were not perfectly correlative, some inequities occurred under the 1970 standards. To be perfectly equitable, actual exhaust volumes should be measured. Amendments to the 1970 regulations included this provision.

The standard for fuel evaporative emissions in 1970 model cars (later postponed to 1971 models) is 6 grams of hydrocarbons per test as described in the regulation. The test simulates fuel tank and carburetor hot soak losses on an average day. Beginning with 1972 models, the standards became 2 grams per test.

1971 and Later Models

Prior to 1970 the state of California was the leader in motor vehicle pollution control and the federal government followed later with similar con-

TABLE 9.5

Federal Exhaust Emission Standards for 1970 Cars as Compared to Standards in Effect for 1968 and 1969 Cars

Name*	Trans-mission	Cylin-ders	Engine displacement (in.³)	Load vehicle weight	Exhaust volume† (ft³/vehicle mi)	Hydrocarbon concentration (ppm) 1968	1970	Hydrocarbon weight (g/vehicle mi) 1968	1970	Carbon monoxide concentration (%) 1968	1970	Carbon monoxide weight (g/vehicle mi) 1968	1970
Renault	M	4	67.6	2030	36.6	410	341	2.6	2.2	2.3	1.90	28	23
Volkswagen	A	4	91.1	2100	40.7	410	307	2.9	2.2	2.3	1.71	31	23
Volvo	M	4	108.6	2900	52.4	350	238	3.2	2.2	2.0	1.33	35	23
Falcon	A	6	170	3130	58.3	275	214	2.8	2.2	1.5	1.19	29	23
Mustang	M	6	200	3152	52.4	275	238	2.5	2.2	1.5	1.33	26	23
Dart	M	6	275	3258	59.0	275	212	2.9	2.2	1.5	1.18	29	23
Chevrolet		6	250	3450	65.6	275	190	3.2	2.2	1.5	1.06	33	23
Cougar	A	8	390	3463	65.6	275	190	3.2	2.2	1.5	1.06	33	23
Fury	M	8	318	4093	64.6	275	193	3.1	2.2	1.5	1.07	32	23
Ford	A	8	302	4270	77.3	275	162	3.7	2.2	1.5	0.90	38	23
Chevrolet	A	8	307	4448	77.3	275	162	3.7	2.2	1.5	0.90	38	23
Thunderbird	A	8	429	4806	81.6	275	153	3.9	2.2	1.5	0.83	41	23
Cadillac	A	8	472	5100	81.6	275	153	3.9	2.2	1.5	0.83	41	23
Imperial	A	8	440	5450	84.9	275	147	4.1	2.2	1.5	0.82	42	23
Lincoln	A	8	460	5654	84.9	275	147	4.1	2.2	1.5	0.82	42	23

* All cars listed are 1968 models.
† Calculated from vehicle weight.

trols. Even though the field of motor vehicle pollution control was pre-empted by the federal government, waivers were granted California to en-force more restrictive regulations. Thus the California regulations were fairly good indications of later federal regulations. In early 1970, California established standards for nitrogen oxide emissions for 1971 model cars, and progressively more restrictive standards for 1972 and 1974 models. More restrictive regulations for CO and hydrocarbon emissions also applied to 1972 models (see Table 9.6).

TABLE 9.6

Evolution of California Exhaust Standards for Light-Duty Motor Vehicles*

Model year	Hydrocarbons, g/mi (equiv ppm)		Carbon monoxide, g/mi (equiv %)		NO_x, g/mi (equiv ppm)	
1966–1968		(275)		(2.3)		—
1970	2.2	(180)	23	(1.0)		—
1971	2.2	(180)	23	(1.0)	4.0	(1075)
1972–1973	1.5	(125)	23	(1.0)	3.0	(800)
1974	1.5	(125)	23	(1.0)	1.3	(350)
1975–1976	See Federal standards, p. 271					

* 1966–1968 standards were expressed in parts per million and per cent; since 1970, mass emissions (g/mi) are specified.

Note: Because of changes in Federal test procedures (see text) California also changed its procedures so that 1972-model standards became 3.2 g/mi hydrocarbons, 39 g/mi carbon monoxide, and 3.2 g/mi NO_x. Standards for 1973 were changed to 3.2, 39, and 3.0 respectively; for 1974 they will be 3.2, 39, and 2.0 respectively.

In late 1970, however, the federal government published new standards and test procedures [31] that confused the situation somewhat. Not only were the standards for 1972 and later model cars changed, but so were the test procedures and the analytical method for determining hydrocarbons. Prior to this time both California and the federal agency used the so-called California test cycle, a 7-mode dynamometer cycle starting with a hot engine. Both also used a nondispersive infrared spectrometric method for determining hydrocarbons. The new federal regulation specified a different cycle starting with a cold engine. The test was said "to simulate an average trip in an urban area of 7.5 miles from a cold start." There are numerous other changes in the regulation, but those mentioned affected the standards to the greatest degree. The exhaust standards now became 3.4 grams per mile hydrocarbons and 39 grams per mile carbon monoxide. They were said to be only slightly more stringent but more realistic than the 1970 standard, but no pat relationship exists.

More confusion arose in late December, 1970 when President Nixon signed the new Clean Air Act of 1970. Title II of the Act directed the Ad-ministrator of the Environmental Protection Agency to prescribe standards for 1975 and later model cars "which require a reduction of at least 90 percentum from emissions of carbon monoxide and hydrocarbons allowable . . . in model year 1970," and to prescribe standards for NO_x for 1976 models which will be 90% of the average of nitrogen oxides emissions in 1971.

Accordingly, EPA announced in July, 1971 that the 1975-model standards will be 0.41 g/mi hydrocarbons, 3.4 g/mi carbon monoxide, and 3.0 g/mi nitrogen oxides (calculated as NO_2). The NO_x standard for exhaust emissions for 1973–1974 models will be 3.0 g/mi. The standard for 1976-models has been set at 0.4 g/mi NO_x. Even though the automobile industry says this value is not attainable, the EPA Administrator, in May 1972, refused to change it.

Heavy-Duty Vehicles

Large gasoline-powered motor vehicles, i.e., those weighing 6001 lb or more, are subjected to specialized emission standards in California. Total *combined* emissions of hydrocarbons and nitrogen oxides are limited to 16 grams per brake horse power hour (BHP hr) in 1973 and 1974 models and 5 g/BHP hr for 1975 and subsequent models. Carbon monoxide emissions are restricted to 40 g/BHP hr and 25 g/BHP hr respectively.

Federal standards as of mid-1972 are still stated in terms of the concentration in exhaust gases, i.e., 275 ppm hydrocarbons and 1.5% carbon monoxide.

Oxides of Nitrogen in Exhaust Gases

Nitric oxide formation in internal-combustion engines is a function of temperature in the cylinder and the rate at which the exhaust gases are quenched when they leave the cylinder. Inasmuch as the highest temperatures and fastest quench rates take place during acceleration and cruise, nitric oxide content is much higher in these cycles than in idling or deceleration. The general range of reported values is shown for pre-1966 model cars in Table 9.3. The average NO_x concentration in exhaust gases from an engine operated on a typical California (7-mode) cycle was 700 ppm. When hydrocarbon and carbon monoxide controls were applied to 1966 and later models, they included an increase in air-fuel ratio (leaner mixture) with the result that oxides of nitrogen increased to 1100 ppm and more as an average value. A few models showed average NO_x concentrations as high as 2000 ppm.

Aldehydes and Acetylene

Various investigations of the aldehyde and acetylene content of exhaust have been made, probably because simple analytical methods were available.

In most recent work, however, these materials have been determined as "hydrocarbons" and referred to the common base, hexane. Published values of both the aldehyde and the acetylene content of exhaust gases vary so widely that no conclusions may be drawn. Generally, however, their concentrations are very low compared to total organic values.

Sulfur Dioxide

Most of the sulfur in gasoline probably appears in auto exhaust as sulfur dioxide. Even when gasoline with maximum sulfur content (0.25%) is used, this would rarely be more than 150 ppm. In practice, it will normally be less than 100 ppm.

Lead and Aerosols

Few comprehensive studies have been made of the particulate matter in auto exhaust. Stanford Research Institute [17] studied the exhaust particulates from five makes of automobiles. They were sampled after dilution with a large volume of air for idling, cruising, and combined acceleration and deceleration. All of the automobile exhaust products sampled in these studies contained large quantities of very small (0.02 to 0.06 μ diameter) particles or larger agglomerates of these small particles. The quantity of material produced varied between 0.22 mg per g and 3.2 mg per g of gasoline burned, the average being 0.78 mg per g gasoline burned.

The results of various examinations indicated that the particulate material contained organic compounds of high molecular weight. Some of these compounds were aldehydes. Oxidants and strong organic acids or bases were not present. Once formed, the particles or droplets were nonvolatile and only slightly hygroscopic.

A comprehensive study was made by the Ethyl Corporation [20] on the amount, composition, and particle size distribution of the lead compounds in exhaust of automobiles operated on gasoline containing tetraethyllead.

Approximately 20 to 30% of the lead burned in the fuel was retained in passenger car exhaust system deposits and lubricating oil. The remainder was evolved at various rates, depending on how the car was operated. Generally, the lead emission increased as car speed increased.

The exhaust particles varied from 0.01 μ to several millimeters in diameter. Although the smaller particles (less than 1 μ) were the most numerous, they accounted for less than 5% by weight of the exhausted lead. The heavier particles (5 μ and larger) represented 27% of the exhausted lead under city-type driving and 39% under accelerating conditions.

The lead was exhausted principally as mixtures of $PbCl \cdot Br$, alpha and beta forms of $NH_4Cl \cdot Br$, and $2NH_4Cl \cdot PbCl \cdot Br$. When phosphorus was present in the fuel, about one-fifth of the exhausted lead was $3Pb_3(PO_4)_2 \cdot PbCl \cdot Br$.

Particle size distribution and composition were independent of the concentration of TEL in the gasoline. Further, neither sulfur concentration nor the presence of phosphorus-containing gasoline additives affected the amount and size of the exhausted lead compounds, although phosphorus altered their composition.

Habibi, et al [19] measured total particulate emissions, including "tars," from two 1970 model vehicles operated under a variety of conditions. Emissions were 0.3 g/mi filterable particulate matter and 1.0 g/mi "tar." The total amount of "nontar" particulates varied with type of operation. When all starts were made using full-choke, total particulates increased.

At present there are no emission standards for particulate matter in motor vehicle exhaust, although a value of 0.1 g/mi has been proposed. If the 90% reduction specified in the Clean Air Act of 1970 is appropriate, then the standard, when issued, may well be 0.03 g/mi.

ANALYTICAL METHODS AND TESTING PROCEDURES

Many engine operating procedures, sampling devices, and analytical methods have been applied to the determination of the constituents of automobile exhaust. The proliferation of procedures stems from the difficulties encountered. Some of the problems arise from (1) differences in exhaust composition resulting from engine variables, such as air-fuel ratio, manifold vacuum, compression ratio, rapid change in phase of operation, and mechanical condition of the engine; (2) the effect on exhaust composition of the type of fuel used; and (3) the influence of water vapor and elevated temperatures of automobile exhaust on subsequent analyses. In the past few years considerable attention has been given to these matters by several committees of the Automobile Manufacturers Association as well as other agencies concerned with air pollution resulting from automobile emissions.

Now that the federal government is responsible for enforcing emission standards, the procedures specified in the reguation (q.v.) have become standard methods for determining motor vehicle emissions.

Engine Operation for Exhaust Sampling

For purposes of exhaust sampling, engines mounted on dynamometers may be used, or automobiles may be used on chassis dynamometers or on the road. Two types of engine dynamometers and one chassis dynamometer are briefly described here. These and other types are discussed by Judge [25].

Engine Dynamometer. Two types of engine dynamometers are commonly used. One of these is the all-electric dynamometer; the other is an eddy-current absorption type. Other types, such as the water brake and fan brake, are not satisfactory for the purpose at hand.

The all-electric dynamometer has an electric generator with rotor and stator mounted so that the turning moment or torque on the stator can be measured. The load is absorbed by means of air-cooled resistances. A separate d-c generator is provided to supply the desired current to the field of the generator. To simulate road operating conditions, it is necessary to use an inertia flywheel or an electronic controller to reproduce the effect of the inertia of the chassis of the automobile.

The eddy-current absorption dynamometer is similar to the all-electric type, but the electrical energy is dissipated within the machine itself. Eddy currents are induced in the stator of the electric generator and the heat is dissipated with cooling water. A separate d-c generator is provided to excite the field coils. The entire installation is more compact and less expensive than the all-electric type. However, deceleration cannot be simulated with a magnetic brake, and it is again necessary to provide an inertia flywheel.

Chassis Dynamometer. This device permits the simulated operation of an automobile under road conditions. It is essentially a treadmill, with rollers on which the rear wheels of the automobile are placed. The dynamometer is equipped with a hydraulic power-absorption unit and, for the purposes described here, with a flywheel so the automobile may be driven under conditions encountered in road service. It may be used to provide exhaust samples from a car that idles, accelerates, cruises, and decelerates in a manner similar to that expected from a driving pattern on city streets. An illustration of a car on a chassis dynamometer appears in Fig. 9.2.

Federal Test Procedures

The test procedures for vehicle and engine exhaust and fuel evaporative emissions for light-duty vehicles are specified in the Code of Federal Regulations (CFR) Title 45, Subtitle A, Part 1201. The following is taken from that document:

> The procedures described . . . will be the test program to determine the conformity of gasoline-fueled light-duty vehicles with the applicable standards set forth in this part.
>
> (a) The test consists of prescribed sequences of fueling, parking, and operating conditions. The exhaust gases generated during vehicle operation are diluted with air and sampled continuously for subsequent analysis of specific components by prescribed analytical techniques. The fuel evaporative emissions are collected for subsequent

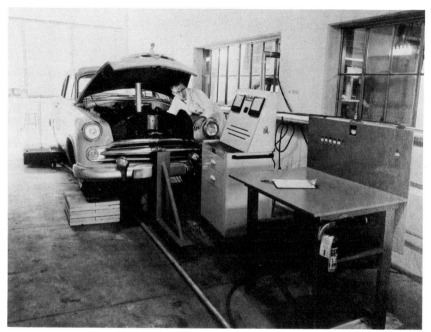

FIGURE 9.2.
Automobile mounted on chassis dynamometer.

weighing during both vehicle parking and operating events. The test applies to vehicles equipped with catalytic or direct-flame afterburners, induction system modifications, or other systems or to uncontrolled vehicles and engines.

(b) The exhaust emission test is designed to determine hydrocarbon and carbon monoxide [Author's note: Insert the words "and oxides of nitrogen" here beginning with 1973 models] mass emissions while simulating an average trip in an urban area of 7.5 miles from a cold start. The test consists of engine startup and vehicle operation on a chassis dynamometer through a specified driving schedule, as described in Appendix A to this part. A proportional part of the diluted exhaust emissions is collected continuously, for subsequent analysis, using a constant volume (variable dilution) sampler.

(c) The fuel evaporative emission test is designed to determine fuel hydrocarbon evaporative emissions to the atmosphere as a consequence of urban driving, and diurnal temperature fluctuations during parking. It is associated with a series of events representative of a motor vehicle's operation, which results in fuel vapor losses directly from the fuel tank and carburetor. Activated carbon traps are employed in collecting the vaporized fuel. The test procedure is specifically aimed at collecting and weighing:

(1) Diurnal breathing losses from the fuel tank and other parts of the fuel system when the fuel tank is subjected to a temperature increase representative of the diurnal range;

(2) Running losses from the fuel tank and carburetor resulting from a simulated trip from a cold start;

(3) Hot soak losses from the fuel tank and carburetor which result when the vehicle is parked and the hot engine is turned off.

Details of the test procedures are specified. They include gasoline fuel specifications, vehicle preparation, collection procedure, dynamometer operating cycle, dynamometer procedures, sampling and analytical systems, and methods of calculating results. Nondispersive infrared analyzers (see Chapter 5) are specified for CO and flame-ionization detectors for hydrocarbon determinations. Procedures of the California Air Resources Board are identical with federal procedures with the exception, of course, of determination of nitrogen oxides.

MOTOR VEHICLE POLLUTION CONTROL METHODS

Control methods for motor vehicle pollution may be separated into direct (emission reduction) and indirect (curtailment of use) means. Although a great deal has been written (and said) about indirect controls, only direct methods are currently in use. Direct methods of control include devices and methods for reduction of hydrocarbon emissions from the exhaust system, crankcase, fuel tank, and carburetor; reduction of carbon monoxide from the exhaust system; and reduction of oxides of nitrogen from the exhaust system. Reduction of exhaust smoke should also be included. Indirect methods of control include a variety of suggestions ranging from abolition of the internal-combustion engine (by substituting electric power, steam power, etc.) to use control (traffic control, no-traffic zones).

Control of Hydrocarbon Emissions

As mentioned previously, hydrocarbon emissions from motor vehicles are divided into three types: (1) crankcase emissions, (2) exhaust emissions, and (3) evaporative losses (from fuel tank and carburetor).

Crankcase (Blowby) Emission Control

Since 1963, the emissions from crankcases of new automobiles have been essentially zero. Prior to that time, any gasoline or exhaust vapors that blew by the piston rings during the compression stroke and into the crankcase were vaporized and emitted along with oil droplets from the road draft tube into the atmosphere. It was estimated that hydrocarbon emissions from this source amounted to 3.15 g/mi or nearly 30% of total hydrocarbon loss.

In the basic blowby control system, the road draft tube is replaced by a tube leading to a flow-modulating or positive crankcase ventilation valve which is connected to the intake manifold below the carburetor. The in-take-manifold vacuum draws the blowby gases and crankcase ventilation air (from the air cleaner) into the engine induction system. The control valve is spring-loaded against manifold vacuum, so that the flow increases as the manifold vacuum decreases. Maximum flow occurs at low-vacuum, high-speed, high-power engine conditions. If abnormally high volumes of blowby gases occur they are conducted to the air cleaner through the intake air hose. In time the control valve may become partially clogged with de-posits. Periodic cleaning or replacement is therefore necessary. A diagram of a closed positive crankcase ventilation system is shown in Fig. 9.3.

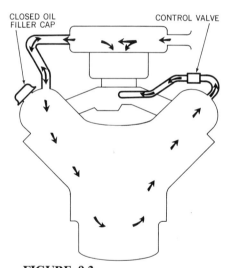

FIGURE 9.3.
Diagram of closed positive crankcase ventilation system.

Exhaust Hydrocarbon Control

Devices and methods to control exhaust hydrocarbon emissions fall into three classes:

1. Devices that modify engine operating conditions,
2. Devices that "treat" exhaust gases, and
3. Use of modified or alternate fuels.

Induction Devices.   Devices that modify engine operating conditions (usually called induction devices) have as their goal improvement of combustion during all or a portion of the driving cycle. They may be generally classified as follows:

1. Fuel cutoff during deceleration.
2. Intake manifold vacuum breakers.
3. Exhaust system vacuum breakers.
4. Throttle retarders (dash-pots).
5. Vacuum control throttle openers.
6. Carburetion mixture improvers.
7. Dual-intake manifolds for better fuel distribution.
8. Fuel injection.

Operating and design changes that may also reduce hydrocarbon emissions are:

1. Faster idling speed.
2. Improved combustion chamber design.
3. Improved electrical system.

In fact, engine modification based on combinations of the items listed have allowed many cars to meet hydrocarbon (and CO) emission specifications even for 1970 models. Leaning the air-fuel mixture by carburetor modifications and combining this with a change in engine spark advance to provide partial spark retard at idle and fuel spark advance during deceleration have been responsible for most of the hydrocarbon and CO reduction attained to date. An increase in idling speed has allowed better combustion at idle and further reduced CO emissions.

The smaller engines in foreign cars have yielded to similar engine modifications. Several manufacturers use fuel injection to improve fuel distribution to the various cylinders and thus promote more uniform combustion. Several manufacturers use dual-intake manifolds. At idle and low speeds, the air-fuel mixture passes to the cylinders through a small-diameter heated manifold. At wide-open throttle the larger, unheated manifold is used. One American laboratory has incorporated this system with modified carburetor design to reduce hydrocarbon emissions markedly [21].

Oxidation of Exhaust Gases.   An obvious and appealing means of reducing hydrocarbon and carbon monoxide emissions is to burn or otherwise oxidize the combustibles in the exhaust gases before they are emitted to the atmosphere. Two types of devices, afterburners and catalytic converters, offer possibilities in this regard, but commercial development has been fraught with technical difficulties.

The principle of the *afterburner* is simple enough: the hydrocarbons and carbon monoxide in the exhaust gases are ignited and burned to carbon dioxide and water. Problems arise because of the tremendous variations in the gases entering the afterburner. Flow rates, temperatures, and concentration show wide variations over short intervals of time. Low-temperature and low-concentration gases need added heat before they will ignite. Gases at high temperatures or high concentrations burn readily but produce so much heat that exotic construction materials are needed. Air must be added under some conditions but not under others. Other constraints that must be met by the afterburner include: no additional fire hazard or excessive heating of other vehicle components, design to fail safely, low back-pressure on engine, suitability for all types of weather and road conditions, i.e., flooding, and no excessive noise.

An efficient afterburner is certainly technically possible, but to date large size, complicated controls, and expensive materials of construction have slowed commercial achievement.

Nevertheless, a close relative of the afterburner was developed by American automobile manufacturers and applied to some engines to achieve 1968 hydrocarbon and CO requirements. Its use in 1970 models was minimal. The air injection reactor, as it is called, uses an engine-driven air pump with a built-in relief valve to deliver air through a manifold system to the exhaust manifold where air is injected under pressure at the point of hottest exhaust gas adjacent to the exhaust valve. Combustion of hydrocarbons and carbon monoxide continue in the exhaust manifold. The quantity of air injected is controlled so that optimum combustion may be attained. When the air injection reactor is used in conjunction with engine modifications, it allows combustion to proceed sufficiently for the exhaust gases to meet regulatory requirements. It is particularly useful in reduction of emissions of carbon monoxide, which is more difficult to control by engine modifications than are hydrocarbons. A schematic drawing of one type of air injection reactor is shown in Fig. 9.4.

In spite of the general discarding of the reactor from 1970 model cars, there is considerable optimism that a more reliable and more effective thermal reactor can be developed eventually and perhaps reduce exhaust emissions to less than 50 ppm hydrocarbons and 0.5% CO. One such reactor is said to be ready for commercial use [19] (see Fig. 9.5).

Two of the inherent problems of the afterburner, flame maintenance and difficulty of low-temperature ignition, are overcome by the *catalytic convertor*. On the other hand, the presence of a catalyst imposes other requirements on the system, particularly high-temperature limitations and absence of materials which might foul or poison the catalyst. Although catalysts are available which are not adversely affected by high temperatures, they gener-

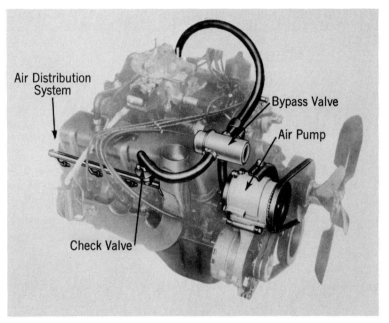

FIGURE 9.4.

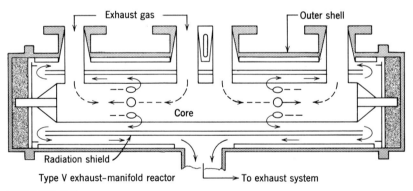

Type V exhaust-manifold reactor

FIGURE 9.5.
Exhaust-manifold thermal reactor system. (From Ref. 19).

ally are not active at low temperatures, and therefore one of the advantages of the catalytic system is lost. Active catalysts may also be susceptible to fouling (with rapid loss of activity) by lead, sulfur, barium, calcium, and phosphates which may be present in exhaust gases. Indications are that catalysts may be developed that are resistant to lead and other contaminants, at least for operation under restricted conditions. There are also indications that small amounts of lead-resistant catalysts can be used near the engine where exhaust temperatures are still high and the catalyst is most effective for reducing hydrocarbons and carbon monoxide (perhaps nitrogen oxides). Such a catalyst would be used only during the warm-up period when afterburners or thermal reactors are ineffective.

Alternate and Modified Fuels

Several suggestions have been made that use of fuels other than leaded gasoline might help alleviate the photochemical smog problem. The most persistent suggestion is replacement of gasoline with liquefied petroleum gases (LPG), which are essentially propane and butane. Use of LPG as fuel not only allows better distribution of fuel to the cylinders, but the photochemical reactivity of the exhaust hydrocarbon emissions is half that of the same engines powered with gasoline. On the other hand the limited supply of LPG and problems in distribution will probably restrict its use to fleets.

Both liquefied natural gas (LNG) and pressurized natural gas have also been suggested as fuel for internal-combustion engines. Use of LNG requires addition of a cryogenic fuel tank as well as a specially designed carburetor. Pressurized natural gas use requires a high-pressure fuel tank. Although exhaust hydrocarbons have very low reactivity and CO and NO_x emissions are low [12, 29], distribution problems probably limit natural gas use to fleets only.

Other suggestions made from time to time include use of nonleaded gasoline and reduction of the volatility and olefin content of gasoline. Elimination of alkyl-lead antiknock compounds from gasoline supposedly would allow catalytic convertors to be used in the exhaust system. Although this solution could be worked out, the economic problems are formidable [5].

Other suggestions for modifying gasoline include reduction of gasoline volatility by 25 and 50% and replacement of C_4 and C_5 olefins by saturates of comparable volatility. None of these modifications reduce exhaust emissions, but they would reduce evaporation losses and relative evaporative-loss reactivity respectively [1].

Evaporative-Loss Control

Beginning with 1971 models, motor vehicles had to be equipped with means to reduce evaporative emissions from fuel tank and carburetor to 6 grams per day as measured by an arbitrary test [11]. Vehicles without evaporative controls are estimated to lose 30 grams hydrocarbon per day from fuel tank filling and breathing (vapor discharge induced by ambient temperature rise). Another 40 grams per day is lost by evaporation from the carburetor (hot soak loss) when the vehicle is parked after being operated. On this basis, evaporative losses from an average car are estimated to be 2.77 g/mi, or 23% of total hydrocarbon emissions.

At present, the device used to control evaporative emissions is a canister filled with activated charcoal which absorbs hydrocarbon vapors led to the canister through tubing connected to the fuel tank and carburetor (see Fig. 9.6). The absorbed vapors are desorbed and fed back to the intake

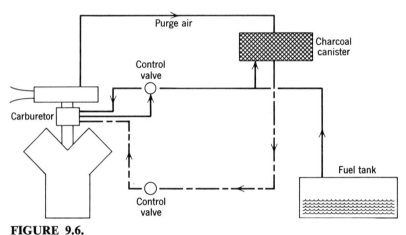

FIGURE 9.6.
Evaporative emission control.

manifold during high-power operating conditions, thus affecting exhaust emissions and engine performance very little. Rather elaborate baffling of the fuel tank is required to keep liquid gasoline from entering the tube to the canister by liquid expansion at high ambient temperatures or splashing on curves and nonlevel roads.

CONTROL OF CARBON MONOXIDE EMISSIONS

Carbon monoxide has been the object of exhaust control methods for many years, chiefly because of its extreme toxicity in concentrations easily attainable in enclosed spaces. In the open atmosphere the problem is less important, but still a potential threat. Another problem, the cumulation of toxic concentrations in passenger compartments of automotive vehicles, has been studied thoroughly by automobile manufacturers [16].

As the carbon monoxide content of exhaust gas is also a measure of inefficient power production, reduction in the amount formed has been the objective of engine designers over a period of years. Considerable reduction has been achieved [9], but more will be required.

Because carbon monoxide emissions are confined to the exhaust gases they are subjected to the control systems used for reducing hydrocarbon emissions in the exhaust. Fortunately, control systems which reduce hydrocarbons also effectively reduce carbon monoxide emissions, albeit to less extent. However, if eventual standards for hydrocarbons (50 ppm is said to be practicable) are met by engine modifications, some type of afterburner or its equivalent will still be required to bring CO to suggested levels. The estimated emissions from pre-1963 model cars are 68 g/mi.

CONTROL OF NITROGEN OXIDE EMISSIONS

When the initial motor vehicle pollution control regulations were being considered in California, it was believed that complete control of either organics (hydrocarbons) or oxides of nitrogen alone would be adequate to eliminate photochemical smog [8].

Since exhaust hydrocarbons could be destroyed by burning and means for eliminating nitrogen oxides were not known, control regulations were developed on the basis of drastically reducing hydrocarbon concentrations. Whether or not nitrogen oxide concentrations should also be drastically reduced is still controversial, inasmuch as nitric oxide inhibits photo-oxidation of hydrocarbons [18]. Some believe, however, that rigorous control of both hydrocarbons and nitrogen oxides will be necessary if proposed oxidant standards are to be met. On the other hand, the arguments are becoming academic now that air quality standards for NO_2 have been adopted based on effects other than photochemical smog (see Chapter 5). To meet these standards, emissions of oxides of nitrogen from motor vehicles must be reduced. In fact, the federal government has followed the lead of the California Air Resources Board and has published standards calling for a reduction of NO_x emissions from motor vehicles to 0.4 g/mi by 1976. Although considerable work has been done on the development of NO_x controls for motor vehicles [4, 14, 38], suitable control systems have not been devised. The two most promising approaches are exhaust gas recirculation and catalytic reduction. It has been shown that addition of an inert gas, e.g., partial exhaust recycle, to the air-fuel mixture entering the combustion chambers markedly reduces NO_x formation with only slight loss in power. The problem of varying the amount of recycle at different operating modes so as to achieve maximum NO_x reduction and not affect vehicle drivability still remains to be solved. It is believed that partial recycle in conjunction with retarded spark timing may be effective.

Another approach is known as the dual-catalyst concept. Exhaust gases flow through a catalytic convertor in a reducing atmosphere to reduce NO to nitrogen or to nitrous oxide (N_2O). An air injector then injects air into the system before the exhaust gases pass through a hydrocarbon and carbon monoxide catalytic oxidation chamber.

CONTROL OF SMOKE EMISSIONS

Smoke is caused largely by improper maintenance of the automotive engine, although faulty engine design is occasionally responsible. The public soon takes care of poor design by refusing to buy "chronic smokers," but the diligence of the same people in properly maintaining a well-designed car leaves much to be desired. The origin and intensity of smoke from automotive vehicles has been discussed by Bowditch [6].

Many cities have regulations prohibiting "excessive smoke" from motor vehicles, but the definition of "excessive" is often inadequate for successful prosecution of violators. In California, smoke from vehicles may not exceed a No. 2 Ringelmann value for more than three minutes.

EFFECT OF CONTROLS

The principal purpose of motor vehicle pollution control regulations is to eliminate or greatly reduce photochemical smog, and to reduce carbon monoxide concentration in urban atmospheres. Since controls are now uniform across the country, those communities with the least photochemical and CO pollution will notice the effects of controls first. In fact, many rural communities have no need of motor vehicle pollution controls, but the mobility of cars has made uniform nationwide regulations desirable. The last area in which photochemical smog will be eliminated is the Los Angeles Basin where smog is most frequent and most intense. Even so, a noticeable change in both frequency and intensity is already apparent according to the Los Angeles County Air Pollution Control District. The rate of smog abatement is determined by the rate at which the percentage of controlled cars in the community increases and the degree of control imposed on succeeding models. When motor vehicle pollution control laws were first adopted in California, it was thought that controls could be developed for all motor vehicles, old and new, but public resistance to the high cost of controlling used cars forced controls to be limited to new models only. As a consequence, quite a few years must elapse before a significant number of uncontrolled cars are scrapped and an adequate percentage of operating motor vehicles are equipped with controls. It may be that Los Angeles may also need to control hydrocarbon emissions from stationary sources in order to eliminate completely adverse smog days (see Chapter 8). In other parts of the country, motor vehicle pollution control will be adequate. If hydro-carbons from stationary sources are to be controlled, it will have to be for other reasons (see Chapter 5).

As for carbon monoxide, the motor vehicle is almost the sole source. Planned automotive controls should bring atmospheric levels well below those which are undesirable.

INDIRECT CONTROLS OF MOTOR VEHICLE POLLUTION

Public disappointment with the inability of government and industry to solve motor vehicle pollution problems "today" have led people in both high and low places to suggest a variety of indirect means of solving the problem. Generally these suggestions are for new nonpolluting engines or for traffic control of one sort or another.

Three types of engines have received serious consideration as nonpolluting alternatives to the conventional internal-combustion engine. They are:

1. Electric motor vehicles
2. Steam-powered vehicles
3. Gas turbine powered vehicles

Prototypes of several *electric cars* are available. All have top-speed limits of 35 to 50 mph and a severely restricted operating range because of need for battery recharging [28]. Even if all other problems are solved, a breakthrough in battery or fuel-cell technology will be required before an electric vehicle aceptable to the public can be produced.

Steam-powered vehicles have been used from time to time since the invention of the automobile. Pollutant emissions are minor (hydrocarbons —20 ppm, CO—less than 0.1%, oxides of nitrogen—40 ppm) [34], but no one has been able, with current technology, to produce an economically competitive vehicle.

Gas turbines have been available in experimental cars and trucks for several years [33], but they have been ruled out for passenger car use because of cost, high fuel consumption, and poor response in stop-and-go driving. Emissions are said to be minor at full power but may be excessive under part load.

Substitution of a different type of engine for the internal-combustion engine also presents formidable problems in retooling factory and maintenance facilities. If a suitable replacement for the internal-combustion engine were available today, full-scale manufacture would still be many years away.

The unlikelihood of early replacement of the internal-combustion engine with a low-emission engine has led to suggestions to curtail present emissions by traffic control methods including more freeways and better synchronized signals (to reduce stop-and-go driving), establishment of central city no-traffic zones, banning traffic during periods when smog is likely, promotion of car pools (to reduce the number of operating vehicles), staggering of work hours (to reduce congestion), and enlargement of urban mass-transit systems, all in the name of reduced pollution. Few of the suggestions have ever been implemented, even partially.

Diesel Engine Exhaust

The contribution of the diesel truck and bus to air pollution has been the subject of considerable investigation both in this country and in Europe. These studies were motivated chiefly by the use of diesel-powered equipment in mines and other enclosed spaces, and the obvious smoke and odor emissions.

The diesel engine presents an entirely different air pollution problem from that of the gasoline engine. Unlike the latter, the diesel engine operates with excess air. The fuel is injected into the cylinder and ignited by the temperature and pressure in the cylinder rather than by a spark. The fuel used is heavier than gasoline and has a higher boiling range. Because of the excess air used, very little of the fuel is normally exhausted unburned. Under certain conditions of operation, however, both smoke and aldehydes are evolved. High concentrations of aldehydes are largely responsible for the pungent odor and eye-irritating characteristics of diesel exhaust.

Two distinctly different types of diesel engines are in general use. Each produces a characteristic and different exhaust. The diesel bus, for instance, commonly uses a blower-scavenged, two-stroke cycle diesel engine with a torque-converter drive. The fuel used is a high-cetane,* high-speed automotive virgin diesel fuel resembling kerosene. The maximum amount of fuel is injected when the engine is first started so that it will start easily. This low air-fuel ratio causes a large puff of black smoke, which disappears as the engine picks up speed and the air-fuel ratio is increased. When the bus is slowed down, fuel injection is cut off but the engine continues to turn because of the momentum of the bus. Because of the partial oxidation of the oil film in the cylinder under these conditions, the exhaust has a highly irritating aldehyde odor.

The large highway diesel transport trucks, on the other hand, generally

* The cetane number of diesel fuels is analogous to the octane number of motor fuels. The reference fuel in this case is a blend of normal hexadecane (100 cetane number) and alpha-methyl-naphthalene (0 cetane number).

use a large 6-cylinder, 4-stroke cycle engine, sometimes supercharged (air added to the cylinder at super atmospheric pressure). These units use a heavier grade diesel fuel than is used by city buses. If the engines are kept in good adjustment, very little visible smoke and only a barely noticeable odor is produced. However, when the engine is heavily loaded and "lugged down" to low speed, maximum fuel is injected, poor mixing of fuel and air results, and black smoke is exhausted. The intensity of the smoke may be accentuated or minimized by the quality of the fuel. Generally, smoke increases in proportion to the end point of the fuel and inversely with the cetane number. Smoke discharge is also greater at high altitudes where air is insufficient for the amount of fuel injected.

Composition of Diesel Exhaust Gases

Emission data on diesel engines have been reviewed by Kinosian, et al [26]. Hydrocarbon and carbon monoxide emissions varied considerably but were well below California standards for motor vehicles. Oxides of nitrogen concentrations varied from 80 to 1370 ppm. The National Air Pollution Control Administration reports that current emissions from diesel vehicles range from 4 to 10 grams NO_x per mile [35].

Abatement of Diesel Exhaust

Obviously, the normal air pollution effects of diesel exhaust are primarily black smoke and aldehyde odor. In mines and other enclosed spaces, all emissions are a problem. Because of the wide use of diesel-powered vehicles in such areas, the minimization of diesel exhaust emissions has been the subject of continual research. The usual solution is good maintenance, proper operating procedures, and adequate ventilation. Even so, some work on removal of aldehydes and oxides of nitrogen has also been done [13]. A variety of afterburners, both catalytic and direct-flame, have been used to reduce hydrocarbons, aldehydes, carbon monoxide, smoke, hydrogen, and other combustibles. The biggest problem here is the low temperature and low combustible concentration of the exhaust. Both factors limit the effectiveness of any practical device.

Little hazard exists in the open atmosphere, so the only motivating force to diesel exhaust control has been public indignation at black smoke, odorous aldehydes, noise, and traffic congestion. On many occasions the latter two items have been the real basis of nuisance complaints. Nevertheless, there is little excuse for 70% of the smoking diesel exhausts one sees on our highways. Good maintenance, quality fuel, and reasonable operating methods are adequate to control these emissions. In addition to these methods, additives are available which will reduce smoke emissions even under adverse operating conditions [39]. The aldehyde problem, particularly with the 2-stroke-cycle engine, is more difficult to solve; however, further dilution with air may be a reasonable palliative.

Regulations limiting the emissions of diesel smoke and odor have been adopted in several communities. They vary all the way from limiting idling time on city streets to specific measures putting limitations on smoke density and odor concentration. For years the Los Angeles County APCD has limited diesel smoke emissions greater than No. 2 Ringelmann (40% equivalent opacity) to three minutes per hour (the same limitation as is enforced against smoke from stationary sources).

Beginning with 1972 models, federal standards for exhaust smoke from heavy-duty diesel engines became effective. Smoke opacity may not exceed 40% during the engine acceleration mode or 20% during the engine lugging mode.

Aircraft Exhaust Emissions

Little attention was paid to aircraft as a source of air pollution until the late 1950's, when turbojet aircraft were introduced commercially. The turbojets emitted highly visible exhaust plumes during takeoff and landing, thus instigating considerable protest from residents in the vicinity of airports.

Aircraft engines are of two basic types—reciprocating or piston engines and gas turbines. Exhaust gases from the two types are distinctly different, largely because a turbine engine operates at air-fuel ratios 5 to 20 times greater than those of piston engines. Emissions from each type and those from automobiles (for comparison) are shown in Table 9.7 [37]. Generally only those emissions during the landing-takeoff (LTO) cycle are important in community air pollution. The fuel consumption of various types of aircraft during the LTO-cycle are shown in Table 9.8.

TABLE 9.7

Comparison of Pollutant Emission Indices for Aircraft and Automobile Engines

Engine	Operating mode	Emission index: pound pollutant per 1,000 pounds fuel			
		CO	Hydro-carbons	NO_x	Particu-lates
Turbofan medium range jet {	Idle and taxi	50	9.6	2.0	0.6
	Approach	6.6	1.4	2.7	2.7
	Takeoff	1.2	0.6	4.3	2.5
Radial	Idle	600	160	0	2
Piston	Approach	800	60	5	2
Transport	Takeoff	1,250	190	0	2
Average automobile engine without emission control	Average overall modes	405	71	21	2

TABLE 9.8

Fuel Consumed per Engine during a Landing-Takeoff Cycle

Aircraft type	Ground operations				Flight operations		Total cycle	
	Taxi and idle		Landing and takeoff		Approach and climbout			
	Duration (minutes)	Fuel consumed (pounds per engine)	Duration (minutes)	Fuel consumed (pounds per engine)	Duration (minutes)	Fuel consumed (pounds per engine)	Duration (minutes)	Fuel consumed (pounds per engine)
Long-range jet	13	236	1.4	195	5.8	577	20.2	1,008
Medium-range jet	13	199	1.1	120	4.9	338	19.0	657
Business jet	13	117	.8	35	2.1	69	15.9	221
Turboprop	13	127	.9	25	8.1	165	22.0	317
Piston transport	13	75	1.2	26	9.6	177	23.8	278
Piston utility	13	6	.7	1	6.3	7	20.0	14
Turbine helicopter	7	40	0	0	13.2	136	20.2	176

Note: LTO-cycle includes all normal operations below 3,000 feet above runway. Assumed conditions: (a) runway at sea level; (b) zero runway slope; (c) temperature 60° F.; (d) zero wind; (e) aircraft at maximum load.

Turbine Engines. Four basic types of gas turbine engines are used for aircraft propulsion: turbojet, turboprop, turboshaft, and turbofan. The basic common element to all four is the gas turbine which consists of a compressor, a combustion chamber, and a turbine. Air entering the forward end of the engine is compressed and led to the combustor where fuel is injected. Most of the energy in the hot combustion gases is used for aircraft propulsion. A portion is expended in driving the turbine, which, in turn, drives the compressor. Over-rich fuel-air mixtures near the point of fuel injection lead to the formation of free carbon, particularly during takeoff and landing, which shows up in the atmosphere as a black plume. In the early days of jet aircraft operation, this plume was enhanced by injection of water for increased power during takeoff. This procedure has been abandoned.

Although the present contribution of aircraft emissions to the total atmospheric pollution problem is small, industry and government are cooperating in the development of means of reducing emissions. It appears that minor modifications to the combustor will quickly reduce particulate, CO, and hydrocarbon emissions without compromising safety.

Piston Engines. Aircraft piston engines are similar to automobile engines in their operation and produce much the same type of pollutant emissions. Nearly all aircraft piston engines have two or more cylinders arranged either in opposed or radial configurations. Radial engines are used mostly in large transport aircraft and are gradually being replaced with turbine engines.

Emissions from piston engines are high at idling conditions and particularly during takeoff when rich-mixture conditions lead to large quantities of exhausted hydrocarbons and carbon monoxide. Exhaust gas treatment similar to that suitable for automobile engines should be effective in reducing exhaust emissions.

Railroad Locomotives and Marine Vessels

In the past, the major air pollution problem of locomotives and ships was smoke from coal-fired boilers. Coal-fired locomotives have all but disappeared from the American scene and been replaced by diesel-powered engines. Some coal-fired marine vessels still are in service, but generally the diesel engine is taking over.

These large, slow-speed, multicylinder units are similar to the reciprocating engines used in stationary installations and are not usually considered important from an air pollution standpoint. Recent emphasis on nitrogen oxide emissions has drawn attention to such units, however. Very few emission data are available and methods of control have only recently become a matter of concern [36].

REFERENCES

1. American Petroleum Institute, "Vehicle Emissions vs. Fuel Composition" (May 1968).
2. Anon., "DuPont's Exhaust-Manifold Thermal Reactor System," *Oil and Gas J.* (June 15, 1970).
3. Anon., "Ethyl Unveils Lean Reactor Auto System," *Hydrocarbon Processing,* 17 (May 1970).
4. Benson, J. D., "Reduction of Nitrogen Oxides in Automobile Exhaust," Paper No. 690019, Society of Automotive Engineers. Presented at International Automotive Engineering Congress, Detroit, Michigan (January 13–17, 1969).
5. Blanchard, L. E., Jr., "Taking Another Look before We 'Take the Lead Out,'" *Automotive Engineering,* **78,** No. 12, 24–29 (1970).
6. Bowditch, F. W., et al, "Two Aids Pinpoint Excessive Exhaust Smoke," *S.A.E. Journal,* **65,** 129–131 (October 1957).
7. California Dept. of Public Health, "Motor Vehicle Pollution in California" (Jan. 1967).
8. California Dept. of Public Health, "Technical Report of California Standards for Ambient Air Quality and Motor Vehicle Exhaust," Berkeley, Calif., 96 (1960).
9. Campbell, J. M., "Improved Combustion Held Aid to Air Pollution Abatement," *S.A.E. Journal,* **62,** 54–58 (December 1954).
10. "Control of Air Pollution from New Motor Vehicles and New Motor Vehicle Engines," CFR Title 45A, Part 85, Federal Register 31, No. 61, Part II, Washington, D.C. (March 30, 1966).
11. "Control of Air Pollution from New Motor Vehicles and New Motor Vehicle Engines—Standards for Exhaust Emissions, Fuel Evaporative Emissions, and Smoke Emissions, Applicable to 1970 and Later Vehicles and Engines," CFR Title 45A, Part 85, Federal Register 33, No. 108, Part II, Washington, D.C. (June 4, 1968).
12. Corbeil, R. J., and S. S. Griswold, "Advantages of Natural Gas as a Fuel for Motor Vehicles," Paper EN 10-A, 2nd International Clean Air Congress, Washington, D.C. (Dec. 6–11, 1970).
13. Davis, R. F., and J. C. Holtz, "Exhaust Gases from Diesel Engines," in *Problems and Control of Air Pollution,* Frederick S. Mallette, ed., Reinhold, New York, 74–94 (1955).
14. Deeter, W. F., H. D. Daigh, and O. W. Wallin, "An Approach for Controlling Vehicle Emissions," Paper No. 680400 Society of Automotive Engineers. Presented at SAE Mid-Year Meeting, Detroit, Michigan (May 20, 1968).

15. Ethyl Corporation, "Gasoline Quality Survey," December, 1968 (*with permission*).

16. Fowler, D. G., et al, "Development of a Test Procedure for Measurement of Carbon Monoxide in Automobile Passenger Compartments," Society of Automotive Engineers National West Coast Meeting, Seattle, Wash. (August 1957).

17. Gates, R. W., "Particulate Matter in Automobile Exhaust," Stanford Research Institute, Stanford, Calif., Rept. No. 2 (April 1955).

18. Glasson, W. A., and C. S. Tuesday, "Inhibition of the Atmospheric Photo-oxidation of Hydrocarbons by Nitric Oxide," 148th Meeting Amer. Chem. Soc., Chicago, Ill. (Aug. 30–Sept. 4, 1964).

19. Habibi, K., E. S. Jacobs, W. G. Kunz, Jr., and D. L. Pastell, "Characterization and Control of Gaseous and Particulate Exhaust Emissions from Vehicles," Fifth Technical Meeting, West Coast Section, Air Pollution Control Association, San Francisco, Calif. (October 8–9, 1970).

20. Hirschler, D. A., et al, "Particulate Lead Compounds in Exhaust Gas," *Ind. Eng. Chem.*, **49**, No. 7, 1131 (1957).

21. Hirschler, D. A., and F. J. Marsee, "Meeting Future Automobile Emission Standards," presented at National Petroleum Refiners Association Annual Meeting, San Antonio, Texas (April 6–7, 1970).

22. Horn, S. A., "Mechanical Factors Relating to Vehicle Emissions Diagnosis and Repair," Proceedings Fourth Technical Meeting APCA West Coast Section, San Diego, Calif. (Oct. 20, 1967).

23. Hurn, R. W., "Fuel: A Factor in Internal-Combustion Engine Emissions," Preprint 69-WA/APC 8, American Society of Mechanical Engineers, Winter Annual Meeting, Los Angeles, Calif. (Nov. 16–20, 1969).

24. Hurn, R. W., T. C. Davis, and P. E. Tribble, "Do Automotive Emissions Inherit Fuel Characteristics?" 25th Midyear Meeting, Division Refining, Am. Petrol. Inst., Detroit, Michigan (May 1960).

25. Judge, A. W., *Testing of High-Speed Internal-Combustion Engines,* 4th Ed., Chapman and Hall, London (1955).

26. Kinosian, J. R., J. A. Maga, and J. R. Goldsmith, "The Diesel Vehicle and Its Role in Air Pollution," California Dept. of Public Health, Berkeley, Calif. (1962).

27. Lawson, S. C., et al, "Economics of Changing Volatility and Reducing Light Olefins—U.S. Motor Gasoline—A Report by the Lead/Volatility Economics Task Force." Paper presented during the 33rd Midyear Meeting, Division Refining, Am. Petrol. Inst. (May 16, 1968).

28. Linford, H. B., "Power Sources for Electric Vehicles," *Environ. Sci. and Tech.*, **1**, 394–399 (1967).

29. McJones, R. W., and R. J. Corbeil, "Natural Gas Fueled Vehicles Exhaust Emissions and Operational Characteristics," SAE Preprint 700078, Automotive Engineering Congress, Detroit, Mich. (Jan. 12–16 1970).

30. Maynard, J. B., and W. N. Sanders, "Determination of the Detailed Hydrocarbon Composition and Potential Atmospheric Reactivity of Full-Range Motor Gasolines," *J. Air Poll. Control Assoc.*, **19**, 505–510 (1969).

31. "Motor Vehicle Air Pollution Standards," CFR Title 45A, Part 85, Federal Register 35, No. 219, Part II, Washington, D.C. (Nov. 10, 1970).

32. National Air Pollution Control Administration, "Air Pollution Dangers" (1970).

33. National Air Pollution Control Administration, "Control Techniques for Carbon Monoxide, Nitrogen Oxides, and Hydrocarbon Emissions from Mobile Sources," NAPCA Publ. AP-66, pp. 7–2 to 7–5 (1970).

34. National Air Pollution Control Administration, "Control Techniques for Carbon Monoxide, Nitrogen Oxides, and Hydrocarbons Emissions from Mobile Sources," NAPCA Publ. AP-66, p. 7–10 (1970).

35. Ibid., p. 5–19.

36. National Air Pollution Control Administration, "Control Techniques for Nitrogen Oxide Emissions from Stationary Sources," NAPCA Publ. No. AP-67, pp. 5–5 to 5–10, U.S. Govt. Printing Office, Washington, D.C. (1970).

37. "Nature and Control of Aircraft Engine Exhaust Emissions," U.S. Senate Document No. 91-9. A Report of the Secretary of Health, Education, and Welfare to the United States Congress, U.S. Govt. Printing Office, Washington, D.C. (1969).

38. Nicholls, J. E., I. A. El-Messiri, and H. K. Newhall, "Inlet Manifold Water Injection for Control of Nitrogen Oxides—Theory and Experiment," Paper No. 690018, Society of Automotive Engineers. Presented at International Automotive Engineering Congress, Detroit, Michigan (January 13–17, 1969).

39. Norman, G. R., "New Approach to Diesel Smoke Suppression," *Transactions, Soc. Automotive Engineers,* 105 (1967).

10

SOCIAL ORIGINS OF AIR POLLUTION

Until the late 1960's, air pollution was viewed almost entirely as a scientific and economic problem whose control was possible only through the exacting pace of technological development as applied to source control systems. But in a few short years public awareness of ecologic systems and of complex man-environment relationships has become an important factor in a new social revolution. At first, the "ecology movement" had all the earmarks of a fad. Although it is still marked by occasional misdirected ardor, it can hardly now be dismissed as a temporary diversion for the dilettante.

Indeed, mature examination of man-environment relationships indicates that undesirable air pollution conditions have a social origin and may be corrected by changes in our way of living. It is therefore appropriate to examine the circumstances and factors within our society that have given rise to these conditions and that support their continued existence. Among these factors are: population growth, rising per capita demands for energy and manufactured goods, changing patterns of fuel usage, urbanization, defects in the design and operation of urban communities, economic disincentives to pollution control, constraints on community problem-solving capacities, rising human expectations, and public policy priorities.

Population Growth

Varying in depth from five miles at the poles to ten miles at the equator, the physical dimensions of the troposphere fix a limit on the volume of air available for the dispersal and dilution of planetary pollution discharges. Unfortunately, however, no such natural limit has yet constrained the size of the planet's pollution-creating population.

Some have referred to recent rates of population increase as the *population explosion* and have viewed this phenomenon as a factor of major importance to both global and continental burdens of air pollution. According to Ehrlich [12], global population doubling times have been dropping in recent generations (see Table 10.1). At current rates of increase,

TABLE 10.1

World Population Growth

Year	Estimated world population	Doubling time (years)
1,000,000 B.C.	2,500,000	—
8000 B.C.	5,000,000	1,000,000
1650 A.D.	500,000,000	1,000
1830 A.D.	1,000,000,000	200
1930 A.D.	2,000,000,000	100
1970 A.D.	3,600,000,000	40
2000 A.D.	7,500,000,000	30

Source: Paul R. Ehrlich, *The Population Bomb* (Revised Edition). New York: Ballantine Books, Inc., 1971, p. 4; and *First Annual Report of the Council on Environmental Quality*, August, 1970, p. 150.

the planetary population will more than double again in the next thirty years.

As population levels increase, the per capita availability of air declines. Looking at in another way, each new increment of population leads to an increase in the discharge of pollution into the planet's fixed air volume and thereby increases the burden of air-borne pollutants within the air mass. To the extent that global atmospheric burdens of carbon dioxide and suspended particulate matter are significant to the ecologic health of the globe (see discussion in Chapter 1), continuing population increases become a factor of major importance to the design of effective air pollution control strategies.

Although population increases have been less severe in the United States than in other parts of the world, they have nevertheless been high enough to inspire concern by some groups and public officials. From the date of the first U.S. census in 1790, almost 100 years were required to increase our population by fifty million persons. The next fifty million was added in less than forty years, and the next in approximately thirty years [33, 34]. The most recent increase of this magnitude occurred between 1950 and 1970, when 52.5 million persons were added to the U.S. population [15]. At current rates of increase the U.S. Census Bureau expects the national population to swell by an additional eighty to one hundred twenty-five million persons within the next thirty years [19].

Growth in population is the fundamental reason why southern California counties have been unable to curb severe and pervasive air pollution conditions in that section of the United States. In spite of the success of these counties in mounting sophisticated and aggressive community air pollution control programs, the gains achieved by these efforts have been negated by a spectacular increase in human population, automobiles, industrial plants, and energy usage. From 1940 to 1970 the five southern California counties (Los Angeles, Orange, Riverside, San Bernardino, San Diego) increased their combined populations from 3,472,383 to 10,989,461 persons [31, 32].

Demands for Energy and Consumer Goods

The air pollution consequences of population growth have been accentuated by rising per capita demands for electrical energy, fuels, and manufactured goods.

Although the U.S. population increased only 29% between 1950 and 1966,* total industrial production more than doubled during this same period [34]. Currently, the per capita demand for electrical energy is doubling every thirteen years, while the demand for motor fuels is doubling every twenty-six years (see Table 10.2). Since the end of World War II, the United States has given birth to more automobiles than babies each year.† Annual sales of such products as washing machines, refrigerators, radios, and television sets also have outstripped our baby production.

In terms of electrical energy and motor fuels consumption each current resident of the United States is the equivalent of 5.1 and 2.3 persons, respectively, in our 1940 population.

These trends, rather than population growth, are probably the major factors influencing air pollution occurrences in such areas as Buffalo, N.Y.; Pittsburgh, Pa.; Chicago, Ill.; Detroit, Mich.; Cleveland, Ohio; and New York City. In these places population growth has ended and, in some cases,

---

* The index of industrial production rose from 75 in 1950 to 156 in 1966.
† Automobile factory sales for 1950, 1955, 1960 and 1966 were 8,003, 9,169, 7,869 and 10,329 thousand units, respectively. For the same years, births were 3,645, 4,128, 4,307, and 3,661 thousands. See Tables 3 and 819, *Statistical Abstract of the United States,* 1968 [34].

TABLE 10.2
Annual U.S. Consumption of Energy and Selected Products
per 1000 Population, 1902–1966*

Product	1902	1912	1920	1930	1940	1950	1960	1966
1. Electrical energy (1000 kwh)	76	262	536	940	1432	2554	4660	6340
Doubling time (years)	—	7	8	13	16	13	12	13
2. Motor fuel consumption (1000 gallons)	—	—	32	128	184	264	353	374
Doubling time (years)	—	—	—	7	14	19	21	26
3. Raw steel production (1000 lbs)	423	734	887	741	1010	1270	1100	1360
4. Paper and paper board (1000 lbs)	—	—	134.9	165.2	218.5	320.1	381.2	479.3
5. Factory sales of automobiles	—	2.02†	20.9	27.3	33.7	52.6	43.6	52.5

* Computed from data presented in Tables 2, 756, 817, 819, 988, 1148, *Statistical Abstract of the U.S.*, 1968 and Series Al-3, S81-93, Q310-320, pp. 187–232, L61-71, *Historical Statistics of the U.S.*
† Data for 1910.

population levels have even declined.* Nevertheless, rising consumer demands have continued to press the capacities of these communities to reduce gross rates of air pollution emissions.

Although U.S. per capita requirements for energy and basic goods will rise somewhat more slowly in the future than in the past, projected increases are nevertheless substantial. (See Table 10.3). To the extent that air pollution emissions are associated with these indices of consumption, it appears that rising consumer demands for energy, goods, and services are of more importance to air pollution conditions than the general growth in population.

* Populations in these cities were as follows in 1950 and 1960:

Place	1950 population	1960 population
Buffalo, N.Y.	580,132	532,759
Pittsburgh, Pa.	676,806	604,332
Chicago, Ill.	3,620,962	3,550,404
Detroit, Mich.	1,849,568	1,670,144
Cleveland, Ohio	914,808	876,050
New York City	7,891,957	7,781,984

TABLE 10.3

U.S. per Capita Demands for Electrical Energy, Raw Steel, and
Motor Fuels, 1970–2000

	1970	1985	2000
Electrical energy production (kwh)	7,204 (est)	14,408	28,816
Assumed doubling time (yrs)	15	15	15
Motor fuel consumption (gallons)	432 (est)	617	864
Assumed doubling time (yrs)	26	30	30
Raw steel production (lbs)	1,400	—	1,960*

* Assumes the same rate (40%) of increase between 1970 and 2000 as between 1940 and 1970.

Rising consumer demands also will escalate pollutant emissions in other sections of the world. With only 7% of the planetary land area and 6% of its population, the United States currently accounts for 40% of the net world output of goods and services and one-third of its total electric power producing capacity [30]. As underdeveloped nations of the world expand their economies, industrialize, and provide a more abundant supply of goods, services, and energy to their expanding populations, global pollution discharges will increase at rates far in excess of the general rate of population growth. If, by the year 2000, the non-U.S. fraction of the world population enjoys the same per capita usage of electrical energy as the 1950 U.S. population, global energy consumption will reach 18.5 billion kwh per year, a fivefold increase over 1967 levels.

Patterns of Fuel Usage

Population growth and rising per capita demands for energy have led to substantial changes in fuel usage patterns within the United States and to developing shortages of low-pollution fuels.

In 1940, hydroelectric projects provided more than one-third of the nation's electrical energy, but by 1967 only 18% of our needs were being satisfied by such projects [34]. Although much of the new energy requirement during this period was met by natural gas, growing shortages of this fuel are now leading to reductions in its percentage share of the electrical power generation market. Thus, future power generation needs will rest even more heavily than in the past on such fuels as coal, lignite, liquid fuels, and nuclear energy, including fast-breeding reactors and—in the far future—fusion reactors.

Changes in fuel requirements for electrical generating plants have been estimated by Spaite and Hangebrauck [27] as shown in Table 10.4.

Current shortages in low-pollution fuels, coupled with the technologic difficulties involved in removal of sulfur compounds, particulate material,

TABLE 10.4
Projected Power Generating Capacity of Electric Utilities
(with Breeder Reactor Capability)*

Year	Total power capacity†	Nuclear	Coal	Natural gas	Oil	Hydroelectric
1970	330	11	215	45	22	38
1975	430	40	265	52	27	40
1980	560	106	330	59	33	43
1990	1,000	350	480	760	48	48
2000	1,600	740	690	88	68	52

* Spaite, P. W., and R. P. Hangebrauck, "Sulfur Oxide Pollution: An Environmental Quality Problem Requiring Responsible Resource Management," paper presented at 19th Canadian Chemical Engineering Conference, Canadian Society of Chemical Engineers, Edmonton, Alberta (October 19–22, 1969).
† Thousand megawatts.

and other contaminants from power plant stacks have led to official pessimism over the nation's near-future capacity to reduce pollution yields from steam-electric generating plants.

In a staff report to the U.S. Senate Committee on Public Works, Walter Planet concluded that [36]:

. . . allowing for the projected introduction and availability of "clean" breeder reactors for electric power generation and applying current technically feasible pollutant control technology and equipment, the projected emissions of specific gaseous pollutants (i.e., SO_2 and NO_2) will not be reduced to the 1967 level by the year 2000.

URBANIZATION AND METROPOLITANIZATION

The emergence of modern air pollution conditions has closely paralleled the urbanization of the human population and the development of the sprawling metropolis. Since the pollution burden of any air mass is related to the mass rate of pollution emissions from the land areas over which that mass has traveled, the aggregation of people and pollution-causing activities on comparatively small and densely populated land areas is of obvious importance to the creation of air pollution conditions.

In a 1965 report the Air Conservation Commission of the American Association for the Advancement of Science stated that [2]:

Air pollution, other than from nuclear explosions, is primarily a big city problem. Or more accurately, it is a metropolitan or regional problem, since small communities in the same region will be affected if they are downwind of the industries, power plants, automotive exhaust, and other pollutants of large cities. Air pollution is created and experienced principally in the large metropolitan region, and all forecasts indicate that the overwhelming proportion of the nation's population will be living in metropolitan areas in the future. Thus, the changes that are taking place in metropolitan areas must be understood in shaping air pollution strategies.

The relationship between the size of cities and atmospheric burdens of air pollution is suggested by Table 10.5 which relates measured burdens

TABLE 10.5

Relationship between Size of Urban Area and Ambient Particulate
Pollution Levels (1957–1965)*

Population size	No. of areas in group	Annual average level of particulate matter (micrograms per cubic meter)
1,000,000+	5	170
700,000–999,999	7	127
400,000–699,999	19	126
100,000–399,999	95	113
50,000– 99,999	92	113
2,500– 49,999	118	85
Rural areas	51	34

* U.S. Bureau of the Census, *Statistical Abstract of the United States: 1968* (89th edition), Washington, D.C.: U.S. Govt. Printing Office, 1968. Tables 257, 259.

of suspended particulate matter to the size of the city in which the measurements were performed.

Although cities have existed throughout man's recorded history, until the last century urban dwellers accounted for only a small fraction of the total population of any nation.

In the United States only twenty-four urban settlements were identified during the first census in 1790 and only 5.1% of the nation's population were recorded as urban dwellers in that year [33]. Since then, however, the urban fraction of the population has grown steadily larger. *Urbanization* has been taking place.

By the close of the Civil War, U.S. urban settlements were recording larger total population gains than all rural areas, and nine cities were boasting populations of 100,000 or more. Not until 1920, however, did the U.S. cease to be a rural nation. In that year the urban fraction of the population surpassed the rural in size [33]. By 1970, 73.5% of the U.S. population were living in cities and towns of 2,500 population or more and urban communities were swallowing up the entire U.S. population increase [15].

In contrast to the urban growth which can take place as the result of a simple excess of births over deaths in urban population segments, urbanization requires a flow of migrants from rural to urban places. Constituting the largest mass migration in human history, massive migrant flows of this sort have been observed throughout the world and are now referred to as the *population implosion* [11].

In countries like the United States, these migrant movements have involved much more than the simple flow of people from rural to urban settlements. In addition, substantial numbers of migrants have moved from rural and small urban places into large metropolitan areas, each containing a population of 50,000 or more persons and consisting of a set of interrelated urban settlements oriented around a focal core city.

For example, seven contiguous counties in the states of Missouri and Illinois environ the city of St. Louis. In 1960 these seven counties covered 4,119 square miles of land area and contained a combined population of 2,105,000 persons distributed among eighty urban places, several lightly populated suburban areas, and a farming belt containing nearly 2,500 square miles.

The largest city, St. Louis, contained a population of only 750,000 persons in 1960 and was declining in population even though the six-county area was growing at the rate of 20% per decade. Viewed as a whole, the area's urban places occupied only 289 square miles, contained almost 1,630,000 persons, and exhibited a combined population density of 5,626 persons per square mile. In contrast, the 1,378 square miles of semiagricultural suburban territory contained only 447,000 persons and exhibited a population density of 424 persons per square mile. The agricultural belt sprawled over 2,453 square miles and showed a population density of only 12.9 persons per square mile. Taken as a whole, the six county area showed a 1960 population density of 511 persons per square mile, as contrasted to a density of 28.3 persons per square mile in those areas of Missouri lying outside the St. Louis and Kansas City metropolitan areas [31, 32].

Some metropolitan regions sprawl over several states, contain hundreds of urban places, and boast populations numbering many millions of persons. The first of these to emerge within the United States has been examined by Gottman [14] and extends from north of Boston to south of Washington, D.C.

Sometimes referred to as BOS-WASH, this metropolitan region sprawls over all or part of eleven states (New Hampshire, Massachusetts, Rhode

Island, Connecticut, New York, New Jersey, Pennsylvania, Delaware, Maryland, the District of Columbia, and Virginia) and ninety-eight counties. In 1960 its boundaries embraced nearly 45,000 square miles of land area and a total population of 36,538,000 persons, more than half of which (19,-226,000) were residents of eighty cities individually containing 50,000 or more population and collectively representing 1801 square miles of land area. With a 1960 population density of 817 persons per square mile, this region accounted for almost 19% of the 1950–1960 mainland population increase by adding 115.2 persons per square mile to its total population [32].

To the extent that population densities are an indicator of the pollution stresses exerted on the natural environment, the areas listed in Table 10.6 are of particular interest to environmental analysts and air quality managers.

Viewing the nation as a whole, eighty-six metropolitan centers and regions now contain 70.1% of the nation's population and account for 98.5% of our continuing population increase. Covering 527 counties and 13% of the mainland area, these places received nearly eight million migrants from the nonmetropolitan areas of the nation between 1950 and 1960 and boosted their combined population from 98,266,000 to 125,-232,000 persons [20].

Thirty of these areas individually claim 500,000 population or more, cover only 417 counties and 10% of the national land area, but account for 62.8% of our current mainland population and 87.4% of our continuing mainland population growth. Data pertinent to these areas are presented in Tables 10.7 and 10.8.

The data suggest the extent to which the factors of population growth, urbanization, rising consumer demands, and fuel usage patterns combine to produce situations in which high air pollution emissions and atmospheric pollution burdens can be expected.

However productive it may have been in the past to concentrate air pollution control efforts on the abatement of emissions from single sources of pollution, these data suggest the need for considering somewhat broader and more comprehensive air pollution control strategies in the future.

Urban Design

Simply because cities do aggregate large numbers of people and pollution-causing activities on limited land areas, their very existence is a factor in the genesis of air pollution problems. However, since cities seem to differ greatly from each other in the extent to which they generate pollution problems, recent interest has focused on the relationship between urban factors and such environmental problems as air pollution.

TABLE 10.6

U.S. Metropolitan Centers and Regions with Population of 1,000,000 or More (1960)*

Area	1960 population	1960 land area (mile²)	Population density 1960	Persons added per mile², 1950–1960
1. Boston-Washington Metropolitan Region	36,537,527	44,729	816.9	115.2
2. Great Lakes Metropolitan Region	24,118,707	54,580	441.9	78.3
3. California Coastal Metropolitan Region	13,933,750	33,881	411.3	140.8
4. East Florida Metropolitan Region	2,838,054	15,248	186.1	91.6
5. Gulf Coast Metropolitan Region	5,693,235	31,659	179.8	52.2
6. N.E. Great Lakes Metropolitan Center	2,323,562	4,555	510.1	N.A.
7. Seattle-Tacoma Metropolitan Center	1,639,016	8,959	183.0	N.A.
8. Central Colorado Metropolitan Center	1,249,991	11,677	107.5	N.A.
9. Central Indiana Metropolitan Center	1,358,472	4,695	289.3	N.A.
10. St. Louis Metropolitan Center	2,104,669	4,119	511.0	N.A.
11. Atlanta Metropolitan Center	1,094,240	2,436	449.2	N.A.
12. Dallas-Ft. Worth Metropolitan Center	1,490,022	1,743	854.9	N.A.
13. Central Ohio-Kentucky Metropolitan Center	3,271,255	7,730	423.2	N.A.
14. Minneapolis-St. Paul Metropolitan Center	1,516,800	2,891	524.7	N.A.
15. Kansas City Metropolitan Center	1,719,441	4,469	384.8	N.A.

* Arthur A. Atkisson, *U.S. Population Settlement and Migration Patterns*. Houston, Texas: University of Texas School of Public Health, 1971 (mimeo.).

In March, 1971, the American Institute of Planners issued a policy statement in which this professional society of city planners averred [4]:

Planning, including plan implementation, can and should be used to abate and prevent pollution of all kinds, to assure balances between

TABLE 10.7

Population Trends in Thirty U.S. Metropolitan Centers and Regions (1950–2000)*

	Total	Five metropolitan regions	Ten metropolitan centers (1,000,000+ pop., 1960)	Fifteen metropolitan centers (500,000– 999,999 pop., 1960)
Number of counties	417	263	86	68
Land area	269,466	180,097	53,274	63,095
1960 population	112,090,074	83,121,273	17,767,468	11,201,333
1960 population density (mile²)	415.97	461.54	331.51	177.5
1950 population density	325.89	364.94	249.21	137.99
Persons added per mile², 1950–1960	90.08	96.6	82.3	39.51
% of mainland population increase, 1950–1960	87.4%	62.67%	15.79%	8.98%
Estimated in-migration, 1950–1960	9,613,332	6,425,249	2,149,320	1,038,763
Estimated population, 2000 AD†	200,875,290	—	—	—
Estimated population density, 2000 AD	745.5	—	—	—

* Arthur A. Atkisson, *U.S. Population Settlement and Migration Patterns*. Houston, Texas: University of Texas School of Public Health, 1971 (mimeo.).
† Assumes that the U.S. mainland population will increase to a level of 280 million persons by 2000 AD and that these areas will account for the same fraction of the national population increase as they tallied between 1950 and 1960.

land use and density of development and the capacity of supporting systems, to provide and promote suitable development standards for planning in relation to existing and desired environmental quality, and to minimize the impact of unavoidable pollution on populations through the appropriate arrangement of land uses.

In adopting the federal Clean Air Act of 1970, Congress clearly assumed that there is a relationship between urban factors and air pollution occurrences. The award of a planning grant to an air pollution control agency was conditioned, in part, on the capability of the agency to develop "a comprehensive air quality plan . . . which plan shall include . . . [consideration of] the concentration of industries, other commercial estab-

TABLE 10.8

Factors Influencing Potential Air Pollution Emissions in Thirty
U.S. Metropolitan Centers and Regions (1950–2000)*

Factor	1950	2000	Increase
1. Population	87,815,370	200,875,290	113,059,920
2. Population density	325.9	745.5	419.6
3. Density increase due to excess of births over deaths + immigration from extra-mainland source	—	—	279.6
4. Density increase due to in-migration (metropolitanization)	—	—	140.0
5. Electrical energy consumption per mile² (kwh)	832,323.0	21,482,328	20,650,005
6. Motor fuel consumption per mile² (1,000 gals per yr.)	109,816.0	644,112	534,296
7. Increase in electrical energy consumption per mile² due to:†			
a. Rising per capita demands	—	—	8,558,785.8
b. Population growth	—	—	
(1) Excess of births over deaths	—	—	8,056,953.6
(2) In-migration	—	—	4,034,240.0

* Calculations assume:
         1950 total mainland population = 150,700,000
         1960 total mainland population = 178,464,000
         2000 total mainland population = 280,000,000
These areas will account for same proportion of 1950–2000 mainland population increase as between 1950–1960.

†   7a = increase resulting from projected change in per capita energy demand and *no* increase in population.
    7b(1) = increase resulting from projected change in per capita energy plus general population increase.
    7b(2) = increase resulting from projected change in per capita energy demand plus in-migration from non-metropolitan areas.

lishments, population and naturally occurring factors . . ." Congress also declared that federal approval of state-developed *implementation plans* (see Chapter 11) would be based, in part, on the extent to which they include provisions for "land-use and transportation controls."

Nevertheless, little scientific progress has yet been made in determining the quantitative relationship of urban design, land-use and other community factors to air pollution emissions and problems. The bulk of the literature dealing with such matters is either broadly descriptive in tone or philosophic, speculative, and legalistic in its orientation [22]. One urban planner has noted this fact and commented that few studies have yet been performed on the relationship between air pollution emissions and "the form of the community where the emission occurs" [16]. Clearly, however, such relationships do exist.

In terms of highly localized air pollution episodes the physical proximity of industrial, commercial and residential properties complicates the task of air pollution control. Mixed land uses result in situations where people reside in zones where highly localized emissions may occur as a result of equipment failures, operator errors, or inadequate source control measures. Residential intrusions into manufacturing areas and manufacturing intrusions into residential areas simply heighten the probability that substantial numbers of people will be annoyed or adversely affected by local air pollution emissions. Air pollution control officials in one large metropolitan community have testified that, under some conditions, *plume-looping* (see Chapter 2, Meteorology) could result in toxic ground-level concentrations of sulfur compounds within populous subdivisions adjacent to a large steam-electric generating station and to a complex of petroleum refineries [3].

Similarly, heavy concentrations of pollution-creating sources in the same land area can negate the results otherwise attainable through use of air pollution control equipment on these sources. Since few source control systems can achieve the complete elimination of pollution emissions, the proliferation of controlled sources within a limited area can result in the same *area rate* of pollution loss as a smaller number of completely uncontrolled sources.

Because of this, air pollution control analysts and planners now are careful to distinguish between point sources (a single industrial plant), area sources (a manufacturing complex or entire city), and linear sources (an inter-urban freeway).

Metropolitan communities differ widely in their dependence on gasoline-powered motor vehicles, in the extent to which high, pollution-yielding traffic conditions are present, and therefore in their capacities for the creation of photochemical and other vehicular-related air pollution conditions. Alan Voorhees has examined "work trips" in thirty-four U.S. cities and found that average trip length varies from 2.0 to 8.7 miles, that average trip duration varies from 6.1 to 20.1 minutes, and that average trip speeds vary from 19.2 to 35.0 miles per hour [6]. Since hydrocarbon and carbon monoxide emissions from motor vehicles are inversely proportional to the speed of travel, and directly proportional to the amount of stop-and-go driving (see Chapter 9), the relevance of these widely varying community travel characteristics is obvious. Atkisson has shown the wide variations between metropolitan areas in terms of the number of vehicles per capita, gasoline consumption per square mile, and gasoline consumption per unit of annual air volume [5].

In terms of other types of sources, additional factors are also important. These include the energy requirements of the community, the energy sources available for meeting these requirements (viz. hydro, coal, liquid fuels, nuclear reactors, geothermal sources, etc.), and the location and number of power-generating sources. Some communities, such as Los Angeles, have banned any further construction of large power-generating facilities within their borders and have supported power plant locations at sites remote from heavily populated metropolitan areas.

Some analysts have suggested that urban communities seek ways of reducing per capita energy requirements and have argued in favor of limitations on energy usage for frivolous or socially undesirable purposes. Since use of electrical energy for space-heating consumes more fuel than other heating methods, these analysts have proposed restraints on electrical energy usage for such purposes [30]. Similarly, some have proposed that inefficient, high fuel-using home furnaces and air conditioning equipment be replaced by the direct delivery of heat and cold to dwelling units and other places from more efficient neighborhood or community facilities [1, 13].

Currently, Spilhaus and his associates are designing an "experimental city" in which greater attention is given to these and other variables [28].

Economic Disincentives to Pollution Control

Economists view air pollution as the result of powerful incentives which lead producers and consumers of goods to seek ways to *externalize* the costs associated with the production and/or consumption of such goods [17]. For example, producers of soft drinks may wish to keep production costs low so as to capture a large market; consumers of such drinks may similarly wish to keep purchase prices low so that they can consume a greater quantity of the product. Both groups may therefore be motivated to pass the costs associated with the collection and disposal of the soft drink container onto the general public through public tax payments for the operation of refuse collection and disposal systems. Were these waste disposal costs to be *internalized* (viz. paid for by the producers and consumers of the product) the producers might experience a smaller market for their products and the consumers might be unable to purchase as much of the product as they desire to consume.

Considering the subject of "external effects," "external diseconomies," or "externalities" of production operations, Walter Heller has observed that [23]:

> . . . indirect costs [may be] inflicted on society which do not enter the private producer's costs and therefore do not influence market supply. Classic examples are the costs of smog, water pollution, denuding of forests, and the like. In these areas private output will exceed the optimum level unless government corrects the situation either by regulation or by a combination of expenditure and chargebacks to the private producers involved.

Since most commercial organizations are in competition with other organizations and since the price of a product may influence its popularity within a market, strong incentives are operative within our economic system to reduce the price of the goods and services dispensed by our many economic enterprises. Improved operational efficiency is only one way of securing such reductions and is, of course, a technique available to all competitors. Other methods of price reduction include subsidization of the costs of production by outside parties (such as federal subsidies to agricultural, air transport, rail, and shipping enterprises) or the transfer of production costs to persons other than the producers and purchasers of the product. When the latter occurs, *externalization* of cost is said to have occurred and the result of the shift is an "external diseconomy." In the case of air pollution, the cost which is shifted to "external" parties is the annoyance, ill health, or economic loss associated with the pollution occurrence, while the cost savings to the producer is measured by the capital and operating costs he avoids by not investing in air pollution control equipment.

The relationship between the capital cost of control devices and the value of the production equipment whose emissions these devices are intended to control has been documented by Robert L. Chass from permit records of the Los Angeles County Air Pollution District [9]. Among the sources listed by Chass in which control device costs are a significant fraction of the cost of the basic equipment are those shown in Table 10.9.

The significance to private business of capital investments in air pollution control devices is not completely revealed by these data. The total cost burden to business includes these costs plus the annual maintenance, operating, interest, and "opportunity" costs associated with the investment.*

The significance of these matters to large industrial and business enterprises was well stated by the pollution coordinator of a large industrial firm during the 1962 National Conference on Air Pollution [25].

> For the most part, air control devices represent added costs to business. Not only does the capital cost of purchase and installation divert

---

* "Opportunity" costs represent the income that has been lost as a result of not investing this sum in an income-producing venture.

TABLE 10.9

Relative Costs of Basic and Control Equipment for Selected Operations

Source	Size of equipment	Cost of basic equipment	Type of control equipment	Cost of control equipment
Airblown asphalt system	500 bbl/batch	$10,000	Afterburner	$3,000
Asphalt saturator	6 × 65 × 8 ft	40,000	Scrubber and electrostatic precipitator	50,000
Bulk gasoline loading rack	667,000 gal/day	88,000	Vapor control system	50,000
Carbon black plant	2,000 gal/day	5,000	Baghouse	5,000
Gray iron cupola	48 in. ID	40,000	Baghouse and quench tank	67,000
	27 in. ID	25,000	Baghouse and quench tank	32,000
Steel electric-arc furnace	18 ton/heat	75,000	Baghouse	45,000
Open-hearth steel furnace	60 ton/heat	200,000	Electrostatic precipitator	150,000
Oil-water separator	300,000 bbl/day	170,000	Floating roof	80,000

huge sums of money, but also the cost of operation adds to continuing overhead expenses in production. Higher costs mean less profit, lower profit means reduced income to government, increased cost of consumer items, and decreased return on investments. Lower profit also influences the competitive position of industry. Management is not inclined to devote the capital funds for improvements or expansion to plants which are in a poor competitive position; the growth of the plant is slowed or stopped and the opportunities for growth of the community are similarly affected.

The same author also stated that "in this day and age no responsible group of businessmen would knowingly submit employees or neighbors to conditions which would inevitably lead to death or disability" [25]. This conclusion, however, does not mean that risks are not taken. Substantial numbers of the U.S. population are exposed to disabling environments daily.

For example, occupation-related diseases and premature deaths have been widely documented for many years. Both employers and employees are aware of the hazardous circumstances associated with certain occupations and certain types of work environments. Yet the hazardous circumstances of these occupations and environments continue because of the economic consequences associated with their discontinuance, rather than because of any malign intent on the part of the owners and operators of the businesses or the self-destructive impulses of the workers. By isolating specific activities within American society, Chauncey Starr has been able to show that individuals voluntarily enter into high-risk situations when such risks are proportional to the cube of the benefits [26]. Stated another way, people will not accept a 50% increase in risk unless the risk-taking activity is accompanied by a three to fourfold increase in the benefits they receive. When such benefits are provided, the risks are accepted.

Thus, workers, neighbors and whole communities may accept the annoyance of pollution or overt risks to their health if a substantial number of them believe that the benefits of the risk-taking are worthwhile. In single company towns, environmental pollution is accepted frequently by the community if control of the pollution-causing activity would produce competitive disadvantages leading to a closure of the plant or a reduction in its activities.

Similarly, knowledge that pollution emissions may lead to ill health within a community does not in itself inevitably lead an organization to abate its pollution emissions. Frequently, reductions in emissions by a single source will not produce significant improvements in air quality unless similar action is taken by all other pollution contributors. Absent assurance that all others will take such measures, the operator of a single source who voluntarily assumes the cost burden of controlling his air pollution emissions may achieve little more than the destruction of his own competitive posture within the community without even securing the improvement in community air quality that was his initial objective.

The impact of control costs on economic enterprises has been examined by Seymour Schwartz who states [24]:

> The competitive structure within an industry or among industries may be changed by expenditures for pollution control. For example, the domestic steel industry which is faced with severe competition from foreign producers may not be able to pass along much of the cost of control to consumers unless it receives added tariff protection from foreign suppliers. A change in tariff policy may have serious international political implications. A significant change in profitability of a large company may result in employment cutbacks and economic loss to an entire area.

> To meet oxides of nitrogen emission standards in some regions it may be necessary for power plants to shift from fossil fuel to nuclear fuel.

Since utility companies are the major purchasers of coal, the impact on the coal industry of such a shift will be great. The regional consequences will be to worsen a situation which is already desperate in coal-producing regions.

Thus, substantial economic disincentives operate within our society to discourage voluntary, self-initiated action to curb pollution emissions. Except where the costs of control are trivial, and for most important pollution sources they are not, voluntary abatement action is unlikely. Without effective governmental action to control pollution emissions it therefore seems more likely that publicly unwanted pollution conditions may grow worse rather than better. In short, public measures are the only means to overcome the economic disincentives to pollution control.

Community Problem-Solving Capacities

However much metropolitan communities may be plagued by unwanted pollution conditions, their capacities for coping with these conditions have been severely impeded by several important factors. At the top of the list is *powerlessness*.

As we have seen, air pollution problems have their genesis in a complex network of forces operative within our society, including population growth, rising energy demands, rural-urban migration, modes of power generation and transportation, and economic disincentives to pollution control. The first four factors are all but immune to regulation by local governmental entities.

Population growth and rural-urban migration are phenomena of national scope. Population flows where it will, responding to such incentives as employment opportunities, natural and community amenities, and the mix of public services provided by competing receptor areas. Constitutionally, the free flow of people, goods, money, and businesses across state, county, and municipal boundary lines may be impeded by no agency short of the federal government itself. Even the authority of the federal government to limit the flow of interstate commerce is strictly circumscribed by constitutional safeguards.

Similarly, the primary authority over the circumstances which give rise to pollution emissions from power-generation and transportation sources is vested principally in federal and state governments. Federal agencies, such as the Federal Power Commission, are solely responsible for decisions concerning the interstate shipments of such low-pollution fuels as natural gas; the Atomic Energy Commission determines the location of nuclear facilities and controls the use of nuclear fuels.

After Los Angeles County officials denied utilities the right to construct power-generating facilities fueled by other than natural gas or nuclear energy, the area's largest private electrical utility joined with the municipal department of water and power to secure out-of-state supplies of this low-pollution fuel. Following several years of hearings and litigation, the application was denied by the Federal Power Commission. Without regard to the technical merits of the case, it serves as an example of the extent to which local authorities are impotent to cope with the larger issues involved in control of pollution emissions from power-generating sources.

Clearly, no single metropolitan area has the fiscal capacity to mount the massive research and development efforts necessary to the development of low-pollution fuels. Each area is dependent on the actions of federal and state governments for the development of such solutions to pollution problems.

State agencies, such as Public Utility Commissions, control the bulk of the activities associated with intrastate location and fueling of steam-electric generating stations. Plant expansions, rates, and location of generating facilities and transmission lines are matters of ascendant concern to such bodies.

Granted, local authorities may exert some control over such enterprises through their powers to award franchises and licenses to utilities for the use of city streets and property; they may control site location through their power over land use; and they may even impose and enforce emission standards on power-generating equipment. However, cities finally face the ultimate imperative that their populations must be served by electrical energy, and their regulatory acts, therefore, must give way to this need and the *superior* judgments of state and federal authorities as to *how* these needs will be satisfied.

Much the same situation prevails in respect to control of pollution emissions from transportation sources. Interstate trucks are licensed by the Federal Interstate Commerce Commission and travel the nation on highways subsidized by federal monies. Jet aircraft are built under standards imposed by federal authorities, fly routes awarded by the Federal Aviation Agency, and take off and land at facilities whose construction was heavily subsidized by federal funds. The manufacture and use of gasoline-powered motor vehicles involves a classic expression of our governmental philosophy concerning the free flow of interstate commerce. Under current United States law, with the exception of the state of California, only the federal government has the power to impose emission standards on such contrivances.

Looking at the needs of metropolitan areas for public services commensurate with the scope and severity of the problems they face, Ostrum and his associates have commented [21]:

> The first standard applicable to the scale of public organization for the production of public services requires that the boundary conditions of a political jurisdiction include the relevant set of events to be controlled.

Events, such as air pollution occurrences, may result from complex networks of causal factors beyond the statutory, jurisdictional, or territorial control of a local government or a single-purpose federal, state, or local agency.

For example, the Attorney General of California has ruled that the California Air Resources Board (a state agency) does not have the power to adopt a regulation prohibiting the use of lead additives in gasoline; the Texas Air Control Act denies the Air Control Board the power to adopt regulations specifying the *type* of control systems to be utilized in the abatement of a pollution emission; and federal pollution control agencies are denied the power to, themselves, make binding decisions concerning the distribution and use of low-pollution fuels within the United States.

Thus, the root causes of many of the social, health, and environmental problems of metropolitan communities are beyond control by state agencies and by the decisions and activities of local jurisdictions of government.

Even the powers of local government to take direct, on-site action to deal with pollution emissions from individual sources operating within their boundaries may be strictly limited. Cities and counties are legal creatures of state government and ordinarily are viewed to have only such powers as may have been conferred upon them by the constitutions and statutes of their states. Counties, for example, have traditionally been viewed as little more than administrative subdivisions of state government and have only such powers as have been expressly granted them by state constitutions and statutes. Although cities usually claim a somewhat broader scope of authority to enact and enforce ordinances dealing with local matters even their powers are strictly limited [20].

Following "Dillon's Rule," the courts traditionally have held that the powers of municipalities (municipal corporations) are strictly limited to "those granted in *express words;* . . . those necessarily or fairly *implied* in or *incident* to the powers expressly granted; . . . those *essential* to the accomplishment of the declared objects and purposes of the corporation, not simply convenient, but indispensable. Any fair, reasonable, substantial doubt concerning the existence of power is resolved by the courts against the corporation, and the power is denied" [20].

Thus, the legal capability of a city or county to adopt and enforce air pollution control measures may depend on the enactment of appropriate "enabling legislation" by the state legislature. Even in those states where cities and counties have been authorized to adopt "home rule" charters, the superior law-making powers of state legislatures are protected by the courts. Where state legislatures have "seen fit to adopt a general scheme for the regulation of a particular subject, the entire control over whatever phases of the subject are covered by state legislation ceases as far as local legislation is concerned" [20].

Thus, in states like Texas, the power of local governments to adopt air quality and source emission standards has been pre-empted by the state government.

The general powerlessness of local governments to cope with some of the more important problems besetting their populations is compounded in metropolitan communities by the jurisdictional pluralism characteristic of these areas. As we have seen in respect to St. Louis and BOS-WASH, large metropolitan centers and regions slop over the boundary lines of numerous jurisdictions of government. Pollution generated in one city, state, or county may produce formidable problems in cities and towns outside the originating jurisdiction and may therefore be immune to control by local governments within the receptor areas.

Recognition of this fact has led the California legislature to authorize the creation of both single-county and multi-county air pollution control districts. Single-county districts, such as the Los Angeles County Air Pollution Control District, are governed by an Air Pollution Control Board consisting of the several members of the County Board of Supervisors and are empowered to enact and enforce rules and regulations intended to control air pollution emissions. These regulations have the full force and effect of state law in both the unincorporated and incorporated areas of the district.

Multi-county districts, such as the Bay Area Air Pollution Control District, are governed by boards consisting of elected officials from cities and counties within the district and have powers comparable to those of single-county districts.

In California, the adoption of state legislation and the creation of state agencies to cope with air pollution conditions from stationary sources has been carefully designed to protect the powers of cities, counties, and districts to adopt more stringent standards than those enacted by the state. In other areas, however, state legislation has completely pre-empted the power of local governments to adopt standards compatible with locally-perceived needs.

Adding to the legal powerlessness of many local governments to cope with pollution conditions are the fiscal and administrative incapacities of these jurisdictions. Tied to property tax leves for the bulk of their funds, many local units of government have been hard pressed to pay for publicly desired pollution control services. The conduct of an effective community air pollution control program is dependent on recruitment of a qualified staff containing representatives from many scientific disciplines, including engineering, chemistry and meteorology. However, raising funds to support such staffs has proven to be beyond the capacity of many lightly populated jurisdictions of local government. In other jurisdictions poor pay scales and other factors have prevented the recruitment and continued retention of technically qualified air pollution control staffs.

Although these constraints on the problem-solving capacities of local governments could be removed through appropriate action by state legislatures, such action has been infrequent. Until recent rulings by the United States Supreme Court, the legislatures of all states were dominated by rural interests. As recently as 1960, legislators representing "urban" districts were in a minority in each of the state legislatures of the nation.

Even now, state legislatures are widely viewed as unresponsive to the needs of metropolitan communities and unsympathetic to legislative proposals enabling state and local agencies to take more effective action to curb pollution problems.

Human Expectations

One of the more important forces that has been operating within our society has been referred to as "the revolution of rising expectations." The revolution involves continuing escalations in the wants, needs, and demands of individuals and groups within our society.

This "revolution" is of importance to community air pollution conditions in at least two ways: (1) it fuels rising consumer demands for goods and services produced by pollution-causing activities, and (2) it fuels public demands for higher levels of life and environmental quality.

The escalating nature of human motivations has been discussed by A. H. Maslow who has suggested that man's motivations fall into an ascending hierarchy beginning with those oriented around his needs for survival, safety, and security and ending with those related to self-actualization [18]. According to Maslow, the satisfaction of lower-level needs leads to a loss or reduction in their motivational power and to a reorientation of motives around the achievement of higher-level satisfactions. Thus, as the simple requirements for survival, safety, and security are satisfied, higher-level satisfactions are increasingly valued and become the basis of an individual's value and behavioral orientation.

Within this context, air quality objectives may be viewed as moving targets whose definition is subject to the changing motivations and attitudes of the public. In these terms, an air pollution problem may be measured by the gap between the air quality aspirations of a community and the reality of pollution conditions within that community. How much of the current public reaction to air pollution conditions is a result of a worsening of such conditions as contrasted to rising air quality aspirations is not known.

Frank Stead has suggested that environmental quality be judged by a scale involving four criteria [29]. First, does it insure survival? Second, does it prevent disease and accidents? Third, does it promote efficient or unusual human performance? Fourth, does it promote comfort, pleasure, or joy? Stead's fourth criterion can be viewed as an "amenity scale" along which can be ordered the several types of responses man makes to those circumstances in his environment which generate feelings of comfort, pleasure, or joy. Some of those responses may lead to heightened productivity or to such a relaxation of pressure, tension, and stress that they may be found to increase man's longevity or health. Other responses may be said to have economic value since they motivate people to purchase private amenity goods (such as homes in pollution-free neighborhoods) or to support community investments in public amenity goods (such as community air pollution control programs). Still other higher-order responses may be psychologically valued by the individual but result in no clearly discernible behavioral manifestations, either economic or political [7].

Politics and Public Policy Priorities

The occurrence and continued existence of unwanted air pollution conditions within U.S. communities also may be viewed as the result of our system of public policy priorities. The mere fact that a governmental jurisdiction or agency possesses the legal power to cope with a public problem does not in itself guarantee that the power will be utilized.

Practical political considerations, institutional inertia, or greater concern for other matters may impede community action to control air pollution problems. It is quite likely that all three of these factors have operated to slacken the pace of community air pollution control efforts within the United States.

Until the recent national surge of public interest in environmental matters, few metropolitan communities exhibited much political interest in air pollution conditions [8]. Among the notable exceptions, of course, were the counties of southern California where "smog" has been a political issue since the late 1940's. In the county of Los Angeles a recent governmental survey of public attitudes concerning community conditions revealed that "air pollution" was viewed as the number one problem of the area by all income and ethnic groups and by all geographic sections of the county [10]. Few communities can demonstrate this intensity of public concern for air pollution conditions.

In American politics it is the "squeaking wheel that gets the grease," and the absence of public "squeaks" concerning pollution conditions has probably influenced the slow pace of pollution control in many communities.

Now, however, growing public concern and political activity concerning environmental matters has resulted in a veritable eruption of legislative pronouncements concerning environmental quality and air pollution conditions. It is not unlikely that this wave of legislative interest in pollution conditions may soon result in a re-ordering of our public policy priorities.

Clearly, many of the factors contributing to current pollution conditions are within our social and technical capacity to correct. That they have not been corrected may be more a function of our public policy priorities than of any other factor.

For example, in spite of the well-documented public concern over traffic congestion, vehicular-related pollution conditions, and the stressful circumstances of the journey to work, little of our national wealth has yet been allocated to improvements in ground transportation systems. In recent years more of the federal budget has been spent annually for space research and technology than for all ground transportation services, including the construction of streets and highways. The cumulative expenditures of the federal government on air pollution control activities over the past decade do not equal the sums expended in one year for income subsidies to agricultural enterprises. Annual federal expenditures for economic and financial assistance to foreign governments have been almost twice as great as the federal sums allocated for urban renewal and community facilities programs. For the 1971 fiscal year, total projected federal expenditures for research, development, and demonstration projects related to the prevention and control of pollution from transportation, electric power production, industrial energy, and other industrial sources totalled approximately 122.1 million dollars [36]. Projected expenditures for fuel resources development and production projects related to pollution control totalled an additional 4.7 million dollars [36]. In contrast, 1969 expenditures for subsidies to agricultural enterprises totalled more than 3.3 billion dollars, subsidies to ship operators totalled 211 million dollars, and expenditures for space research and technology exceeded 4.6 billion dollars (see Table 10.10).

Whatever the merits of these subsidies (space research and international

TABLE 10.10
Federal Subsidy Payments and Expenditures for Selected
Programs and Purposes (1960–1970)* †

Program/Purpose	Year	Expenditure (millions of $)
A. Federal Subsidies (Selected programs):		
1. Price support and related programs for agricultural enterprises	1969	$3,313
2. Shipping enterprises	1969	211
3. Payments to air carriers	1969	54
B. Federal Expenditures for Selected Programs:		
1. Space research and technology	1969	4,574
2. Urban renewal and community facilities	1969	1,432
3. Foreign economic and financial assistance	1969	2,564
4. Ground transportation	1969	4,420
5. Air pollution control and abatement (all agencies)	1970	163

* U.S. Bureau of the Census. *Statistical Abstract of the United States: 1968* (89th edition). Washington, D.C., 1968. Tables 538, 539, 547.
† Executive Office of the President. *Environmental Quality: The First Annual Report of the Council on Environmental Quality.* Washington, D.C., 1970. Table 2, p. 320.

assistance expenditures) may be, it is clear that expenditures related to pollution control have been assigned a lower public policy priority. To that extent the continuation of air pollution conditions within the United States may be ascribed to our national expenditure priorities.

Implications for Air Quality Management Policies

In 1965, one responsible scientific group concluded that "air pollution is a social and economic problem. Its cause and its solution may be found in man's relationship to man, and in his relationship to his tools and his economic life" [2].

In this chapter we have seen how population growth, rising consumer demands for energy and manufactured goods, migration to metropolitan communities, and patterns of fuel usage contribute to the occurrence of air pollution problems. Similarly, we've seen how economic factors, public policy priorities, and statutory constraints on community problem-solving capacities may contribute to the continued occurrence of such problems.

Each of these factors is of importance to the design of strategies having a high probability for achieving air quality objectives consistent with public wants and demands. The extent to which current federal, state, and local programs are responsive to these factors is discussed in Chapters 11 and 12.

REFERENCES

1. Abrahamson, D. E., "Environmental Cost of Electric Power," Scientists Institute for Public Information (1970).

2. Air Conservation Commission, *Air Conservation.* Washington, D.C.: American Association for the Advancement of Science, 272, 307 (1965).

3. Air Pollution Control District, County of Los Angeles. *Defendant's Trial Brief, G.W.A., Incorporated vs. Air Pollution Control District of Los Angeles County,* Los Angeles, Calif.: Office of the County Counsel (Two volumes) (1965).

4. American Institute of Planners, "Environmental Policies," *Journal of the American Institute of Planners, 37,* No. 4, 209 (July, 1971).

5. Atkisson, A. A., "National Motor Vehicle Contaminant Control Requirements," *Journal of the Air Pollution Control Association, 12,* No. 5 (May 1962).

6. Atkisson, A. A., and R. Gaines, eds., *Development of Air Quality Standards,* Columbus, Ohio: Charles E. Merrill Publishing Company, 148 (1970).

7. Atkisson, A. A., and I. M. Robinson, "Amenity Resources for Urban Living," *in* Harvey S. Perloff, ed., *The Quality of the Urban Environment.* Baltimore, Md.: The Johns Hopkins Press, 179–201 (1969).

8. Caldwell, L. K., *Environment: A Challenge for Modern Society.* Garden City, New York: The Natural History Press (1970).

9. Chass, R. L., "The Status of Engineering Knowledge for the Control of Air Pollution," *in:* U.S. Dept. of Health, Education, and Welfare, *Proceedings: National Conference on Air Pollution,* U.S. Govt. Printing Office, Washington, D.C., 272–280 (1963).

10. City of Los Angeles Planning Department, *Summary Report of the Los Angeles Goals Council,* Los Angeles, California, Los Angeles Goals Council (November 1969).

11. Davis, K., "The Urbanization of the Human Population," in: *Cities: A Scientific American Book,* New York: Alfred A. Knopf, 3–24 (1966).

12. Ehrlich, P. R., *The Population Bomb* (Revised Edition), New York: Ballantine Books, Inc., 4 (1971).

13. Fabricant, N., and R. M. Hallman. *Toward a Rational Power Policy: Energy, Politics, and Pollution.* New York: George Braziller (1971).

14. Gottman, J., *Megalopolis: The Urbanized Seaboard of the United States,* Cambridge, Mass.: The M.I.T. Press (1961).

15. Hauser, P. M., "The Census of 1970," *Scientific American, 225,* No. 1, 18 (July, 1971).

16. Jammal, I. M., "Vehicular Air Pollution: Variables Influencing the Urban Transportation System," *in* Atkisson, Arthur A., and Richard Gaines, eds.,

Development of Air Quality Standards, Columbus, Ohio, Charles E. Merrill Publishing Company, 127 (1970).

17. Kapp, K. W., *The Social Costs of Private Enterprise,* New York: Shocken Books (1971).
18. Maslow, A. H., *Motivation and Personality.* New York: Harper (1954).
19. Mayer, L. A., "U.S. Population Growth: Would Slower be Better?" *Fortune,* 81–82 (June 1970).
20. Michelman, F. I. and T. Sandalow, *Materials on Government in Urban Areas: Cases—Comments—Questions,* St. Paul, Minn.: 'West Publishing Company, 179–185, 252–253, 332–339, 382–383 (1970).
21. Ostrum, V., et al, "The Organization of Government in Metropolitan Areas," *American Political Science Review,* **55,** 831–842 (Dec. 1961).
22. Pelle, W. J., *Annotated Bibliography on the Planning Aspects of Air Pollution Control,* Study for Northwestern Illinois Planning Commission and U.S. Public Health Service, March 1965.
23. Phelps, E. S., ed., *Private Wants and Public Needs* (Revised Edition), New York: W. W. Norton and Company, Inc., 161–162 (1965).
24. Schwartz, S., *Models for Decision-Making in Air Quality Management,* Los Angeles: University of Southern California, Dept. of Industrial and Systems Engineering, Technical Report 70-4, 103 (August 1970).
25. Schwegmann, J. C., "Prepared Discussion: Economic Considerations in Air Pollution Control." *In:* U.S. Dept. of Health, Education, and Welfare, *Proceedings: National Conference on Air Pollution,* U.S. Govt. Printing Office, Washington, D.C., 187–188 (1963).
26. Starr, C., "Social Benefits versus Technological Risk—What is Our Society Willing to Pay for Safety?" *Science,* **165,** 1128–1133 (Sept. 1969).
27. Spaite, P. W., and R. P. Hangebrauck, "Sulfur Oxide Pollution: An Environmental Quality Problem Requiring Responsible Resource Management," paper presented at 19th Canadian Chemical Engineering Conference, Canadian Society of Chemical Engineers, Edmonton, Alberta (October 19–22, 1969).
28. Spilhaus, A., "The Experimental City," *Daedalus,* **96,** No. 4, 1129–1141 (Fall, 1967).
29. Stead, F., "Levels of Environmental Health," *American Journal of Public Health,* **50,** No. 3 (1960).
30. Task Force for Nuclear Power Policy, *Nuclear Power in the South,* Atlanta, Georgia: Southern Governor's Conference, Interstate Nuclear Board, 21 (1970).
31. U.S. Bureau of the Census, *County and City Data Book, 1956,* U.S. Govt. Printing Office, Washington, D.C. (1956).
32. U.S. Bureau of the Census, *County and City Data Book, 1967,* U.S. Govt. Printing Office, Washington, D.C. (1967).

33. U.S. Bureau of the Census, *Historical Statistics of the United States, Colonial Times to 1957,* U.S. Govt. Printing Office, Washington, D.C. (1960).

34. U.S. Bureau of the Census, *Statistical Abstract of the United States: 1968* (89th edition), U.S. Govt. Printing Office, Washington, D.C. (1968). (Tables 3, 20, 756, 819, 1105, 1158)

35. U.S. Government, Executive Office of the President, *First Annual Report of the Council on Environmental Quality,* U.S. Govt. Printing Office, Washington, D.C., 150 (August 1970).

36. U.S. Senate, Committee on Public Works, *Some Environmental Implications of National Fuel Policies,* U.S. Govt. Printing Office, Washington, D.C. (December 7, 1970).

11

AIR QUALITY
MANAGEMENT

Within the past decade, the broad outlines of a coherent program for the management of air quality has emerged within the United States. Involving a mix of mutually supporting activities by federal, state, and local governments, the program exhibits several significant attributes. These include: (1) an increased emphasis on statutory and administrative measures for the abatement of pollution emissions; (2) a commitment to rational processes for the making of air quality management decisions; (3) an orientation to the prevention of air pollution problems; (4) increased reliance on a mix of policy strategies for dealing with the social origins of pollution conditions, including techniques for control of land use; (5) expanded opportunities for citizen participation in air quality management and emission abatement decisions; (6) a complex network of intergovernmental relations; (7) expanded federal support for state and local air pollution control activities.

The Legal Bases for Pollution Control

One of the most significant developments in air pollution control has involved a shift from reliance on common law remedies for control of problem-causing pollution emissions to more direct and effective statutory prescriptions and administrative regulations. This shift in the legal means for dealing with pollution emissions signals a developing awareness that air pollution conditions are a *public* rather than a *private* problem and that the standards to be observed in controlling pollution emissions should be explicitly stated as a matter of public policy rather than be left to the vicissitudes of litigation before a court.

322

As a former colony of Great Britain, the United States has inherited and made use of the English common law. As contrasted to statutes enacted by a legislative body, or regulations issued by an administrative board or official granted such power by a legislative body, the *common law* is judge-made law and grows out of the established customs of the community. It represents the set of rules and principles articulated by countless courts in adjudicating disputes between private parties [23].

In England, the first legal actions to abate air pollution emissions were brought under the common law doctrine of nuisance [1]. Under this doctrine, "one must use his own property so that his neighbor's comfortable and reasonable use and enjoyment of his estate will not be unreasonably interfered with or disturbed . . ." [40]. The doctrine holds that the use of private property is unrestricted only so long as it does not injure the person or property of another. However, if the action causes physical discomfort or injury to a person of "ordinary sensibilities" the injured party may seek assistance from the courts which are empowered to order a halt to the offending action or to compel the payment of "damages" to the injured party, or both.

In a private nuisance suit the injured party (plaintiff) must provide proof that damage or injury has occurred and that the defendant's activity is "unreasonable." The simple fact that some party has been annoyed, or even injured, by the actions of another will not alone sustain a nuisance action. For example, a New York court has stated [12]:

> The compromises exacted by the necessities of the social state, and the fact that some inconvenience to others must by necessity often attend the ordinary use of property, without permitting which there could in many cases be no valuable use at all, have compelled the recognition, in all systems of jurisprudence, of the principle that each member of society must submit to annoyances consequent upon the ordinary and common use of property, provided such use is reasonable both as respects the owner of the property, and those immediately affected by the use, in view of time, place, and other circumstances.

On the basis of this view one court concluded that air pollution is "indispensable to progress" and, in 1931, classified the pollution from fifty coke ovens in Buffalo, New York, as only a "petty annoyance" [8]. Twenty years later the city of Buffalo took legal action against the same pollution source and, like the plant's neighbor, met with little success [36].

Unlike private common law nuisances, where the action is between private parties, a "public nuisance" involves "damage to the *community* in the exercise of its common rights . . ." [37]. In bringing a public nuisance suit against an offending party, the state is obligated to provide proof "that a large number of persons actually . . . [suffer] some impairment to their enjoyment of life. . . ." [37].

Because of the difficulties and uncertainties associated with litigation based on the common law doctrine of "public nuisance," most authorities agree that this approach is of only limited utility in providing a foundation for community air pollution control efforts. For example, the Air Conservation Commission has stated that "attempts to cope with the problem of air pollution on a community-wide basis through legislation will probably be more effective than litigation" [1]. Similarly, John Hanlon has observed that [23]:

> It is perhaps unfortunate that a large proportion of public health officials still consider the use of the law of nuisance as one of their most important if not their chief legal recourse. The pursuit of this point of view eventually leads to many difficulties and dissatisfactions. . . . Decision by a law court that a nuisance exists may result in damages being paid to the plaintiff but does not necessarily effect the solution or abatement of the noisome circumstances . . . a rule of equity may provide a way out for the defendant in that if he can demonstrate to the court's satisfaction his intention or, better yet, partial action to abate the nuisance, the case in all probability will be dismissed from court.

Government adoption and enforcement of air pollution control laws and regulations can avoid the vagaries, delays, expense and uncertainty associated with litigation based on the common law doctrine of nuisance. When the latter method is employed a judge ordinarily must determine, on the basis of evidence submitted, whether an air pollution emission has occurred; the source from which the emission derived; whether the emission caused injury or harm to the plaintiff; whether the harm-causing emission resulted from an "unreasonable" activity; and whether or not the plaintiff is entitled to relief in the form of monetary damages or the imposition of some judicial constraint on the harm-causing source, or both.

Usually, such actions lead only to damage payments. However, the court may order the closure of the source or some pattern of activity intended to control the emission. When this occurs, a nontechnically trained judge is thrust into the role of determining what emission standard should be imposed on the source and what type of control method should be employed. Other judges in other courts within the community may be making similar decisions. Clearly, reliance on such a process is rich in opportunity for the production of many difficulties, uncertainties, and inequities in control measures for comparable sources within a community. One can imagine the even more formidable difficulties that would be associated with court action to prevent or to control the construction of a source capable of producing *future* injury, harm, or annoyance.

On the other hand, air pollution control laws and regulations can deal with whole classes of sources, including those now in operation as well as those which may be built in the future. For each such class, precise emission standards may be stated and requirements may even be imposed governing the type of fuel and equipment which may be utilized. Routine procedures may be established for the inspection of existing sources and for reviewing plans for the construction of new sources. As we shall see, such laws and regulations can be enacted to accomplish a wide range of public purposes, extending from elimination of public health emergencies to the achievement of beautiful and highly productive environments. Moreover, if such laws and regulations are properly drafted, any litigation arising from their enforcement will be limited to far fewer issues. In such litigation, proof of injury-causing emissions is not necessary and the judicial inquiry is confined to such questions as whether or not the terms of the statute or regulation were violated and whether or not the legislative or administrative body acted "irrationally" in enacting such terms. The courts operate on the presumption that the statute or regulation is valid and that the enacting body did properly exercise its authority. In contrast, in a common law nuisance action the court presumes the defendant to be innocent of any harmful activity and the full burden of proof falls upon the plaintiff.

Air pollution control laws and regulations are based on the "police power" of governmental jurisdictions. Although a vague and imprecisely defined term, the "police power" of governments confer upon them the authority to deal with a wide variety of ills and to achieve a wide variety of public purposes. In one case, a court defined the "police power" as "that inherent sovereignty which the government exercises whenever regulations are demanded by public policy for the benefit of the society at large in order to guard its morals, safety, health, order, and the like in accordance with the needs of civilization" [31].

In a case testing the validity of a Sacramento (California) air pollution control ordinance, the court stated [25]:

> That the police power is an inherent attribute of every state or commonwealth in the Union is a proposition which will be readily conceded. It is not only a power which inheres in the sovereignty of the states, but is a power the exercise of which by the states is indispensably essential to the health, peace, comfort and welfare generally of the inhabitants thereof. . . .

> This power embraces the right to regulate any class of business, the operation of which, unless regulated, may in the judgment of the appropriate local authority, interfere with the rights of others, for, as is said in *Dobbins v. City of Los Angeles,* 139 Cal. 179, 96 Am. St. Rep. 95, 72 Pac. 1970, "all property is subject to police power!" In other words, the proposition cannot be maintained that the exercise of this power is confined to the regulation only of such interferences with the public welfare and comfort as come strictly within the common law definition of a "nuisance."

In another case, the Chief Judge of New Jersey held [7]:

> The reason for a municipality making unlawful the emission of smoke is readily apparent. The issuance of dense smoke from a single chimney, in and of itself, may be altogether harmless and cause no inconvenience or damage to the public, but if smoke of like density issued from hundreds of chimneys, the contamination of the atmosphere would be substantial and the injury to the public considerable, yet for lack of the requisite elements of a public nuisance at common law, the municipality could obtain no relief by way of indictment. Ordinances making unlawful the emission of smoke are therefore obviously necessary and reasonable and a valid exercise of the local police power.

In a 1954 opinion the United States Supreme Court said that any attempt to define the reach of the police power or [6]

> . . . its outer limits is fruitless, for each case must turn on its own facts. . . . Public safety, public health, morality, peace and quiet, law and order—these are some of the more conspicuous examples of the traditional application of the police power. . . . Yet they merely illustrate the scope of the power and do not delimit it. . . . *It is within the power of the legislature to determine that the community should be beautiful as well as healthy, spacious as well as clean, well-balanced as well as carefully patrolled.*

Thus, the adoption and enforcement of air pollution control laws and regulations is a proper exercise of the police power. However, Lawrence W. Pollack has cautioned against basing such enactments on legislative objectives not related to the protection of the public from the harm, injury, and health disturbances occasioned by appropriate air pollution conditions [37]. Under some conditions, the utilization of the police power to achieve legitimate public purposes may require the payment of "just compensation" to persons whose property rights have been impaired [37]:

At least one of the frequently used tests of whether police power regulations have gone so far as to be a compensable government "taking" is whether the legislation simply restrains conduct harmful to others or whether its purpose is positive enrichment of the public at the expense of private property.

However, when statutes and regulations are based on public health considerations and on protection of the public from other harmful consequences of air pollution, they may impose considerable economic burdens on the regulated parties and even lead to the closure of otherwise lawful businesses. The leading case of this sort was decided by the United States Supreme Court in 1916. The court held [34]:

> So far as the federal Constitution is concerned, we have no doubt the State may by itself or through authorized municipalities declare the emission of dense smoke in cities or populous neighborhoods a nuisance and subject to restraint as such; and that the harshness of such legislation, or its effect upon business interests, short of a merely arbitrary enactment, are not valid constitutional objections. Nor is there any valid federal constitutional objection in the fact that the regulation may require the discontinuance of the use of property or subject the occupant to large expense in complying with the terms of the law or ordinance.

In an excellent review of the legal foundations for air pollution control rules and regulations, Harold Kennedy has observed that such enactments may: (1) declare air contaminants to be a public nuisance even though such contaminants were not considered by the common law to be a nuisance per se; (2) secure the abatement of air pollution emissions on grounds of discomfort, annoyance, and inconvenience without showing impairment of health; (3) prohibit emissions on the basis of their darkness or opacity, the effects they produce, their qualitative characteristics as measured by the percentage of an offensive contaminant in an emission, and the quantity of a contaminant released; (4) prohibit specific acts (such as open burning), specific fuels (such as soft coal), or specific equipment (such as single-chamber incinerators); (5) require the use of specific types of equipment (such as vapor recovery systems on open skimming ponds) [27].

In hearing litigation concerning pollution control laws and regulations, courts have refused to substitute their judgments as to the objective merits of the regulations for those of the legislative body. Unlike common law nuisance actions the scope of judicial inquiry is strictly limited in cases concerning violations of legislative standards. The major issues before the courts in such cases involve determinations as to whether or not the alleged violation occurred and whether or not the legislature acted *irrationally* in adopting the standard.

In respect to the power of the judiciary to weigh the merits of legislative enactments, a California court has held [26]:

> A legislative body, in the exercise of its police power, has a broad discretion to determine both what public interests are and the measures necessary for the protection of such interests. The determination of the need for a mode of exercising the power is primarily for the legislative body and the courts will not hold enactments invalid unless they are palpably unreasonable, arbitrary or capricious, having no tendency to promote the public welfare, safety, morals, or general welfare. Every presumption is in favor of the reasonableness of the law and its validity. A court is not concerned with the wisdom or policy of the law and cannot substitute its judgment for that of the legislative body. If reasonable minds might differ as to the reasonableness of the regulation, the law must be upheld.

Similarly, the New York Court of Appeals has stated that "it is not for the courts to determine which scientific view is correct in ruling upon whether the police power has been properly exercised [13]. In upholding the validity of a local law regulating automatic and coin operated laundromats, the New York Supreme Court stated [39]:

> This court is unable to say that the city legislative authorities are wholly unreasonable to the point of irrationality in believing that the police and fire problems will be somewhat alleviated by not permitting laundromats . . . to be open between midnight and 6 a.m., and by requiring that between 6 p.m. and midnight laundromats have someone in attendance. Even if the municipal legislative authorities are mistaken in this view, that does not invalidate the legislation. Courts do not sit as an additional house of either state or municipal legislature.

In deciding the constitutionality of the St. Louis air pollution law, the Missouri Supreme Court [5] drew from an earlier case the following principle [33]:

> The methods, regulations, and restrictions to be imposed to attain, so far as may be, results consistent with the public welfare, are purely of legislative cognizance. The courts have no power to determine the merits of conflicting theories, nor to declare that a particular method of advancing and protecting the public is superior or likely to insure

greater safety or better protection than others. The legislative determination of the methods, restrictions, and regulations is final, except
when so arbitrary as to be violative of the constitutional rights of the
citizens.

Administrative regulations issued pursuant to the terms of a statute are
treated as legislation insofar as judicial inquiry into the merits of the regulation are concerned. Thus, in upholding the regulations at issue in a case
the California Court of Appeals observed [16]:

> An administrative rule which is legislative in character is subject to
> the same test with reference to its validity as is an act of the legisla-
> ture. . . . The same principles govern administrative rules as govern
> statutes. . . .

In still another case a California court ruled [9]:

> . . . it must be held that the *de novo* type of review does not apply
> to quasi-legislative acts of administrative officers and that the judicial
> review is limited . . . to determine whether his [the officer's] action
> has been arbitrary, capricious, or entirely lacking in evidentiary sup-
> port. . . .

Thus, it seems clear, the legislative prerogatives to utilize police power
for regulation of air quality will go unchallenged by the courts, that the
factual merits of the controls will not be probed by the courts unless palpably irrational, and that these same circumstances apply to administrative
regulations issued pursuant to legislative enactments.

Thus, Professor James E. Krier has observed that there is little in the
way of "legal" constraints on the exercise of police power by a legislature
in choosing and enacting an air pollution strategy [3]. Distinguishing between "legal" and "extralegal" constraints, such as what is practical, economic, political, and desirable, Krier states [3]:

> The legal . . . constraints which limit the authority of government to
> act and thus make the police power something less than absolute are
> primarily of constitutional origin—the familiar concepts of due proc-
> ess and equal protection of the laws. Due process requires both "rea-
> sonable" and "certain" legislation, but these are broad and dynamic
> terms. Reasonableness requires only that the legislative body not act
> arbitrarily nor more broadly than necessary to accomplish its purpose.
> Since the courts recognize the right of the legislature to exercise wide
> judgment both as to what the facts of a problem are and what action
> might best be taken to meet it, this is a loose-waisted constraint in-
> deed. So, too, with the requirement of "certainty." The utmost

mathematical precision is by no means required; the essential question is whether the meaning of the statute can be reasonably ascertained; whether persons within the scope of the statutory requirements have been put on reasonable notice of the actions required of them. Moreover, the proscription against vagueness has little impact outside the realm of criminal prosecutions.

In terms of the legislature's power to delegate to administrative bodies the authority to issue air pollution control regulations, the courts have insisted only that such delegations be consistent with a legislative declaration of policy, a primary standard fixed by the legislature, and that the legislature itself establish the penalties to be invoked against violators of the regulations.

For example, the California Supreme Court has held [19]:

> The legislature may, after declaring a policy and fixing a primary standard, confer upon executive or administrative officers the "power to fill up the details" by prescribing administrative rules and regulations to promote the purposes of the legislation and to carry it into effect, and provision by the Legislature that such rules and regulation shall have the force, effect, and sanction of law does not violate the constitutional inhibition against delegating the legislative function.

In yet another case, the same court stated [21]:

> . . . it has become increasingly imperative that many quasi-legislative and quasi-judicial functions, which in smaller communities and under more primitive conditions were performed directly by the legislative or judicial branches of the government, are entrusted to departments, boards, commissions, and agents. No sound objection can longer be successfully advanced to this growing method of transacting public business. These things must be done in this way or they cannot be done at all, and their doing, in a very real sense, makes for the safety of the republic and is thus sanctioned by the highest law. . . . And while inferior boards or tribunals cannot be invested with power to compel obedience to their orders or regulations by fine or imprisonment (citing cases), nevertheless it is within the power of Congress to authorize them to make such reasonable regulations and orders and itself to declare that a violation of them shall be punishable as a misdemeanor.

Similarly, the United States Supreme Court has stated [43]:

> From the beginning of the government various acts have been passed conferring upon executive officers powers to make rules and regula-

tions—not for the government of their departments, but for administering the laws which did govern. None of these statutes could confer legislative power. But when Congress had legislated and indicated its will, it could give to those who were to act under such general provisions "power to fill up the details" by the establishment of administrative rules and regulations, the violation of which could be punished by fine or imprisonment fixed by Congress. . . .

In terms of the penalties which should be fixed by the legislature to punish violators of air pollution control laws, rules, and regulations, there has been considerable disagreement. Some few have argued that no penalties should be fixed and that government should rely completely on voluntary cooperation from those whom it seeks to regulate. Others have argued that criminal sanctions, civil penalties, and the power of injunction should be authorized to secure compliance with such enactments. The latter view now seems more common and much of the recent air pollution control legislation therefore includes such provisions. Moreover, federal regulations issued under the authority of the Clean Air Act of 1970 now require that state-developed *Implementation Plans* include provisions demonstrating that "the state has legal authority to carry out the plan, including authority to: (1) Adopt emission standards . . . ; (2) Enforce applicable laws, regulations and standards, and *seek injunctive relief*" [17].

Rationality and Air Quality Management

Even though legislative bodies may be comparatively free of legal constraints in developing air quality management programs, there is now a formidable body of opinion as to the criteria that *should* be observed in their design. Moreover, the Clean Air Act of 1970 outlines several rationally oriented policies and procedures to be observed by federal agencies, as well as by state and local governments, in the design and implementation of air quality management programs.

Among the criteria which have been suggested to govern the design and operation of such programs are the following: (1) that the programs should be oriented around the achievement of specified air quality objectives; (2) that such objectives should be established on the basis of scientific information concerning the relationship between atmospheric pollution concentrations and effects judged to be undesirable by the public; (3) that final decisions on air quality objectives should be influenced by our understanding of the benefits associated with alternative levels of air quality in relationship to the costs associated with achieving such levels; (4) that air pollution emission standards and other control measures should be developed in consonance with scientific information concerning the mete-orology of the target area, the existing and projected emissions within that area, and the relationship between these factors and established air quality objectives; (5) that selection of the mix of measures to be employed in achieving the air quality objectives for an area should be influenced by the "least cost" mix of available control measures. ·

Although an oversimplification of a highly complex and technical web of argument and professional thought, the list suggests the general prop-erties of a "rational" system of air quality management. In a way, it rep-resents nothing more than a sophisticated application of "common sense" to the development of public policies in an area deemed important by the general public.

Clearly, one would not wish to control dusts from a desert gravel pit if his focal interest were the amelioration of eye irritation occurrences in a distant metropolis. If a specified burden of pollution within a community's air supply has the likelihood of producing damage to the public health one would wish to control community emissions to whatever point is necessary to maintain atmospheric burdens below the health-damaging level. Con-versely, if burdens are well below the level at which health damage can be expected, and if no other overt symptoms of pollution problems are in evi-dence, it would be foolish to undertake a costly emission-reduction pro-gram. If a set of air pollution problems are extant or expected as a conse-quence of anticipated future growth, and if several equally effective control measures are at hand for curbing these problems, then most prudent indi-viduals would wish to select the least costly measure for implementation. If $1,000 in damage to ornamental vegetation is being experienced each year as a result of pollution conditions, one might not wish to spend $1,000,000 per year to control the emissions producing such damage.

In real life, of course, choices are rarely this simple. Frequently one is confronted with several means-ends alternatives each of which may exhibit its own mix of costs and benefits, its own lag time from start of effort to time of probable goal accomplishment, and its own set of probabilities con-cerning the likelihood that the course of action associated with the alterna-tive will lead to accomplishment of the desired objective. Under such cir-cumstances there is no "natural law of rationality" which reveals the "best" choice for all persons, in all places, and at all times. Instead, there are a set of decision-assisting tools which a rational man may use in resolving the decision problem. In short, the primary characteristic of a "rational" system has to do with the process by which decisions are made rather than with the "efficacy" of the decisions as measured by some "value-laden" criteria stated by some superior, all-knowing, hyper-rational man.

THE FEDERAL PROCESS

Several features of the Clean Air Act of 1970 are of particular significance in terms of the opportunities they provide for a rational process of decision-making in the field of air quality management.

Air Quality Criteria

Building on the foundation provided by the Clean Air Act of 1963 and the Air Quality Act of 1967, the Clean Air Act of 1970 requires the administrator of the federal Environmental Protection Agency to continue the issuance of air quality criteria and requires that [11]:

> Air quality criteria for an air pollutant shall accurately reflect the latest scientific knowledge useful in indicating the kind and extent of all identifiable effects on public health or welfare which may be expected from the presence of such pollutant in the ambient air, in varying quantities. The criteria for an air pollutant, to the extent practicable, shall include information on
> a. Those variable factors (including atmospheric conditions) which of themselves or in combination with other factors may alter the effects on public health or welfare of such air pollutant;
> b. The types of air pollutants which, when present in the atmosphere, may interact with such pollutant to produce an adverse effect on public health or welfare; and
> c. Any known or anticipated adverse effects on welfare.

This statutory charge makes clear the fact that air quality criteria *are not* to be viewed as normative statements. In other words, they are not policy judgments which express value choices as to how pure or polluted the atmosphere should be. Instead, as stated in testimony before the U.S. Senate Subcommittee on Air and Water Pollution, "air quality criteria are descriptive—that is, they describe the effects that can be expected to occur whenever and wherever the ambient air level of a pollutant reaches or exceeds a specific figure for a specific time period" [3].

In the purest meaning of the term, as well as in the language of the statutory charge, *air quality criteria* are products of scientific inquiry, not the results of a process in which personal values, barter, personal advantage, and compromise are the guiding considerations. In these terms, *criteria* are an expression of *existing knowledge* and may be only feeble approximations of the truth in circumstances where knowledge is incomplete. (Whether or not currently published criteria conform to these standards will be left to the judgment of the individual reader!)

Thus, if properly prepared, *air quality criteria* should be useful to legislators, policy-makers, and citizens as scientifically defensible, unbiased sources of information concerning *what is now known* about the relationship between atmospheric concentrations of pollutants and a broad spectrum of effects. They can serve as a focus for continuing scientific inquiry aimed both at those substances for which *criteria* have not yet been published as well as at those for which *criteria* have already been issued. Clearly, currently published *criteria* represent only a first effort and many gaps in scientific understanding of pollution effects remain still to be filled in.

State of Technology Documents

For each substance for which *air quality criteria* have been published, the Clean Air Act of 1970 also charges the Administrator of EPA with responsibility for issuing [11]:

> . . . to the states and appropriate air pollution control agencies information on air pollution control techniques, which information shall include data relating to the technology and costs of emission control. Such information shall include such data as are available on available technology and alternative methods of prevention and control of air pollution. Such information shall also include data on alternative fuels, processes, and operating methods which will result in elimination of (sic) significant reduction of emissions.

Like criteria, these documents clearly are intended to be unbiased, objective, and scientifically defensible documents concerning the state of the "control arts" and the costs associated with alternative source control methods. Thus, when coupled with other locally developed information, the availability of both *criteria* and *state of technology* documents should provide decision-makers with such information as may be available to assist them in weighing the costs and benefits associated with alternative air quality objectives and with alternative strategies for achieving such objectives.

Air Quality Standards

Air quality standards are unabashedly presumed to be the product of policy and political considerations. Unlike *air quality criteria,* which are descriptive or predictive, "air quality *standards are prescriptive*—they prescribe pollutant levels that cannot legally be exceeded during a specific time period in a specific geographic area" [3].

Within the United States, *air quality standards* probably had their genesis in actions taken during the 1959 session of the California legislature. By that date, photochemical smog conditions in southern California had a fifteen year history during which aggressive and highly sophisticated local control efforts had been unable to produce desired improvements in air quality. Plagued by uncontrolled hydrocarbon and other emissions from motor vehicles, southern California officials had been pressing for state action to curb vehicle contaminant losses. Also, each severe smog attack in Los Angeles produced new waves of public concern over the relationship between such conditions and possible health disturbances. Clearly some new action was necessary, but of what type? Automobile manufacturers were arguing that control systems could not be delivered to the California automobile population unless California officials could provide the performance specifications for such systems. What levels of pollution loss from the motor vehicle population would be acceptable? To answer this question, one obviously had to answer still larger questions: What level of smog would be tolerable to the population? What ambient levels of specific contaminants led to varying intensities and frequencies of eye irritation, visibility impairment, and other conditions?

On February 10, 1959 newly elected Governor Edmund G. Brown delivered a special message to the California legislature. He said [22]:

> The time has now come for California to take the lead and establish standards for the purity of the air. Standards for safe air will give local control officials and health officers a measuring stick for smog. It is essential that we know what level of pollution threatens death or illness, or impairs the health of our people. Unless we establish these guideposts, we risk our happiness, and indeed our lives.

Before the session had drawn to a close the legislature enacted an amendment to the Health and Safety Code which stated [22]:

> The State Department of Public Health shall, before February 1, 1960, develop and publish standards for the quality of the air of this state. The standards shall be so developed as to reflect the relationship between the intensity and composition of air pollution and the health, illness, including irritation to the senses, and death of human beings, as well as damage to vegetation and interference with visibility.

Eight years later the federal Air Quality Act of 1967 imposed on the several states of the union the requirement that they "adopt, after public hearings, ambient air quality standards applicable to any designated air quality control region or portions thereof within" their states [2]. Other sections of the statute required the Secretary of the Department of Health, Education, and Welfare to fix the boundaries of "air quality control regions," and to publish *air quality criteria* and state of technology documents. The timetable for state adoption of *air quality standards* was based on the dates on which the Secretary issued *air quality criteria.*

Also required of each state by the 1967 act and discussed below, was "a plan for the implementation, maintenance and enforcement of such standards of air quality" [2] as the state may have adopted.

During the public hearings called in the several states to adopt *air quality standards,* literally thousands of citizens jammed the hearing rooms. Prompted and prepared through workshops and citizen education programs conducted by the Tuberculosis and Respiratory Diseases Association, groups affiliated with the Conservation Foundation, and other bodies, citizens spoke out across the country in tones not heard before in air pollution hearings. Nationally, the "environmental movement" was born and gathered momentum. Pressures mounted while states and control regions wended their ways through the long, tedious, and complicated procedures and timetables fixed in the act. When 1970 rolled around, many states had adopted *air quality standards,* but few had yet submitted *implementation plans.* In Congress and elsewhere fears mounted that state and local air pollution control efforts were not proceeding quickly enough.

The result? In the closing days of 1970 the U.S. Congress adopted the Clean Air Act of 1970. A principal provision of the act cancelled the timetable provided in the earlier statute and required the Administrator of the newly formed Environmental Protection Agency to, himself, develop and publish "regulations prescribing a *national primary ambient air quality standard* and a *national secondary ambient air quality standard* for each air pollutant for which air quality criteria have been issued" [11]. Air quality criteria developed after the enactment of the statute are to be published simultaneously with "proposed national primary and secondary ambient air quality standards" [11] for the pollutant dealt with in the criteria.

In the language of the statute, *primary standards* are those "which in the judgment of the Administrator, based on such [air quality] criteria and allowing an adequate margin of safety, are requisite to protect the public health" [11].

In contrast, a *secondary standard* is one "which in the judgment of the Administrator, based on such [air quality] criteria is requisite to protect the public welfare from any known or anticipated adverse effects associated with the presence of such air pollutant in the ambient air" [11].

Thus, *primary standards* are those which protect against health effects while *secondary standards* are those which protect against other, but less important, effects. Neither, apparently, pre-empts the right of the states to adopt more stringent air quality standards than those of the federal government.

State Implementation Plans

In the 1967 Air Quality Act, states were required to develop and submit to the federal government plans for the implementation, maintenance, and enforcement of such air quality standards as may have been adopted by the states and approved by the Secretary of Health, Education, and Welfare. In contrast, the 1970 Clean Air Act requires the states, after appropriate public hearings, to develop *"Implementation Plans"* which provide for the implementation, maintenance, and enforcement of *national primary and secondary air quality standards.* Plans for implementation of *primary standards* were to be built around a time schedule providing for the achievement of the primary standard within a three-year period following federal approval of the plan, and within a "reasonable time" for national *secondary standards.* No requirement is imposed on the states for development and federal approval of implementation plans calculated to achieve state-developed air quality standards which may be more rigorous than those of the federal government.

The statute requires that *state implementation plans* specify the control measures the state will employ to achieve the primary and secondary standards within the established time schedule, including emission standards, compliance schedules, land-use and transportation controls, procedures for preconstruction review of the location of new or modified sources of pollution, and such periodic inspection and testing of motor vehicles as may be necessary. The plan also is to define appropriate systems of air monitoring, measures for dealing with "emergency air pollution episodes," and arrangements for such "inter-governmental cooperation" as may be necessary to fully effectuate the plan. It shall include provisions requiring owners or operators of stationary pollution sources to install equipment for monitoring their emissions, to report periodically "on the nature and amounts of such emissions," and provisions providing that such reports, when submitted, "shall be available at reasonable times for public inspection."

Detailed regulations governing the preparation, adoption, and submittal of implementation plans were published on August 14, 1971. These regulations encouraged the states "to identify alternative control strategies, as well as the costs and benefits of each such alternative, for attainment and maintenance of the national standards" [17]. They instructed states to develop emission limitations which take into account both present and projected levels of population, industrial activity, motor vehicle traffic, and other factors which may cause or contribute to increases in pollution emissions.

Under the statute, states wishing to do so could submit plans for "implementing and enforcing" federal emission standards for "new stationary sources" and for "hazardous air pollutants." (Under the 1970 Act, the Administrator of the Environmental Protection Agency was given authority to promulgate such nationally enforceable emission standards.) If such plans are approved by the Administrator, responsibility for enforcing national emission standards may be delegated to the states, subject to stand-by federal enforcement authority in the event of inaction by state and local officials.

Three incentives to state compliance with statutory requirements for development of state implementation plans are implicit in the 1970 Act: (1) federal financial grants for support of state and local air pollution control and planning programs are limited to "air pollution control agencies which have substantial responsibilities for carrying out" implementation plans; (2) failure to develop such a plan will result in action by the Administrator to develop and implement such a plan under federal authority; (3) development of an adequate plan dealing with emissions from "new stationary sources" and with "hazardous air pollutants" may result in delegation of responsibility for enforcing national emission standards to the state.

Source Control Plans

Since that now-forgotten day in history when the first person or governmental unit took the first step to legally compel someone to abate an air pollution emission, the ultimate objective has been to stimulate the owners or operators of air pollution sources to develop and implement plans for the control of their own pollution emissions.

It should be manifestly clear that the control of air pollution emissions is a completely different "beast" than control of other socially rejected acts such as driving a vehicle too fast. In the latter case instant compliance with social prescriptions can be obtained; an alteration in the intent and behavior of the driver will bring about a solution to the problem. Not so with air pollution emissions. Given the best of intentions, the operator of an air-polluting source can rarely comply with socially desired emission limitations simply by altering his routine daily behavior.

Instead, the reduction or elimination of an unwanted pollution emission usually involves one or more modifications in the physical system which gives rise to the emission. Unless fuel substitutions or modification in product input will, alone, solve the problem some physical alteration in the plant usually must be made; basic production equipment must be modified; effluent control equipment must be designed, purchased, installed, and tested. In short, a plan for the solution of the problem must be developed and implemented in accordance with an acceptable time schedule.

This is the end objective of legislative enactments, enforcement systems, plant inspections, cajoling, litigation, injunctions, and court-imposed fines and penalties—to stimulate the owner or operator of a source to develop and implement an acceptable plan for an appropriate reduction in his pollution emissions.

Explicit recognition of the need for such plans and, where possible, their approval by appropriate government officers, is provided in federal regulations concerning development of *state implementation plans*. The regulations authorize state compliance schedules built around individually negotiated "compliance schedules" for specific sources, each of which is presumably based on an approved *source control plan*. Requirements for state programs to deal with "emergency air pollution episodes" specifically state that, for large sources, the episode contingency plan "shall include, or provide for preparation of, a specific legally enforceable emission control action program and shall show that the owner or operator of such stationary source has been notified of the requirements of such emission control action program" [17].

Similarly, requirements for state procedures to provide for the review of plans for construction of new sources or modification of existing sources specifically state [17]:

> Such procedures shall provide for the submission, by the owner or operator of a new stationary source, or existing source which is to be modified, of such information on the nature and amounts of emissions, locations, design, construction, and operation of such sources as may be necessary to permit the state agency to make the determinations required. . . .

Also, of course, the new federal statute requires individual sources to monitor and report on their emissions to appropriate state or local authorities. This suggests some prior knowledge on the part of these authorities of details concerning the plant, its points of pollution emission, and the points at which monitoring equipment has been installed.

The minimal information necessary to the development of a *source control plan* is generally comparable to that which federal regulations now require state agencies to maintain for selected point sources (see Table 11.1).

Thus, in respect to *existing pollution conditions* within the United States the 1970 Clean Air Act provides procedures, policies, and assistance congenial to the development of rational decision processes in the field of air quality management. Scientific information concerning relationships between ambient pollution burdens and specific effects is provided; minimal national air quality standards are stated and are categorized as to those which are related to health and those which are related to other effects; information on the cost and effectiveness of alternative source control tech-

TABLE 11.1

Information Requirements for Development of a Point Source Control Plan*

I. GENERAL SOURCE INFORMATION

A. Establishment name and address.

B. Person to contact on air pollution matters and telephone number.

C. Operating schedule:
1. Percent of annual production by season.
2. Days of week normally in operation.
3. Shifts or hours of day normally in operation.
4. Number of days per year in operation.

D. Year in which data are recorded.

E. Future activities, if available (e.g., addition of new or expansion of existing facilities, changes in production rate, installation of control equipment, phasing out of equipment, fuel change, etc.).

F. Map or general layout of large complex plants showing locations of various facilities, if available.

II. FUEL COMBUSTION

A. Number of boilers.

B. Type of fuel burning equipment for each boiler.

C. Rated and/or maximum capacity of each boiler, 10^6 B.t.u./hr. or kcal/hr.

D. Types of fuel burned, quantities, and characteristics:
1. Type of each fuel used and place of origin.
2. Maximum and average quantity per hour.
3. Quantity per year.
4. Sulfur content (as received), percent.
5. Ash content (as received), percent.
6. Heat content (as received), B.t.u. or kcal/unit of measure.
7. Estimate of future usage, if available.

E. Percent used for space heating and process heat.

F. Air pollution control equipment (existing and proposed):
1. Type.
2. Collection efficiency (design and actual), percent.

G. Stack data:
1. List stacks by boilers served.
2. Location of stacks by grid coordinates (Universal Transverse Mercator, UTM, or equivalent).
3. Stack height, feet or meters.
4. Stack diameter (inside, top), feet or meters.
5. Exit gas temperature, °F. or °C.
6. Exit gas velocity, feet/sec. or meters/sec.

H. Emission data:
1. Based on emission factors.
2. Estimate of emissions by the source.
3. Results of any stack tests conducted.

III. MANUFACTURING ACTIVITIES (PROCESS LOSSES)

A. Process name or description of each product.

B. Quantity of raw materials used and handled for each product, maximum quantity per hour, and average quantity per year.

C. Quantity of each product manufactured, maximum quantity per hour, and average quantity per year.

D. Description of annual, seasonal, monthly, weekly, and daily operating cycle including downtime for maintenance and repairs.

TABLE 11.1 (*continued*)

E. Air pollution control equipment in use (existing and proposed):
  1. Type.
  2. Collection efficiency (design and actual), percent.
F. Stack data:
  1. List of stacks by equipment served.
  2. Location of stacks by grid location (UTM or equivalent).
  3. Stack height, feet or meters.
  4. Stack diameter (inside, top), feet or meters.
  5. Exit gas temperature, °F. or °C.
  6. Exit gas velocity, feet/sec. or meters/sec.
G. Emission data:
  1. Based on emission factors.
  2. Estimate of emissions by the source.
  3. Results of any stack tests conducted.

IV. SOLID WASTE DISPOSAL

A. Amount and description of solid waste generated, quantity per year.
B. Percent of total that is combustible.
C. Method of disposal (on-site or off-site).
D. Description of on-site disposal method, if applicable (incineration, open burning, landfill, etc.) including maximum quantities disposed per hour and average quantities disposed per year and actual operating schedule.
  1. Location of the source by a grid system (UTM or equivalent).
  2. If method of disposal is by an incinerator, include the following information:
    a. Auxiliary fuel used.
    b. Air pollution control equipment (existing and proposed):
      (1) Type.
      (2) Collection efficiency (actual and design), percent.
    c. Stack data:
      (1) List stacks by furnaces served.
      (2) Stack height, feet or meters.
      (3) Stack diameter (inside, top), feet or meters.
      (4) Exit gas temperature, °F. or °C.
      (5) Exit gas velocity, feet/sec. or meters/sec.
      (6) Exit gas moisture content, percent if available.
  3. Emission data:
    a. Based on emission factors.
    b. Estimate of emissions by the source.
    c. Results of any stack tests conducted.

* Appendix E, "Requirements for Preparation, Adoption, and Submittal of Implementation Plans," 36 Federal Register 158, Part II, p. 15499 (Aug. 14, 1971).

niques is provided to state and local authorities; procedures are outlined for the development of information concerning pollution sources, emissions, and ambient pollution burdens within an area; and requirements are imposed on the states for the synthesis of this information into a coherent and comprehensive plan for the achievement of stipulated air quality objectives. Requirements are imposed on the states to include in such plans proposed emission limitations to be enforced in respect to specific sources, and the plans further must show the relationship of such limitations to probable future ambient levels in terms of the projected growth of the area and the probable effectiveness of the control measures in reducing ambient pollution burdens.

Moreover, implicit requirements are suggested for the development of source control plans for individual sources and for the reporting of information concerning individual sources. All of this is to be accomplished in the "public eye," through plans and documents to which the public has full access and under conditions where maximum opportunity is provided for nongovernmental review of area needs, objectives, probable present and future conditions, and proposed control strategies.

Decision Models and Cost Analysis

A burgeoning literature now is available concerning decision-assisting tools of possible utility to air quality planners and managers. An excellent presentation of decision problems and requirements in air quality management has been presented by Schwartz [41], and approaches to the application of cost benefit and cost effectiveness analysis to air pollution control have been suggested by both Schwartz and Kohn [41, 28]. A summary exposition of systems analysis theory and practice in terms of air pollution control requirements has been presented by Fleischer, Atkisson, and Kreditor [20].

Prevention of Pollution Problems

Another attribute of the emerging system for the management of air quality within the United States is its orientation to prevention of pollution problems. In an earlier chapter we saw how population growth, metropolitanization, and rising consumer demands for goods and services can lead to rapid escalations in pollution emissions within a community. Under such circumstances control agencies must "run fast to stay in the same place."

However, the system now emerging is addressed solidly to the problems posed by continuing growth.

Under the 1970 Clean Air Act the Administrator of the federal Environmental Protection Agency has been given authority to establish and enforce emission standards for "new stationary sources." Under the statute he is obligated to: (1) publish a list of categories of stationary sources which "contribute significantly to air pollution which causes or contributes to the endangerment of public health or welfare" [11]; (2) issue regulations establishing "standards of performance" to govern the modification of existing sources or the construction of new sources included on such list; (3) develop such standards on the basis of "the best system of emission reduction . . . (taking into account the cost of achieving such reduction)" [11]; (4) establish procedures and issue regulations appropriate to the enforcement of these standards. The first five performance standards issued by EPA are shown in Table 7.1.

In addition, regulations for *implementation* plans require the states to develop procedures for review of the location of new pollution sources and to take such measures as are necessary to prevent the development of undesirable concentrations of pollution sources in limited areas. Federal authorities also have recommended that states adopt (or authorize their local governments to adopt) laws and regulations requiring that permits be obtained from appropriate air pollution authorities prior to the construction and operation of equipment having a potential to pollute the air.

A system of this sort has been in operation in Los Angeles County since the birth of the County Air Pollution Control District and has been called the "big muscle" in the Los Angeles control program [44].

Under this system, related units of equipment (such as a series of brass furnaces whose emissions are vented through a common collection system) are grouped together into a coherent equipment system for which a permit is required. Prior to the construction or modification of such a system, the plans and specifications must be reviewed by district engineers. If approved, construction may proceed. If not, continuation of construction results in criminal prosecution and, if necessary, injunctive action. Following construction, the operation of the system is tested and, if found in compliance with all air pollution laws, rules, and regulations, a permit to operate is issued. The operating permit may, itself, contain specific conditions which must be honored if the permit is to be valid. For example, such a condition might specify the throughput of water required in a scrubber during various operating conditions.

Aside from the control this system provides in terms of policing the construction of new sources, it also is a powerful enforcement tool. Violation of the terms of an operating permit can result in criminal prosecution, with each day of violation constituting a separate offense [27]. Moreover, upon proper cause, the Air Pollution Control Officer may revoke the operating permit and each day of operation without the permit also is an offense subject to criminal prosecution.

Control of the Social Origins of Pollution Conditions

Although no coherent, well-integrated, and properly funded federal effort yet has been undertaken to deal with the social origins of pollution conditions discussed in Chapter 10, a beginning of such effort is emerging.

In terms of limiting population growth, several groups have been organized to promote reductions in family size and the achievement of a stable national population. These include Zero Population Growth, Inc.; Planned Parenthood; and the Association for Voluntary Sterilization. In

1970 the U.S. Congress enacted the Family Planning Services and Population Research Act which authorized the creation of an Office of Population Affairs in the Department of Health, Education, and Welfare and expansion of programs to limit population growth. Other measures intended to promote smaller family sizes have been incorporated into legislative proposals at both the federal and state level.

Pollution emissions occasioned by fuel usage patterns are attacked in the 1970 Clean Air Act through a charge to the Environmental Protection Agency to "give special emphasis to research and development into new and improved methods, having industry-wide application, for the prevention and control of air pollution resulting from the combustion of fuels" [11]. The agency was specifically charged to pursue projects directed at controlling the combustion by-products of fuel, removal of potential air pollutants from fuel prior to combustion, improving the efficiency of fuel combustion techniques, and producing "synthetic or new fuels which, when used, result in decreased atmospheric emissions" [11].

Under its own statute, the Atomic Energy Commission is proceeding with development of fast-breeding reactors and is considering future efforts pointed toward the development of fusion reactors. Also, during 1971 the President announced a program to develop a "fuels policy and program" for the nation.

The economic disincentives to pollution control have been partially attacked through those provisions of the 1970 Clean Air Act which authorize the enactment and enforcement of "performance standards" for "new stationary pollution sources." These performance standards eliminate any competitive advantage "no control" communities might enjoy in attracting new industrial construction. Federal authorities have also considered the feasibility of "emission fees" for sulfur oxide emissions in order to "internalize" some of the costs now borne by the national community in terms of pollution damage arising from emission of sulfur compounds. The concept has apparently been abandoned, at least for the time being, although some economists believe that "emission fees" might stimulate more rapid action to control emissions as a result of the cost incentives the system would provide. It is argued that control delays occasioned by prolonged litigation contesting control activity would not be worthwhile.

Similarly, comparatively modest federal funding has been provided to assist in the planning and construction of mass transportation systems.

A legislative proposal has been developed by the executive branch of the federal government for a new program of "revenue sharing" with state and local governments. Even though modest fractions of federal income tax receipts are to be shared with local units in the initial phases of the proposed program, the proposal nevertheless is a pioneering attempt to solve the fiscal crises of metropolitan communities and, thereby, to enable them better to meet some of the physical and service demands now contributing to pollution and other problems.

Land Use and Site Location Controls

In theory, there are at least six purposes which *might* be served through the application of land-use controls to the field of air quality management: (1) to provide *complete* protection from pollution of all sorts to highly desired environments, such as scenic shorelines, recreational areas, or agricultural areas devoted to the growing of highly sensitive crops; (2) to prevent the entry of a particularly large or noxious source of pollution emissions into certain types of land areas, such as those occupied by metropolitan communities; (3) to prevent the development of significant "area" sources of pollution emission as a consequence of the proliferation of "controlled" sources within a limited land area; (4) to prevent the mixing of incompatible land uses within a limited area, such as the intermingling of sensitive air pollution receptors (nursing home patients) and sources of highly noxious pollutants; (5) to influence the ultimate design configuration of a community so as to minimize those community conditions which lead to high yields of unwanted pollution emissions, such as the relationship between distance to work and likely exhaust emissions from motor vehicles; (6) to provide areas remote from populous or highly valued environmental settings for high-environment impacting or hazardous operations (viz., the absolute reverse of a "smokeless zone").

Whether or not each of these possible applications of land-use controls to air quality management has been seriously considered by appropriate authorities is not evidenced by the available literature. Indeed, it would appear that only scanty attention has been given to this entire subject, except on a broadly philosophic level.

One otherwise excellent and standard text in the field of urban land-use planning contains no mention of air pollution and offers no edification concerning the conduct of pollution-related site location studies [10]. However, a bibliography of planning and zoning literature related to air pollution has been assembled and published by Pelle [35]; the legal aspects of land-use planning and zoning have been examined by Mandelker [30]; Holland [24] has examined some applications of zoning to air pollution situations; while Taylor and Hasegawa [42] have examined some possibilities for the control of air pollution through site selection and zoning.

Three facts seem clear: (1) there is new and developing interest in this entire subject; (2) in terms of existing statutory authorizations, there are few administrative agencies in the country which are able to deal with many of the six purposes outlined; (3) much more research must be performed and a great deal more hard data acquired if land-use controls are to have much early utility to the field of air quality management.

Among the new developments and ideas, a few are of particular interest at this time.

First, of course, is the extent to which activities unleashed by the 1970 Clean Air Act seem pointed toward the utilization of land-use controls in comprehensive air quality management plans. References to controls over location of sources and to the utilization of land-use controls lace through the current regulations.

The second is the substantial policy interest which now is focusing on formal means for controlling the site location of particularly large pollution sources, with special emphasis on both nuclear and fossil fuel-fired power plants.

In 1970 only thirty-eight steam power generating units with a capacity of over 600,000 kilowatts were located within the United States. Five of these were nuclear plants and the balance were fossil fueled. However, the Federal Power Commission has projected that by 1980 this number will jump to 255 of which 131 will be fossil fueled. By 1990 the total number will increase to 631 of which an estimated 266 will be fired with fossil fuels. Under the National Environmental Policy Act, which created the new Council of Environmental Quality, federal agencies must prepare "environmental impact" statements for projects and programs in which they are involved. Those dealing with power plants—particularly nuclear plants —are being reviewed and attacked by environmental groups [14, 15].

In 1968, the Office of Science and Technology established an interdepartmental committee on power plant siting [14]. Two reports now have been issued by the group, the last of which recommended new federal legislation to deal with siting problems [14]. The administration has introduced such legislation and congressional action on the proposals is pending (1972). Bills are under consideration which would require utilities to develop and publicly disclose long-term plans for plant expansion, including alternative sites under consideration, public participation in hearings related to such sites, and provisions requiring states to establish power plant siting boards or commissions. The state of Washington already has established such a body [14, 15].

The third matter of particular interest is the extent to which land-use planning now has become an important issue at state and federal levels of government, where land-use matters traditionally have been ignored. Legislation is pending in Congress which would stimulate and financially support state-level planning activities. Vermont has adopted a comprehensive state-level planning law which imposes severe restrictions on entry of pollution-causing sources into most areas of the state. Viewed as a whole, these developments could have significant implications to air quality management programs.

The last idea or issue is one which has been suggested and discussed by Seymour Schwartz—the possibility and desirability of developing *area emission standards* as additional underpinning to community control programs. Under Schwartz's scheme, metropolitan communities would be divided into "subregions" for each of which an air pollution emission budget would be established. When the limits of the budget were met, no further growth would be permitted, or tighter emission standards would be invoked [41].

Citizen Participation

One of the most distinctive attributes of the emerging system for air quality management is the extent to which it provides and promotes expanded opportunities for citizen participation in air pollution control decisions.

New federal statutes have built in opportunities for citizen participation at three stages: (1) development of air quality standards; (2) development of implementation plans; (3) enforcement of air pollution standards.

A key feature of the 1967 Clean Air Act was its insistence on state-conducted public hearings prior to adoption of air quality standards and a showing that the state had considered citizen views in making its final standards decisions. This requirement has been extended by the 1970 Act to include procedures related to the adoption of implementation plans. So committed to this view are federal authorities that the Environmental Protection Agency made substantial financial grants to citizen and conservation groups to conduct educational workshops to prepare citizens for participation in these hearings.

The concept of citizen participation now has been dramatically extended by the Clean Air Act and other legislative enactments into the area of litigation. Section 404 of the 1970 act gives citizens the standing to sue in the federal courts to secure the enforcement of the several terms of the act. The section states: "Any person may commence a civil action on his own behalf—(1) against any person (including (i) the United States, and (ii) any other governmental instrumentality or agency to the extent permitted by the Eleventh Amendment to the Constitution) who is alleged to be in violation of (A) an emission standard or limitation under this Act or (B) an order issued by the Administrator or a state with respect to such a standard or limitation, or (2) against the Administrator where there is alleged a failure of the Administrator to perform any act or duty under this Act which is not discretionary with the Administrator" [11].

The theory behind citizen suits in environmental matters has been covered in two recent books [4, 38], one by the principal architect of the much-discussed Michigan act which declared the environment a "public trust" and gave citizens the standing to sue to bring about control of environment-impairing activities.

Inter-governmental Relations

After viewing the American "federal" system of government, an anonymous wit once commented: "The cities have all the problems, the states have all the power, and the federal government has all the money."

To a considerable extent, the charge is true, and the emerging system for air quality management can be viewed as an attempt to more appropriately distribute both *power* and *money* among the several levels of government. In the new system, the federal government has assumed a larger share of responsibility for funding control programs at state and local levels. Considerable functional specialization of responsibilities has occurred at the several levels.

Thus, with but minor exceptions, the federal government now has *exclusive* responsibility for dealing with pollution emissions from aircraft and from motor vehicles. It has *primary* responsibility for basic and applied research and for development of air quality criteria. It also appears to have primary responsibility for matters dealing with pollution derived from fuels.

All three levels share responsibility for enforcement of air pollution control standards and limitations, though this seems to be a *primary* responsibility of local government. Similarly, such direct operational functions as inspection of sources, air quality surveillance, and monitoring of pollution effects seems to be a primary responsibility of local government, with the responsibility being shared by the states.

On the other hand, quite unlike the historic circumstances in California and other pioneering states, there now seems to be a drift away from vesting local governments with primary responsibility for development of air quality and source emission standards. These now appear to be primary responsibilities of the states, with some sharing of this power by federal and local governments.

As was observed in an earlier chapter, cities and counties are little more than legal creatures of the states and have only such powers as state legislatures choose to confer upon them. In many states, such as Texas, the legislatures have chosen to confer little law-making power in the field of air pollution control upon local governments. Protests over the difficulties associated with dealing with distant state agencies and boards now seem to be mounting, however, and some additional realignment in authority distribution may result.

In metropolitan areas the demands for more effective patterns of intergovernmental relations are growing more acute as problems such as air pollution are tackled. Although special districts, such as those in California, still are not the primary mode for achieving concerted control efforts within multi-city and multi-county metropolitan areas, techniques such as this may well become more common in the future (see Chapter 12, Organization and Operations of Air Pollution Agencies).

Federal Assistance

Continuing the pattern set in the Kuchel-Capehart Air Pollution Act of 1955, the 1970 Clean Air Act provides for a broad spectrum of federal assistance to state and local control programs. Monies are provided for direct financial support of state and local planning and control operations. Substantial sums remain authorized for training and manpower development programs, and a well-developed intramural training activity now has been developed in support of local operations. EPA now provides new services intended to facilitate the recruitment and placement of qualified personnel within the air pollution control field, and, under some conditions, persons may be hired by the federal government and assigned to work within state and local agencies.

REFERENCES

1. Air Conservation Commission, *Air Conservation,* Washington, D.C.: American Association for the Advancement of Science, 212–213 (1965).
2. *Air Quality Act of 1967,* P.L. 90–148 (November 21, 1967).
3. Atkisson, A. A., and R. S. Gaines, *Development of Air Quality Standards,* Columbus, Ohio: Charles E. Merrill Publishing Company, 1970, pp. 213–216.
4. Baldwin, M. F., and J. K. Page, Jr., eds., *Law and the Environment,* New York: Walker and Company (1970).
5. *Ballentine v. Nester,* 350 MO. 58, 164 S.W. 2d. 378 (1942).
6. *Berman v. Parker,* 348 U.S. 26, 99 L. Ed. 27, 75 Sup. Ct. 98 (1954).
7. *Board of Health v. New York Central R.R.,* 10 N.J. 294, 306, 90 A. 2d 729, 735 (1952).
8. *Bove v. Donner-Hanna Coke Corporation,* 142 Misc 329, 254 N.Y.S. 403 (1931), aff'd. mem., 236 App. Div. 37, 258 N.Y.S. 229 (1932).
9. *Brock v. Superior Court,* 109 Cal. App. 2d 594, 603, 241 P. 2d 283 (1952).
10. Chapin, F. S., Jr., *Urban Land Use Planning.* Urbana: University of Illinois Press (1965).
11. *Clean Air Act of 1970,* P.L. 91–604 (Dec. 31, 1970).
12. *Cogswell v. New York, New Haven and Hartford R.R.,* 103 N.Y. 10, 13–14 (1886).

13. *Chiropractors Ass'n. of New York v. Hilleboe,* 12 N.Y. 2d 114, 237 N.Y.S. 289, 291, 187 N.E. 2d 756 (1962).

14. Clark, T. B., "Energy Report/Legislation on Power Plant Siting Seeks to Speed Resolution of Environmental Disputes," *National Journal,* 1785–1795 (August 28, 1971).

15. Clark, T. B., "Energy Report/Environmentalists Divide on 'one-stop' Plan for Decisions on Siting Electric Power Facilities," *National Journal,* 1852–1859 (Sept. 4, 1971).

16. *Duke Molnar Wholesale Liquor Co. v. Martin,* 180 Cal. App. 2d 873, 4 Cal Rptr 904 (1960).

17. Environmental Protection Agency, "Requirements for Preparation, Adoption and Submittal of Implementation Plans," *Federal Register,* **36,** No. 158, Part II 15486–15506 (August 14, 1971).

18. *Environmental Quality Improvement Act of 1970,* P.L. 91–224 (April 3, 1970).

19. *First Industrial Loan Co. v. Daugherty,* 26 Cal. 2d 545, 159 P. 2d 921 (1945).

20. Fleischer, G. A., A. Kreditor, and A. A. Atkisson, *Relevance of the Systems Concept to the Development of Motor Vehicle Pollution Control Policy,* Los Angeles: Institute of Urban Ecology, University of Southern California, August, 1968 (mimeo.).

21. *Gaylord v. City of Pasadena,* 175 Cal. 433, 166 P. 348 (1917).

22. Goldsmith, J., "Evolution of Air Quality Criteria and Standards," *In:* Atkisson, A. A., and Richard Gaines, op. cit., pp. 1–18.

23. Hanlon, J. J., *Principles of Public Health Administration* (Fifth Edition), St. Louis: The C. V. Mosby Company, 160–186 (1969).

24. Holland, W. D., et al, "Industrial Zoning as a Means of Controlling Area Source Air Pollution," *J. Air Poll. Control Assoc.,* **10,** No. 2 (April, 1960).

25. *In re Junqua,* 10 Cal. App. 602, 605, 103 Pac. 159.

26. *Justesen's Food Stores v. City of Tulare,* 43 Cal. App. 2d 616, 621, III P. 2d 424 (1941).

27. Kennedy, H. W., "The Mechanics of Legislative and Regulatory Action," *In:* U.S. Department of Health, Education, and Welfare. *Proceedings: National Conference on Air Pollution.* Washington, D.C.: U.S. Govt. Printing Office, 306–314 (1963).

28. Kohn, R. E., "Abatement Strategy and Air Quality Standards," *In:* Atkisson, A. A., and Richard Gaines, op. cit., pp. 103–124.

29. *Leone v. Paris,* 43 Misc. 2d 442, 251 N.Y.S. 2d 277 (1964), *modified,* 25 App. Div. 2d 508, 261 N.Y.S. 2d 656 (1965).

30. Mandelker, D. R., *Managing Our Urban Environment: Cases, Text and Problems,* New York: The Bobbs-Merrill Company, Inc. (1963).

31. *Miami County v. Dayton,* 92 0.5.215.

32. *National Environmental Policy Act of 1969,* P.L. 91–190 (January 1, 1970).

33. *Nelson v. City of Minneapolis,* 112 Minn. 16, 18, 127 N.W. 445, 447 (1910).

34. *Northwestern Laundry v. Des Moines,* 239 U.S. 486, 491–492 (1916).

35. Pelle, W. J., *Annotated Bibliography on the Planning Aspects of Air Pollution Control,* Study for Northwestern Illinois Planning Commission and U.S. Public Health Service, March, 1965 (mimeo.).

36. *People v. Savage,* 1 Misc. 2d 337, 148 N.Y.S. 2d 191, aff'd. mem., 309 N.Y. 941, 32 N.E. 2d (1955).

37. Pollack, L. W., "Legal Boundaries of Air Pollution Control—State and Legal Legislative Purpose and Techniques," *Law and Contemporary Problems* (School of Law, Duke University), **33,** No. 2 (Spring, 1968).

38. Sax, J. L., *Defending the Environment: A Strategy for Citizen Action,* New York: Alfred A. Knopf (1971).

39. *Schact v. City of New York,* 40 N.Y. Misc. 2d 303, 243 N.Y.S. 2d 272, 279 (1963).

40. *Schlotfelt v. Vinton Farmers Supply Co.,* 252 Iowa, 109 N.W. 2d 695 (1961).

41. Schwartz, S., *Models for Decision-Making in Air Quality Management,* Los Angeles: University of Southern California, Dept. of Industrial and Systems Engineering, Technical Report 70-4, 103 (August 1970).

42. Taylor, J. R., and A. Hasegawa, "Control of Air Pollution by Site Selection and Zoning," World Health Organization (1961).

43. *United States v. Grimaud,* 220 U.S. 506 (1911).

44. Weisburd, M. I., *Air Pollution Control Field Operations Manual.* Washington, D.C.: U.S. Govt. Printing Office (1962) (Public Health Service Publication No. 937).

12

ORGANIZATION AND OPERATIONS OF AIR POLLUTION AGENCIES

Activity requirements associated with the web of decisions and actions necessary to modern air quality management programs have led to the development of numerous air pollution and environmental control agencies at federal, state, and local levels of government. By 1970, more than two hundred state, local, and regional air pollution agencies were in operation within the United States, and their combined need for technical personnel was estimated at approximately 8,000 positions [13].

Recognition that air quality management functions are sufficiently large, complex, and demanding as to warrant the creation of special administrative agencies to engage in them has been slow to develop within the United States. As recently as the mid 1950's, the federal government had not yet evolved an organizational unit specializing in air pollution matters, only 82 state and local air pollution agencies were in operation, only 76 of these were spending $5,000 or more per year, and a scant 13 were spending as much as twenty cents per capita per year on air pollution programs [19].

The first air pollution agencies in the nation were developed by local units of government. One of these, the Los Angeles County Air Pollution Control District, was created in 1948 and, until the late 1950's, had a larger staff and annual operating budget than the federal program. Until comparatively recently this single agency accounted for the bulk of all state and local expenditures for air pollution programs.

ORGANIZATION FOR
AIR QUALITY MANAGEMENT

Air quality management programs are carried out at several levels of government, as mentioned above. The specific organizational structure should depend on the various functions performed. In practice, however, the organizational structure of air pollution agencies or air quality management agencies (the two terms are used interchangeably here) reflects the political climate at the time the agency was formed and the changes in political philosophy as the agency developed. Seldom, then, are two agencies alike. One may be certain that the ultimate organization has not yet been formed.

In the United States there is a vast gulf between the federal establishment and local agencies, whether they be state, regional, county, or municipal. Accordingly, in the following discussion, the two levels, federal and local are considered separately.

The Federal Agency

Current federal air pollution activities are authorized by the Clean Air Act of 1970, and are conducted by an Environmental Protection Agency created in late 1970. The agency is the federal body vested with primary responsibility for conducting federally sponsored activities concerned with the direct regulation of environmental pollution, including noise, solid wastes, use of pesticides, and air and water pollution. The agency is one of three federal organizations concerned primarily with environmental matters, the other two being the Council on Environmental Quality and the National Oceanic and Atmospheric Administration [20]. Other federal agencies, such as the U.S. Bureau of Mines and the Department of Health, Education, and Welfare also engage in research and developmental activities related to air pollution matters.

Headed by a presidentially appointed Administrator, the Environmental Protection Agency reports directly to the Executive Office of the President and now consists of ten well-staffed regional offices, each with a regional air pollution director, plus a large national staff which is organized into several functionally specialized organizational entities headed by assistant administrators. These units are concerned with such matters as research and development, enforcement and abatement, public affairs, and media. The latter consists of a series of staffs, each headed by a Deputy Assistant Administrator, one of which is concerned exclusively with air pollution matters.

The Clean Air Act of 1970 also provides for the establishment, within the Environmental Protection Agency, of a sixteen-member Air Quality Advisory Board. The Administrator of the agency serves as chairman of this body, and the other fifteen members are appointed by the President for three-year terms. None of these members may be federal officers or employees. The Board has no regulatory functions and is charged only with the responsibility to "advise and consult with the Administrator on matters of policy relating to the activities and functions of the Administrator under this act and (to) make such recommendations as it deems necessary to the President" [2].

Current federal air pollution activities are traceable to the Kuchel-Capehart Act of 1955 which authorized the Public Health Service of the Department of Health, Education, and Welfare to expend up to $5 million annually for air pollution research, data collection, and technical assistance to state and local governments [1]. In 1960 Congress adopted PL 86-493 which authorized the Department to engage in research related to pollution emissions from motor vehicles [1]. Limited air pollution enforcement authority was given to federal agencies under the terms of the Clean Air Act of 1963. National regulation of air pollution from new motor vehicles was authorized by 1965 amendments to the act, and, in November 1967, Congress adopted the Air Quality Act of 1967 which set in motion a comprehensive federally sponsored attack on air pollution problems. This act was replaced by the Clean Air Act of 1970 which brought the federal government directly into the field of regulating pollution discharges from both transportation and stationary sources of air pollution.

When the Kuchel-Capehart Air Pollution Act of 1955 was adopted the now-abolished Bureau of State Services of the U.S. Public Health Service was designated to administer federal air pollution activities. The Bureau created an air pollution coordinating committee to deal with air pollution matters but no formal air pollution unit was established until a Division of Air Pollution was formed within the Bureau in 1959. Several years later, the division was escalated, first, to the status of a director-headed National Center for Air Pollution Control, then to the status of a commissioner-headed National Air Pollution Control Administration within a newly formed Environmental and Consumer Protection Service of the Department of Health, Education, and Welfare. Finally, in 1970, the Environmental Protection Agency was formed and brought together in a single agency, reporting directly to the President, the range of environmental control activities which earlier had been performed by a number of cabinet-level departments.

Organization of State and Local Air Pollution Agencies

At state, regional, and local levels of government, several major issues have been associated with efforts to create air pollution agencies. These issues have centered on such matters as: the institutional structure of such agencies; their organizational locus within the total governmental structure; and the scope of their jurisdiction over air pollution sources and effects.

1. Institutional Structure of Air Pollution Agencies. As observed in Chapter 11, modern air quality management programs rest firmly on legislative enactments concerning air pollution matters. Legislatures enact laws authorizing administrative bodies to engage in air pollution functions, fix a basic air pollution standard to govern their activities, and empower such agencies to adopt administrative rules and regulations containing precise standards to be observed by regulated parties. Ordinarily, enabling legislation assigns a broad mix of responsibilities to administrative bodies including some of a quasi-legislative and quasi-judicial nature.

Many questions have been raised concerning the most appropriate system for structuring the air pollution agency in terms of several major functions and activities, including those related to: (a) promulgation of rules and regulations which establish air quality and pollution emission standards, permits, and other requirements to be met by owners and/or operators of air pollution sources; (b) quasi-judicial determinations, such as those involving the award of variances, the issuance of abatement orders, the revocation of permits, and appeals concerned with agency action on permit applications, including those dealing with the siting of the source; (c) use of technical advisors and technical advisory committees.

Most school children are taught at an early age that a fundamental characteristic of American government is its institutional separation of legislative, executive, and judicial powers. The prosecutor must never be the judge, the policeman must never also be the lawmaker, and the lawmaker should not administer the products of his labors. The idea that these powers should be separated and not concentrated in the same office has not always, however, guided the structuring of U.S. governmental institutions. Thus, many municipalities—and all but a few county governments—have no chief executive at all. County Commissioners and City Councilmen frequently serve both as lawmakers and as administrators of specific departments within their jurisdictions. Some mayors do not stand for election by the general public and instead are selected by their fellow councilmen to serve only ceremonial functions.

Nevertheless, the idea of separation of powers lies behind some of the organizational issues in the field of air pollution control. The 1947 California Air Pollution Control District Act strictly observed this idea. The act provided that each district would consist of: (a) a five-member rule-making body called the Air Pollution Control Board, (b) a three-member quasi-judicial body, called the Air Pollution Hearing Board, and (c) an administrative-technical organization to be headed by an Air Pollution Control Officer.

Under this act the five elected members of each County Board of Supervisors serve as the Air Pollution Control Board. In that capacity, the Board names two attorneys and one licensed engineer to serve specified terms as members of an independent Air Pollution Hearing Board. In California, such boards act on applications for variances, hear appeals from permit actions taken by the APCO, and determine whether permits to operate shall be revoked when requested by the Air Pollution Control Officer. Each Hearing Board has its own staff and budget, and functions totally independently of the APCO-headed administrative apparatus. The Air Pollution Control Board is the only body which can adopt rules and regulations and it also appoints the Control Officer and approves his requested budget.

Under this system, the Air Pollution Control Officer is the chief executive officer of the district and is charged by law with the responsibility to enforce all air pollution statutes, rules, and regulations. He is not obligated to seek the approval of either the Control Board or the Hearing Board before he institutes court action against an alleged violator of the law.

In contrast, Texas operates under a system where all rule-making power is concentrated at the state level and enforcement power is shared by city, county, and state authorities. The Texas Air Control Board bears both quasi-legislative and quasi-judicial responsibilities. It adopts the rules and regulations which fix limits on air pollution emissions within the state, hears applications for variances, approves the issuance of orders to abate, and is the state body which authorizes the attorney general to begin litigation against a violator of air pollution laws and regulations. At county and city levels in that state, air pollution directors must secure the approval of their city councils or county commissioners before instituting court action against violators.

In contrast to these two types of organizational structure, the National Clean Air Act of 1970 concentrates all rule-making and administrative authority in the Administrator of the Environmental Protection Agency. He issues national emission standards, supervises the staff which enforces these standards, and instigates court action against violators.

Various combinations of these three basic schemes now are being utilized by state and local jurisdictions throughout the nation. Each has its own advocates and opponents, and the most appropriate organizational scheme for accomplishing these three broad institutional functions remains a lively issue. As noted in Chapter 11, further pressures currently are mounting to centralize all power plant siting decisions in a single state-level board so as to provide utilities with one-stop shopping service and thereby avoid the delays associated with the present system which requires clearances from a number of agencies, with one frequently being in conflict with others.

Another issue has been raised concerning the composition of quasi-legislative and quasi-judicial boards. Some state statutes require that positions on such boards be allocated to persons representing specific interests, such as agriculture, industry, local government, industrial medicine, and public health.

Reviewing the schemes employed by the twenty-five states which, by 1966 had adopted air pollution control statutes, Charles Yaffe observed that twenty had established boards, commissions, or similar bodies and had generally specified the composition of such bodies [22]:

> As a rule, the intent seems to be to include representation both from those interests affected by pollution and from those likely to be affected by any regulations.

> On the other hand, one act, passed in 1966, has the novel proviso that "no officer, employee or representative of any industry, county, city, or town which may become subject to the rules and regulations of the board shall be appointed to the board."

In view of the complex, multidisciplinary character of air pollution operations it is not surprising that many agencies have made formal arrangements for securing a broad range of technical advice on emission standards and other matters. Almost from the beginning of its existence, the Los Angeles County Air Pollution Control District retained a number of independent consultants and frequently assembled them as a formal panel to review agency plans and programs. In the mid 1950's the Los Angeles County Air Pollution Control Board adopted Regulation VII which created an emergency episode program and formally established a "Scientific Advisory Committee." Among the specific charges given to that body were several having to do with determining the specific contaminants to be monitored by the district's air monitoring network, the types of monitoring methods to be employed and the specific sites at which air monitoring stations were to be established.

The federal Environmental Protection Agency and its predecessor air pollution agencies have utilized a number of standing advisory committees composed of technical experts, and the federal legislation provides for a presidential air quality advisory board.

Almost since the date of its creation, the California Air Resources Board (not to be confused with the local air pollution control districts) has made heavy use of a technical advisory committee. However, the operations of that committee recently were questioned in a report prepared by the staff of the state's legislative analyst. The staff commented that [17]:

> The preeminence of the TAC (Technical Advisory Committee) membership has been fortunate for California, but it has also resulted in the TAC becoming involved in program and policy matters because the TAC could more clearly delineate and evaluate these problems than many of the board members. . . . Because the TAC has tended to work in private as much as possible, because much of its important work has been through small subcommittees, and because it has frequently reviewed recommendations made by the board's staff to the board (which upon review by TAC are then presented as the latter's recommendations rather than board staff recommendations) the role of the TAC has been obscure and difficult to assess. It can be reasonably clearly stated that there has been an assumption of program and policy guidance and *de facto* responsibility of the TAC which is inappropriate. The technical advice of the TAC and even its strong advice and recommendations on program and policy matters would create no significant problems if the board, itself, fully explored these matters and weighed the public policy issues involved at its public meetings before acting. In too many instances this has not occurred and instead the board has routinely adopted the recommendations of the TAC after listening to witnesses who wished to speak on the proposed action . . . When the board routinely accepts the recommendations of the TAC (whether technical or policy) without undertaking a full exploration of the public policy issues involved, the TAC is exercising *de facto* decision-making authority. This authority is reserved by law exclusively to the board. When this happens, full public and legislative understanding of the significance of the board's actions may not occur.

It would appear that, on matters as sensitive to the public and to regulated parties as air pollution control, considerable caution should be exercised in the assignment of responsibilities to, and in the organization and operations of formal advisory bodies.

2. Organizational Locus of Air Pollution Agencies.   The organizational location of air pollution agencies within the total structure of the sponsoring governmental jurisdiction also has been a matter of considerable debate. In 1966, Schueneman reported that 68 of the then-existing state and local air pollution agencies were located within health departments, 10 within building departments, 10 in safety departments, 15 in other departments, while 27 had the status of being autonomous agencies [19].

More recently, there has been a trend among state agencies to consolidate air pollution operations in new departments with broad environmental responsibilities [8]. Minnesota, Oregon, and Washington have followed the federal pattern and combined *all* pollution control activities (air, water, noise, etc.) into a single department. Wisconsin, New Jersey, New York, and Vermont have placed their air pollution control agencies in new super-departments which consolidate all pollution control functions of the state as well as other functions concerned with conservation and natural resource management. Illinois has consolidated all pollution control operations but functionally specialized them into three departments; one which sets standards and adjudicates cases; another which conducts enforcement and technical-administrative operations; and a third which is responsible for environmental research and planning functions [8].

Governmental jurisdictions with the longest history of air pollution control activities seem to have progressed through a four-stage process of organizational development. In the first stage, official air pollution activities are inaugurated but are carried out by existing organizational entities. In the second stage, specialized air pollution units are organized in existing governmental departments. These may be described as "captive" organizations. As "captive" organizations grow in size and importance they may, in the third stage, be separated from their parent departments and acquire an independent status of their own. At this point they may be described as "autonomous" organizations. The fourth, and current, stage of organizational development is one characterized by the consolidation of environmental control agencies into super-departments charged with broad environmental management responsibilities, usually including control of air and water pollution, noise, solid waste disposal, use of insecticides and pesticides, and other matters.

In the early history of air pollution control in the United States, air quality management activities were conducted primarily by existing organizational entities of Public Health Departments at local, state, and federal governmental levels. Typically, at state and local levels, an Environmental Health Division of a Public Health Department would take on air pollution control duties as an additional assignment. Lumping air pollution functions together with a mix of regulatory activities concerned with such disparate matters as inspection of swimming pools, restaurants, food-processing establishments, water supplies, and waste disposal facilities, usually left much to be desired. Trapped beneath several layers of supervisors, the air pollution control officer in this pattern of organization was impotent to act with much independence, or to hammer out community air pollution policies. Usually he was frustrated in his search for dollars in a limited environmental health budget. Since public agency salaries usually are based on one's hierarchical status within the bureaucracy, this arrangement also placed a severe limit on the salary authorized the community's top air pollution official. Usually, therefore, this arrangement has given way to the creation of a separate air pollution unit within the Environmental Health Division, then to a separate air pollution division or bureau reporting directly to the Health Director.

Even the latter arrangement, however, has not always been a happy one. Tradition and statutory requirements frequently have decreed that Health Directors be physicians. By academic training and disposition, few such individuals have been well equipped to deal with the range of complexities posed by a community air pollution control effort. Overwhelmed by other responsibilities, ranging from care of the medically indigent to mass inoculation programs, few Health Directors have been able to provide the leadership necessary to sustain an effective community air pollution control program. Moreover, the field of public health has labored under dogma initially laid down by its founders which hold that "consultation, communication, and cooperation" is the magic route to program success. Typically, slavish compliance with the dogma has constrained many health departments from the kind of aggressive, firm, and no-nonsense air pollution emission control program which the circumstances of modern urban life require and which the public demands. Autonomous status for the air pollution control function has been the usual result of this situation.

3. Jurisdiction of the Air Pollution Agency. One of the most important issues in the design of air pollution organizations relates to the scope of the agency's jurisdiction. An organization, like an individual human being, is an entity whose survival and effective performance is dependent on continuing satisfactory interactions with its environment. Like a human being, it mines its environment for the resources required to sustain it. In return it must generate products or services sufficiently desirable to its clientele to warrant their continued provision of the resources necessary to sustain the organization. Thus, the first step in designing any organization involves the development of answers to four important questions: What clientele is the organization to serve? What type of service do these clients desire? What mix of input resources is necessary to maintain the organization and to provide the service? What can the clients individually and collectively afford to pay for the service?

In terms of air pollution control organizations, these questions are of particular importance in determining the scope of an agency's jurisdiction. Failure to answer them—if, indeed, they have even been asked—explains much about our past failures to cope with community air pollution conditions. The questions may be restated as the *basic design imperatives* or criteria to be met in the development of an air pollution control organization.

A. *The geographic jurisdiction of an agency should be large enough to cover the total zone in which causal relationships can be observed between air pollution emissions and effects.*

This criterion implies two premises: (1) that the organizational product which is being *purchased* by the general public is *air quality* compatible with the public's desires, and (2) that the organization must therefore be able to deal with the mix of sources and emissions causally linked to qualitatively degraded air received by the organization's clientele.

On the first point, some critics of American government have referred to some governmental operations as being a "triumph of process over purpose." Instead of delivering safe meats to the consuming public, the system may deliver infrequent and ineffectual inspections of meat-packing plants; instead of delivering "safe streets," it may deliver helicopter-borne policemen; instead of delivering air of an acceptable quality, it may deliver air monitoring data and reports on the number of sources inspected. Rationally, these "commodities" are nothing more than the means to certain ends, and it is the "ends" which the public wishes to buy—not the "means." All too frequently governmental organizations are created which are palpably incapable of accomplishing the missions expected of them by their creators. Air pollution control organizations are no exception. Not only have such organizations been created to serve areas far too small in population to financially support the range of professional-technical activities essential to an effective program (see discussion below), but the scope of their geographic jurisdiction frequently is too small to encompass the mix of sources, emissions, and pollution-receptor areas and populations that collectively comprise the "community air pollution problem." From their birth, such organizations ordinarily are doomed to failure.

A situation of this sort once plagued air pollution control efforts in the county of Los Angeles. Although now apparently forgotten by the chroniclers of the U.S. air pollution efforts, significant air pollution problems in the city of Los Angeles date back to the first decade of this century when smoke from space-heating equipment billowed into downtown streets, requiring that street lamps be lighted during the daytime. During that period the area's first smoke control ordinance was adopted by the Los Angeles City Council [10]. In the early years of World War II, both the city and the county of Los Angeles maintained separate smoke-control staffs within their respective Health Departments. However, when photochemical smog first made its presence known within the community (1943–1944) neither staff was able to cope with it. Moreover, the primary victims of the smog attacks were the residents of the then-suburban city of Pasadena whose air supply is imported from the central section of Los Angeles County where the downtown area of the city of Los Angeles is located. After four years of abortive efforts to secure agreement from the county of Los Angeles and the then thirty-six separately incorporated cities of the area to adopt uniform air pollution control laws, disgruntled residents and officials of the county turned to the state legislature. In 1947 a bill introduced by Pasadena's state assemblyman, A. I. Stewart (but largely drafted by the distinguished Los Angeles County Counsel, Harold Kennedy), was adopted by the legislature and signed into law by then-Governor Earl Warren [9]. It provided for the creation of state-sponsored Air Pollution Control Districts within each county of the state and designated the members of the Board of Supervisors (County Commissioners) of each county as the members of the Air Pollution Control Board of each such district. In that capacity, County Supervisors were empowered to enact air pollution control rules and regulations having the full force and effect of state law within both the unincorporated and incorporated areas of their counties and to establish countywide air pollution agencies headed by a chief executive officer carrying the title of Air Pollution Control Officer. The legislation permitted cities and counties to adopt more stringent control ordinances than those adopted either by the State Legislature or District Boards and in no way restrained local units of government (cities and counties) from continuing to maintain their own air pollution control staffs. However, none did so, and the Los Angeles County Air Pollution Control District thereafter emerged as the largest and perhaps the most technically competent and effective local air pollution control agency in the nation. Until comparatively recently its annual appropriations and payroll exceeded that of the federal program.

Thus did Los Angeles County provide a basis for the creation of an air pollution organization with sufficient geographic jurisdiction, and a large enough base of support, to deal effectively with the mix of *stationary* pollution source emissions and adverse pollution effects. (As the experience of later years demonstrated, motor vehicle pollution emissions are beyond the control capacity of local governmental units.)

Morris Neiburger [16] has shown that, unlike the emission from a single source (whose atmospheric concentration decreases as a function of distance from point of emission), emissions from the large mix of community sources result in a progressive increase in atmospheric pollution burdens throughout the community and in the downwind zones immediately proximate to the areas of emission. Considerable distances sometimes are necessary before the air masses leaving a metropolitan community have cleansed themselves sufficiently to restore the quality of the air mass to its original entry state [16]. In view of this, Neiburger suggests "a diameter for the control program region three times the largest dimension of the fully developed metropolitan area." Weisburd [21] has referred to such a region as the "metropolitan pollution zone." In a 1961 survey, Purdom [18] showed that more than half of all responding air pollution control agencies believed that pollution from outside their jurisdictions affected the pollution levels in their areas.

The need for structuring organizations and programs so as to deal with the total system of causes and effects embraced by their missions has long been recognized in other fields. Typical examples are the "river basin" approach of water and hydroelectric resource development programs and the "problem shed" (market area) approach to delivery of health care services which was recommended by the National Commission on Community Health Services.

B. *The legal authority of the agency should extend to the mix of sources, emissions, and emission-causing circumstances which give rise to unwanted air pollution effects.*

Much of the justification for this criterion is provided in Chapters 10 and 11. Granted it may not always be possible for the authority of an air pollution control organization to extend over the total mix of circumstances which give rise to air pollution conditions. Nonetheless, provisions must be made for the agency to impact policies and programs which do extend to the causal circumstances giving rise to such problems, assuming, of course, that the creators of the organization really wish to solve these problems. Thus, short of acquiring total authority to adopt land-use regulations for a metropolitan area, the air pollution control agency may nevertheless be empowered to control the site location of specific air polluting sources. Without becoming the traffic engineering or mass transportation agency for a region, the air pollution control agency may nonetheless be authorized to prepare reports and recommendations on such matters in terms of their air pollution implications. The National Environmental Policy Act currently requires the preparation of environmental impact statements for projects funded by federal agencies [1] and similar statutes have been adopted or are being considered by several states [8]. Air Pollution Control agencies might well be authorized and encouraged to review such statements in terms of the air pollution implications of the projects to which they relate.

Similarly, air pollution agencies bearing primary responsibility for dealing with the air pollution conditions of an area should be vested with legal authority to adopt emission standards respecting the sources causally linked to these conditions and to engage in the range of functions necessary to an adequate community control program (see discussion below on functions and work products of air pollution agencies).

C. *The taxing base within the jurisdiction of an air pollution agency should be large enough to provide financial support for the range of functions required in an effective and comprehensive air quality management program.*

Good intentions will not, alone, solve any human problem nor will an understaffed, inadequately budgeted, and poorly equipped air pollution control agency be able to deal effectively with a community air pollution problem. Yet a large proportion of the nation's air pollution agencies are in precisely this condition. In 1970, 50% of all state air pollution control agencies had fewer than ten full-time positions, and 50% of all local agencies had fewer than seven [13]. For state and local agencies to be able to fully perform the range of functions required in an effective control program, a recent study indicates that their 1970 employment levels would have to be increased from 2,837 to about 8,000 positions [13].

Although some of the observed underbudgeting of air pollution agencies may be due to public miserliness or political considerations, some of it also is due to the fact that small population aggregations simply cannot afford to retain the range of high-paid specialists which any effective program requires. The day when a lone sanitary engineer could be deemed competent to handle every environmental problem of a community has long since passed. Specialists in various source operations, in air monitoring, meteorology, chemistry, and other fields are now required. But the workload and financial capabilities of lightly populated areas is insufficient to justify the development of such multidisciplinary control agency staffs. Even at annual per capita tax burdens of one dollar, which is substantially beyond the present national average, many communities cannot raise enough funds to support such a staff.

D. *The geographic jurisdiction of an agency should be small enough to permit local air quality aspirations to be satisfied.*

Superficially, this criterion appears to be in conflict with the first and third criteria and raises issues which perennially plague the institutions of American government. On the one hand our traditional national commitment to social pluralism and the democratic ethic lead us to support highly responsive and "grass-roots" institutions for policy-making and administration. On the other hand, the imperatives of efficiency and economy and our commitment to protect one person, group, or community from the possibly harmful circumstances created by the activities of another, lead us to configure our decision-making and executive institutions around ever-larger geographic boundaries.

Yet, these apparent conflicts are not insoluble. As observed above, the 1947 California Air Pollution Control District Act found a way out of this dilemma. Similarly, multi-county districts also have been formed, such as that which is operative in the San Francisco Bay Area. Perhaps more in point is the experience of the Los Angeles County Sanitation Districts. In that county, state law has authorized the formation of many legally recognized sanitation districts, each with its own geographic jurisdiction and tax base. However, the same law provides that one sanitation district shall cover the whole of the county and be responsible for regional-level needs, such as construction of ocean out-falls and large sewage treatment plants. Moreover, there is a single administrative-technical staff, headed by a chief engineer, which performs the work of all the districts. In effect, there is one administrative organization, but several boards of directors. In this manner the occupational specialization, efficiency, and economy of a large-scale organization is achieved without sacrificing local autonomy over some types of decisions.

Similar opportunities are provided by such mechanisms as inter-governmental or inter-jurisdictional compacts and contracts and by the more recently developed administrative infrastructures of regional councils of government [12, 14].

Perhaps the most well-developed expression of the former is the "Lakewood Plan," which also was developed in the county of Los Angeles [12]. After the end of World War II and the great flood of inmigration to southern California, new subdivisions grew like weeds throughout this region —sometimes stretching over hundreds of acres. Many of these were developed in unincorporated areas of the county and, once built, their residents wished more than the standard level of services provided to such areas by the County Board of Supervisors. Moreover, many of these local communities wished to gain power over such important regulatory functions as zoning, traffic control, and general community police functions. One of these, the unincorporated community of Lakewood, therefore took action to incorporate itself as a city. However, not wishing to get involved in the hiring and supervision of large numbers of city employees, it contracted for all of its municipal services from Los Angeles County. Under the contract, it purchased the quantity and quality of street maintenance, recreation, police, and other services it desired from the large and well-managed departments of county government. Thereby, it was able to preserve local decisional autonomy while at the same time enjoying the benefits of economy, professionalism, and efficiency that is possible only in large organizations. This same plan now is becoming popular in other areas. Similarly, a great many state air pollution control statutes authorize inter-governmental and inter-jurisdictional contracts for air pollution services, although few jurisdictions have yet taken advantage of this opportunity.

Prompted by federal requirements, states now also have authorized the creation of regional councils of governments. Councils are "voluntary" confederations of local governmental units within an area and typically include among their members the cities, counties, school districts, and other special district boards which operate within their territory. Created primarily to provide regional planning services and to meet the "review and comment" requirements of certain federal grant programs, some states now have authorized these councils to engage in the direct delivery of services on behalf of their member governments [12, 14]. Under such arrangements, the local service-receiving units retain their decisional autonomy over the function (such as air pollution control) but are spared the pain and higher overheads associated with the development of comprehensive and well-staffed departments.

Frequently, the premises implicit in our first and third criteria are cited by proponents of schemes in which state governments are given total regulatory power over air pollution emissions. Typically, proponents of such schemes cite the advantages of large-scale organization and the need for uniform standards in support of their proposals.

On the other hand, opponents of preemptive state legislation hold that, in terms of the democratic ethic and its implications for "grass-roots" control over matters affecting local populations, there appears to be little justification for such preemptive state legislation. On this point, the National League of Cities, which is the official association of municipal officials, has recently stated that "legislation and administrative regulations which establish environmental standards and controls must assure that state and local governments can adopt controls stricter than those applied from the . . ." higher level of government [15].

Those who hold this view argue that the need for minimum national or state standards dealing with air quality and source emissions does not preclude the possibility or desirability of vesting local units of government with the power to adopt more restrictive environmental standards. Opponents to preemptive state legislation argue that so long as local communities wishing to do this fully understand the implications of their actions to their local economy and tax base, there seems little reason to deny them this authority.

Nonetheless, many problems can arise when several jurisdictions of government have concurrent jurisdiction over the same source of pollution emission. In Texas, for example, the State Air Control Board has sole rule-making power, but local units may enforce state standards if they desire. The board's chairman recently noted that some enforcement actions instituted by one county air pollution agency actually have slowed down the implementation of source abatement plans within that community [11]. In that county, the central city, the county itself, and the state of Texas have concurrent enforcement jurisdiction over many of the same sources of pollution emission and one agency may take enforcement action without the prior knowledge of the others. The Chairman of the State Board observed that his agency frequently receives completely opposing abatement recommendations from the two local jurisdictions in respect to the same source, and he has therefore recommended that the state law be amended so that county and state enforcement powers are limited only to territory in which state-approved municipal enforcement programs have not been organized.

Operations of Air Pollution Agencies

The work performed by air pollution agencies may be viewed as consisting of a set of *functions* which are conducted in order to generate certain *work products* necessary to the accomplishment of the agencies' statutory objectives. Although considerable specialization of functions now has occurred between federal, state, and local agencies, the following functions or operations are generally viewed as being essential to any comprehensive air quality management program. The organization structure, budget, and staffing pattern of any single air pollution agency is largely determined by the functions it is to conduct.

1. Air Pollution Effects Research. Concerned with the development of sound scientific understanding of the relationship between ambient air pollution burdens and specific air pollution effects, this function now is performed almost exclusively by agencies of the federal government and by recipients of federal grants and contracts. Until the adoption of the Kuchel-Capehart Act of 1955, almost all research concerning the effects of air pollution was funded by private sponsors or by state and local units of government. At one time, the Los Angeles County Air Pollution Control District was expending approximately $1 million per year on such research, and the privately supported Southern California Air Pollution Foundation was spending a comparable amount. Now, however, the overwhelming proportion of all air pollution effects research within the United States is conducted or funded by agencies of the federal government.

2. Development of Air Quality Criteria. Utilizing the results of air pollution effects research and the scientific judgments of air pollution experts, air quality criteria are developed so as to indicate the effects to be expected when specific burdens of air pollutants are reached within an air mass. Development of such criteria are now the exclusive responsibility of the federal government.

3. Promulgation of Air Quality Standards. Guided by air quality criteria, information concerning the cost of controlling pollution sources, and the air quality aspirations of the public, state and federal agencies now bear principal responsibility for the enactment of air quality standards which specify the levels of air quality to be achieved and maintained within a community, region, or area. Federal primary standards of air quality are intended to protect the public from adverse health effects while secondary standards are intended to prevent other air pollution effects, such as visibility impairment and damage to agricultural crops.

4. Surveillance of Ambient Air Pollution Burdens and Community Air Pollution Effects. Precise measurement of ambient air pollution levels within communities and regions is now a necessary component of an air quality management program. Federal standards issued under the authority of the Clean Air Act of 1970 require a fixed number of air monitoring stations for specific pollutants depending on the population size of a community, region, or state. Since many metropolitan communities sprawl over a considerable territory, a more appropriate determinant of the number of air monitoring stations required in a community might be the number of square miles of land area to be served by each air monitoring station. In any event, the air monitoring network of a community or region should be of sufficient size to adequately record the range of air pollution values to which the population of the area is exposed. Beyond the mere measurement of air pollution levels, however, communities also face the need to determine the specific air pollution effects being experienced within their boundaries. Some communities, such as Los Angeles, have routinely conducted studies to determine the geographic distribution, occurrence, and severity of such air pollution effects as eye irritation, odors, visibility impairment, and damage to growing vegetation. Information of this sort is essential to the development of air pollution control strategies appropriate to the needs of a community. Studies of this sort have been funded in the past by all three levels of government but are few in number.

5. Source Registration. Identification of the many possible sources of air pollution emissions within any area is the primary objective of a source registration system. Registration of sources is necessary to any effort to determine the types, quantities, and sources of pollution emissions within a community and provides the foundation for a community source inspection and abatement program. This function now is a primary responsibility of state and local air pollution agencies.

6. Development of Emission Inventories. Pollution emission inventories are intended to reveal the types and quantities of air pollution emitted within a specific area and usually are conducted so as to reveal emissions derived from the several categories of sources and geographic areas within a community. Inventories may be based on actual measurement of pollution losses from individual sources, or on emission estimates prepared on the basis of information concerning the size and product throughput of specific source categories. Such estimates are based on standard emission factors which suggest the types and quantities of pollutants ordinarily liberated from specific types and sizes of sources.

7. Development of Source Control Strategies. Guided by air quality standards, emission inventories, data on source control costs, information concerning air movement patterns, and the geographic distribution, occurrence, and severity of air pollution effects within an area, this function is concerned with determining the types of sources and pollutants to be subject to governmental regulation and the precise emission standards to be observed by individual sources and whole categories of sources. Development of source control strategies is now a primary function of state and local air pollution agencies. However, the Clean Air Act of 1970 requires that such strategies be presented in state implementation plans and approved by the Administrator of the Environmental Protection Agency.

8. Design and Enactment of Emission Control Standards. The key to any successful community pollution emission control program is to be found in the enactment and enforcement of appropriate emission control standards. These may limit the quantities of pollutants which may be liberated from any source, prescribe the type of equipment or fuel to be utilized in an operation, or prescribe the permissible concentration of a pollutant in any given volume of effluent gas. The development of such standards is a technically demanding task well beyond the capacities of most legislators and therefore comprises a principal function of the technical staff of an air pollution agency. The enactment of the standard itself usually is reserved to the legislature or a quasi-legislative board and may even be subject to prior review by a technical advisory committee.

9. Source Inspection and Surveillance. A primary function of all state and local agencies involves the surveillance of air pollution sources within the community and the periodic inspection of specific sources or source categories. Routine area patrols may be conducted to detect visible sources of pollution, such as open burning, or to detect odors and other air pollution effects within the community. Programmed inspections of specific sources also are necessary; the required frequency and thoroughness of such inspections are determined by the size, complexity, and importance of the source as well as by the type of control system being utilized. For example, sources equipped with baghouses and other filtering devices subject to deterioration may be inspected more frequently than sources equipped with more durable mechanical control devices, such as vapor recovery systems.

10. Source Testing.   A special form of source inspection, source testing, involves the actual collection of effluent samples during various operating cycles of a source. Usually, multimember source testing teams must be organized to conduct this type of operation, and only the largest air pollution agencies and jurisdictions are able to support such an operation.

11. Sample Analysis.   Modern air pollution agencies require well-equipped and competently staffed laboratories in which both wet chemical and instrumental analyses of ambient air and source samples may be performed. The primary determinant of workload associated with this function is the volume of source tests performed and the nature and size of the air monitoring network. Considerable care in the handling of samples and in the recording of analytic results is necessary because of the legal significance of such findings in formal abatement proceedings before the courts.

12. Equipment Development and Maintenance.   Both air monitoring and laboratory operations require the use of expensive and complicated instruments. The accuracy of data derived from these instruments is greatly influenced by the maintenance they receive and by their periodic calibration. Technicians skilled in the maintenance, modification, and calibration of air monitoring and laboratory equipment, therefore, are needed by modern air pollution agencies.

13. Complaint Investigation.   Communities experiencing air pollution problems severe enough to warrant the creation of an air pollution agency usually generate dozens to hundreds of public complaints each year concerning specific air pollution sources or localized air pollution effects. Well-staffed agencies therefore maintain systems which permit the reporting of complaints during most hours of the day and their investigation by trained inspectors and technicians.

14. Review and Approval of Source Abatement Plans.   The development of precise plans for the control of emissions from a specific source or equipment system usually is a responsibility of the owner and/or operator of the equipment. However, consultation with the air pollution agency may be sought during its preparation. Moreover, the review and approval of such plans is a necessary part of any permit system which requires that permits be secured from an air pollution agency prior to the construction and operation of any equipment, device, or contrivance capable of emitting or controlling the emission of any air pollutant. Since 1948, the Los Angeles County Air Pollution Control District has administered such a system and has developed an Engineering Division to perform this function. The division is structured into a number of specialized units conforming to the major industrial operations conducted within the county, and is staffed by a large number of professional engineers in such fields as mechanical and chemical engineering.

15. Review of Site Location Plans. Applications for zone exceptions or for permission to locate an air pollution source at a specific site requires that a competent technical study be made of such matters as current air quality within the area, typical wind movements at the site, current air pollution emissions within the area, probable new emissions to be derived from the source (including emissions resulting from plant breakdowns), and the proximity to the site of sensitive air pollution receptors.

16. Law Enforcement Activities. In spite of the highly technical activities which they conduct, air pollution organizations also are law enforcement agencies and therefore engage in the range of activities associated with such agencies. When a violation of air pollution laws or regulations is detected, a whole series of tasks is then necessary. Usually, a notice of violation must be issued to the owner or operator of the source, evidence of the violation must be collected, a request for institution of a court action must be made to an appropriate prosecutor, the case must be prepared in cooperation with the prosecuting law office, and agency enforcement officers must be scheduled to offer testimony in court or before the appropriate administrative tribunal.

17. Data Processing and Records Management. The combination of functions described above result in the generation of much data concerning a broad variety of matters—ranging from measurements of air quality to files containing plans and data relevant to specific sources. In large agencies, much of this data can be handled only through use of computers, and complex records management and data processing programs therefore must be developed.

18. Public and Inter-Governmental Relations. Like most other important public programs, air quality management requires careful attention to the information needs of the press, other governmental jurisdictions, and the general public. Large agencies, therefore, have organized small staffs which specialize in this function.

Although by no means an exhaustive list of all the activities conducted by a comprehensive air pollution control agency, these eighteen functions represent the operations of principal importance to a community air quality management program. In any specific region or community, such as a large metropolitan area, these operations must be conducted by some agency or combination of agencies if effective and predictable air pollution control results are to be achieved. Formal systems of inter-governmental cooperation should be designed so as to assure that each function is performed within the focal community or region.

Work Products Used or Generated
by Air Pollution Agency Operations

Several major types of documents or work products are associated with the functions described above. Some of these are utilized by personnel engaged in the function while still others are the terminal products of the function. Among the documents and work products of major importance to agency operations are the following:

ABATEMENT ORDER — In some jurisdictions, air pollution statutes authorize air pollution control boards or officers to issue abatement orders to the owners or operators of specific sources. Such orders specify the source or sources to which the order relates; the specific action which is required of the source owner or operator, including any emission standards which are to be met; the time period authorized for compliance with the order; and any reports or procedures required. Under appropriate conditions, violation of such orders can result in criminal, civil, or injunctive action before a court.

AIR POLLUTION EFFECTS ISOMAP — A graphic portrayal of the rate, frequency, prevalence, or incidence of a specific air pollution effect within a specified community, region, or state.

AIR POLLUTION EFFECTS REPORT — A document describing the geographic distribution, frequency, and severity of specific air pollution effects (eye irritation, odors, respiratory disorders, etc.) within a community, region, or state.

AIR POLLUTION
LAW

As used here, an air pollution law is an enact-ment of a legislative body. Enactments of state and federal legislative bodies usually are re-ferred to as statutes, while those of municipal and county legislative bodies are referred to as ordinances. Semilegislative enactments of ad-ministrative boards, commissions, and officers are referred to as rules or regulations. An air pollution statute or ordinance usually defines the legislature's policy in respect to pollution emissions and conditions, contains a basic standard to guide the enactment of administra-tive rules and regulations, assigns an organiza-tion or department responsibility for imple-menting and enforcing the statute, and gives such organizations authority to engage in spe-cific functions and to utilize specific procedures. The penalties for violations of the statute also are defined.

AIR QUALITY
CRITERIA

Descriptive or predictive statements concerning the relationship between ambient burdens of specified pollutants or combinations of pollut-ants and specific air pollution effects, such as eye irritation, visibility impairment, and dam-age to human health.

AIR QUALITY
STANDARDS

Legally defined statements concerning the am-bient burdens of specified pollutants or com-binations of pollutants that will be tolerated in a specific place for specified time periods.

AIR QUALITY
SURVEY REPORT

A document showing the ambient pollution burdens that have been measured within a com-munity, the sites at which such measurements have been made, and the methods used in col-lecting and analyzing the ambient air samples.

AMBIENT AIR POL-
LUTION ISOMAP

A graphic portrayal of the ambient burdens of a specific pollutant throughout various areas of a mapped region. Isolines are like the contour lines on a topographic map, but instead of in-dicating altitude they indicate the ambient con-centrations of a pollutant in the mapped land area to which they relate. Techniques now are available for the computer-mapping of air pollution concentrations.

BUILDING PERMIT

A legal document which authorizes some person or persons to proceed with the construction of some specified object at some specific location. Used by most cities and urban counties to control the structural safety of new buildings, applications for such permits can provide air pollution control agencies with useful information concerning possible new sources of pollution within a community. Moreover, under some conditions building permit ordinances may be modified so as to require approval of building plans by air pollution control agencies prior to the award of the permit.

BUSINESS LICENSE

A legal document, usually issued annually or at some other stipulated time interval, which permits an enterprise to conduct business at some specified location within a community. The award of business licenses usually is conditioned on the payment of a fee and a showing by the applicant that he is in compliance with all laws, rules, and regulations governing the conduct of his business. To operate without such a license usually exposes the operator to criminal or civil prosecution. Licenses are used to assure that proper practices are followed by operators of waste disposal sites, and license provisions may be extended to deal with other types of air pollution problems.

EMERGENCY
EPISODE PLAN

Federal regulations require that each State Implementation Plan include a specific plan for dealing with emergency episodes. The plan must specify the procedures to be followed in forecasting and dealing with such episodes and provide for the closure of sources, when required.

EMISSION
INVENTORY

A quantitative statement of the types and quantities of air pollutants emitted for specified source categories within a specified place or region over a specified period of time. For any pollutant, emissions usually are expressed in terms of the tons per day emitted from specified source categories.

EMISSION ISOMAP

A graphic portrayal of the quantities of pollutants emitted from various areas of a mapped region. Isolines are like the contour lines on a map, but instead of indicating altitude, they indicate the quantity of pollutants emitted per square mile in the mapped land area to which they relate.

EMISSION
STANDARDS

Legally defined and enforced prescriptions which prohibit the emission of more than some specified quantity of a pollutant from some designated source or sources, which prohibit the operation of some source or the use of some type of fuel or product, or which require the use of some type of equipment, fuel, or air pollution control system.

ENVIRONMENTAL
IMPACT STATEMENT

A document required of certain public agencies under the terms of the National Environmental Policy Act of 1970. Such statements are to be filed for each federal agency program, project, or decision having possible implications to environmental quality. Such statements are to assess the probable impacts of the program, project, or decision on environmental quality, the costs of these impacts as contrasted to the benefits of the project, the specific nature and geographic scope of the impacts, the measures relevant to control of such impacts, and the probable costs of each.

IMPLEMENTATION
PLAN

A document which describes a comprehensive plan of action for achieving specified air quality objectives and standards for a particular place or region within a specified time period.

INTER-ORGANZA-
TIONAL WORK PLAN

A formal document which describes, for any community or region, the functions which will be performed by two or more air pollution agencies. The plan may be quite detailed and serve as the basis for appropriate inter-agency or inter-jurisdictional contracts, compacts, or agreements.

NOTICE OF VIOLATION

A document used by field air pollution enforcement personnel to record the facts concerning an observed violation of air pollution laws, rules, or regulations. A copy of the notice may be given to the person committing the violation while other copies may be routed to other agency personnel, and/or to other governmental departments for appropriate review and action.

PERMIT TO CONSTRUCT

A document issued pursuant to an appropriate statute or regulation which authorizes a person or persons to proceed with the construction of some specific equipment, device, contrivance, or system of such units at a specific location subject to specified conditions. Usually such conditions deal with the types of basic and control equipment to be constructed and the performance specifications appropriate to each. Issuance of a permit to construct usually provides no guarantee that a permit to operate will be issued. Usually, construction without such a permit is prohibited by law and violators are subject to appropriate penalties.

PERMIT TO OPERATE

A document issued pursuant to an appropriate statute or regulation which authorizes a person or persons to operate some specific equipment, device, contrivance, or system of such units at a specific location subject to certain conditions. Such conditions may extend to specifications to govern the maintenance of control equipment, the periodic testing of emission discharges from the source, and other matters. Violation of permit conditions, or engaging in operations without a permit, usually are declared by the statute to be a legal offense punishable by certain prescribed penalties.

PROGRAM OF WORK A document required of an air pollution agency when requesting a federal grant. Usually such a document also is necessary as part of the agency's budget request to its sponsoring jurisdiction. Usually such a document shows the organization structure of the agency, the types and numbers of positions currently funded and requested, the functions performed by the agency and to be performed during the period covered by the report, the specific tasks, projects, or program changes to be accomplished during the report period.

PUBLIC USE PERMITS As used here, the term relates to those permits which must be obtained by an individual before he can use certain public facilities, such as water distribution, sewage disposal, and highway facilities. Subdividers may be required to secure such permits before tying on to existing public water distribution and sewage collection lines; builders may have to secure such permits before emptying a driveway into a public street; and industrial plants may require such permits before connecting with an industrial waste water line. Under some conditions, ordinances requiring such permits may be modified so as to require applicants to conform to air pollution requirements and to secure appropriate clearances from the local air pollution agency.

REPORT OF COMPLAINT INVESTIGATION A document which shows the name and address of complainant, date and time of complaint, and a description of the complaint, including the effect observed by the complainant and/or the source to which the complaint relates. An appropriate portion of the document is used to report the results of the investigation including the name of the officer(s) conducting the investigation and the time at which performed.

REPORT OF SOURCE INSPECTION A working document indicating the location and type of source inspected, the name of the inspector, the date and time of the inspection, and the principal findings of the inspection.

REQUEST FOR COM-PLAINT OR COURT ORDER
A document which requests a district attorney, city attorney, State Attorney General, or U.S. attorney to file a complaint in or seek an order from an appropriate court. The format of such requests varies from jurisdiction to jurisdiction. These documents are prepared by air pollution control agencies and signal the passage of a "case" from their hands to the responsibility of an appropriate prosecuting agency.

SITE LOCATION REPORT
A document which assesses the appropriateness of a specific site as the location for a source of air pollution emissions. If comprehensive, such a report should disclose the meteorologic characteristics of the site and cover such matters as: the frequency of occurrence of various wind speeds and directions of wind movement; receptor areas prone to receive pollution generated from the site, including the probable frequency of such reception; ambient pollution burdens now existing at the site and in probable receptor areas which will be affected by on-site pollution emissions; the probable on-site pollution emissions which will occur as a result of controlled and uncontrolled operations of the proposed source; an estimate of the frequency of plant breakdowns that will result in emissions from the source; an assessment of the probable impact of such emissions on air pollution conditions within the site and within potential receptor areas.

SOURCE ABATEMENT PLAN
A document which presents a comprehensive plan for the reduction of air pollution emissions from a specific point source or plant location. Such plans usually depict the operations conducted by the source, the product flows and types of equipment involved in such operations, the points of pollution loss, the types and quantities of pollutants emitted during various phases of each operation, the emission standards programmed for achievement, the type of control equipment or methods to be used in meeting

these standards, the estimated cost of basic and control equipment, and the time period during which various aspects of the plan will be accomplished.

SOURCE FILE

Upon registration, a file is opened for each source of air pollution emissions and thereafter is used as the repository for all reports related to the source. The file is of major importance to abatement proceedings.

SOURCE REGISTER

A document or documents containing a listing of all premises within the community, region, or state from which significant air pollution emissions occur. The register, together with associated reports and plans relevant to each entry therein, should present the following information for each entry: the address and/or location of the premise; the name and address of the owner or operator of facilities located on the premise; the type of operation conducted there; the type and quantity of pollution emissions derived from these operations; specific sources of pollution emissions within the premise.

SOURCE TEST REPORT

A document showing the results of a test conducted on a particular source to determine the types and quantities of pollution emissions resulting from various conditions of source operation. At a minimum the report should show: the names of the person or persons conducting the test; the date on which the source sample was collected and analyzed; the methods used in collecting and analyzing the source samples; the specific conditions under which the sample was collected.

VARIANCE

An order issued pursuant to law which extends to some person or persons the legal right to operate a specific air pollution source or sources in violation of air pollution laws, regulations, and emission standards. Usually *variances* are authorized in order to give the owner or operator of a source sufficient time to comply with an emission standard.

Enabling Legislation

The essential first ingredient in any state or local air pollution control program is an adequate enabling statute. Since the powers of city and county government ordinarily are fixed by state constitutional and statutory provisions, and since state administrative agencies can perform only those functions authorized by state statutes, this observation therefore suggests that the quality of the state air pollution statute is the primary determinant of the quality and effectiveness of state and local air pollution agency operations.

As recently as 1963, only eleven states had adopted an air pollution statute [22]. However, between 1963 and 1966, twelve more states enacted such laws and twenty-three additional states took such action during the next two years [20]. By the end of 1970, all fifty states had adopted air pollution statutes [13], but these were of widely varying quality.

In 1967 the Council of State Governments published a model state air pollution statute [5]. However, the national Environmental Protection Agency has stated, "in view of the provisions of the Clean Air Act, as amended, and due to other reservations, the Environmental Protection Agency does not endorse such model law in its present form" [6].

Both the Council and the Environmental Protection Agency agree that a state air pollution statute should deal with a core set of topics, including: the policy of the legislature in respect to air pollution and the objectives to be achieved by the act; definition of terms employed in the act; any specific emission standards or requirements which the legislature wishes to impose on regulated parties; designation of the state agencies and local jurisdictions of government which are to be vested with responsibility and authority for carrying out various provisions of the act; specification of the structure and organizational locus of any state or regional administrative agencies on special districts which are created or authorized by the act, including designation of the officials to be given authority for the appointment of board members and administrative officers and employees within these bodies; the penalties which may be imposed on violators of the statute and of administrative rules and regulations issued under its authority; the power of designated agencies or boards to issue orders and to enact rules and regulations dealing with specific matters, such as emission standards, air quality standards, requirements for review and approval of construction, operating plans for air pollution sources and emergency episodes, and requirements relating to the site location of air pollution sources.

In terms of the discussion in Chapters 11 and 12, it seems clear that a state air pollution statute should specify the organizations and jurisdictions which are to engage in air pollution operations and should specifically deal with the authority and responsibility of each to engage in each of the several functions discussed in these chapters.

Current federal requirements specify that the state agency responsible for carrying out an approved implementation plan must have the authority to: (1) adopt emission standards and limitations and any other measures for attainment and maintenance of national ambient air quality standards; (2) enforce applicable laws, regulations, and standards through such means as orders, hearings, and court actions which, under authority of the basic statute, can lead to injunctive relief and the imposition of criminal or civil penalties; (3) abate pollutant emissions on an emergency basis during health-threatening air pollution episodes; (4) review and approve plans for the location, construction, and operation of air pollution sources; (5) obtain information from owners and operators of air pollution sources concerning the nature of their operations, the qualitative and quantitative characteristics of their emissions, including the authority to inspect such sources, test their emissions, and require the maintenance of certain types of records and the use of emission-monitoring devices and systems; (6) make public the data concerning emissions from any source; (7) carry out a program of inspection and testing of motor vehicles in order to enforce compliance with applicable emission standards [6].

The Environmental Protection Agency also has recommended certain policies respecting the composition of Air Pollution Boards, the use of variances, judicial review of air pollution agency actions, the legal definition of air pollution and of state air pollution control policy, and the size and character of penalties to be imposed on violators of air pollution laws and regulations [6].

Specifically, EPA has recommended that Air Pollution Control Boards *not* be structured so as to represent the several interest groups concerned with air pollution emission control standards and enforcement policies. As noted earlier, many statutes now require that such boards include representatives of the several categories of business and industry regulated by the agency. Substantial penalties for violations of the law have been recommended, and EPA has approvingly noted that "some states now provide for civil penalties of up to $10,000 per day" for each violation. States are cautioned not to adopt too stringent a definition of "air pollution," and to define the term in such a way as to permit preventive control action unencumbered by overly stringent statutory constraints concerning economic feasibility and reasonability. Although agreeing to the need for variances, EPA cautions that "the law should make abundantly clear that variances are not a mechanism for maintaining the status quo" [6]. They urge that the statute "contain standards or guidelines to restrict the number and duration of variances" [6].

In respect to judicial review of air pollution agency actions, the Environmental Protection Agency has recommended that "de novo judicial review should be avoided" [6]. In short, EPA urges that a court should not sit as an independent judge of the factual merits of a decision made by the air pollution board or agency. Instead, EPA recommends that judicial review be confined essentially to the following: (1) the scope of the agency authority; (2) whether the agency acted within the scope of its authority; (3) whether the agency action, determination, or finding being appealed is supported by *substantial evidence.*

Much controversy still surrounds questions concerning the proper content of state air pollution statutes, the most appropriate means for organizing an air pollution program, the type of standards to be enforced, and the nature of the sanctions which may be invoked against violators of air pollution laws. It seems likely that the controversy will continue, that demands will be voiced from many quarters for changes in existing statutes, and that the legal authorities, organizational structure, and operations of air pollution agencies will remain important public policy issues for some time to come.

REFERENCES

1. Andrews, Richard N. L., "Three Fronts of Federal Environmental Policy," *Journal of the American Institute of Planners,* **37,** No. 4, 258–266 (July 1971).

2. Clean Air Act of 1970 (42 U.S.C. 1857 et seq.)

3. Cluster, R. C., "State and Local Manpower Resources and Requirements for Air Pollution Control," *J. Air Poll. Control Assoc.,* **19,** No. 4, 217–223 (April 1969).

4. Committee for Economic Development, *Modernizing Local Government to Secure a Balanced Federalism,* New York: Committee for Economic Development (July 1966).

5. Committee of State Officials on Suggested State Legislation, *State Air Pollution Control Act,* Chicago, Ill.: Council of State Governments (1967).

6. Environmental Protection Agency, *Necessary Legislative Considerations for Coordinated Local, State, and Federal Air Pollution Control Programs,* Washington, D.C.: Air Pollution Control Office, Environmental Protection Agency (March 1971).

7. Executive Office of the President, *The History of Revenue Sharing,* Washington, D.C.: U.S. Govt. Printing Office (1971).

8. Haskell, E., "New Directions in State Environmental Planning," *Journal of the American Institute of Planners,* **37,** No. 4, 253–257 (July 1971).

9. Kennedy, H. W., *The History, Legal, and Administrative Aspects of Air Pollution Control in the County of Los Angeles,* Report submitted to the Board of Supervisors of Los Angeles County, 83 pp. (May 9, 1954).

10. Koster, B., *History of the Los Angeles County Air Pollution Control Program,* Los Angeles: Los Angeles County Air Pollution Control District, 1956 (mimeo).

11. McKee, H. C., "Testimony Before the Senate Interim Committee on Land Use Planning and Environmental Pollution," Texas State Senate, 1970 (mimeo).

12. Michelman, F. I., and T. Sandalow, *Materials on Government in Urban Areas,* St. Paul, Minn.: West Publishing Co. (1970).

13. National Air Pollution Control Administration, *Manpower and Training Needs for Air Pollution Control: Report to the President and the Congress.* Washington, D.C.: Department of Health, Education, and Welfare (June 1970).

14. National Commission on Urban Problems, *Building the American City: Report of the National Commission on Urban Problems to the Congress and to the President of the United States,* Washington, D.C.: U.S. Govt. Printing Office (1968).

15. National League of Cities, *National Municipal Policy,* Washington, D.C.: National League of Cities, 96 pp. (December 10, 1970).

16. Neiburger, M., "What Factors Determine the Optimum Size Area for an Air Pollution Control Program?" *Proceedings: The Third National Conference on Air Pollution,* Washington, D.C.: U.S. Govt. Printing Office, 442–449 (1966).

17. Office of the Legislative Analyst, *Air Pollution Control in California,* Sacramento: State of California (January 1971).

18. Purdom, P. W., "Interjurisdictional Problems in Air Pollution," *Public Health Reports,* **77** (1962).

19. Schueneman, J. J., "A Roll Call of the Communities—Where Do We Stand in Local or Regional Air Pollution Control?" In: *Proceedings: The Third National Conference on Air Pollution (Dec. 12–14, 1966),* Washington, D.C.: U.S. Govt. Printing Office (1966).

20. U.S. Government, Executive Office of the President, *First Annual Report of the Council on Environmental Quality.* Washington, D.C.: U.S. Govt. Printing Office, 150 (August 1970).

21. Weisburd, M. I., ed., *Air Pollution Control Field Operations Manual: A Guide for Inspection and Enforcement,* Washington, D.C.: U.S. Govt. Printing Office (PHS Publication No. 937) (1962).

22. Yaffe, C. D., "A Roll Call of the States—Where Do We Stand in State and Interstate Air Pollution Control?" In: *Proceedings: Third National Conference on Air Pollution,* op. cit.

APPENDIX

Conversion Factors for Common Air Pollution Measurements

To convert from	To	Multiply by
	DUSTFALL	
Tons/sq mile	Pounds/acre	3.125
	Pounds/1000 sq ft	0.07174
	Grams/sq m	0.3503
	Kilograms/sq km	350.3
	Milligrams/sq m	350.3
	Milligrams/sq cm	0.03503
	Grams/sq ft	0.03254
Pounds/acre	Tons/sq mile	0.32
	Pounds/1000 sq ft	0.023
	Grams/sq m	0.1121
	Kilograms/sq km	112.1
	Milligrams/sq m	112.1
	Milligrams/sq cm	0.01121
	Grams/sq ft	0.0104
Pounds/1000 sq ft	Tons/sq mile	13.94
	Pounds/acre	43.56
	Grams/sq m	4.882
	Kilograms/sq km	4882.4
	Milligrams/sq m	4882.4
	Milligrams/sq cm	0.4882
	Grams/sq ft	0.4536
Grams/sq m	Tons/sq mile	2.855
	Pounds/acre	8.921
	Pounds/1000 sq ft	0.2048
	Kilograms/sq km	1000.
	Milligrams/sq m	1000.
	Milligrams/sq cm	0.1
	Grams/sq ft	0.0929

Conversion Factors for Common Air Pollution Measurements (*continued*)

To convert from	To	Multiply by
Kilograms/sq km	Tons/sq mile	2.855×10^{-3}
	Pounds/acre	8.921×10^{-3}
	Pounds/1000 sq ft	204.8×10^{-6}
	Grams/sq m	0.001
	Milligrams/sq m	1.0
	Milligrams/sq cm	0.0001
	Grams/sq ft	92.9×10^{-6}
Milligrams/sq m	Tons/sq mile	2.855×10^{-3}
	Pounds/acre	8.921×10^{-3}
	Pounds/1000 sq ft	204.8×10^{-6}
	Grams/sq m	0.001
	Kilograms/sq km	1.0
	Milligrams/sq cm	0.0001
	Grams/sq ft	92.9×10^{-6}
Milligrams/sq cm	Tons/sq mile	28.55
	Pounds/acre	89.21
	Pounds/1000 sq ft	2.048
	Grams/sq m	10.0
	Kilograms/sq km	10.0×10^3
	Milligrams/sq m	10.0×10^3
	Grams/sq ft	0.929
Grams/sq ft	Tons/sq mile	30.73
	Pounds/acre	96.154
	Pounds/1000 sq ft	2.204
	Grams/sq m	10.764
	Kilograms/sq km	10.764×10^3
	Milligrams/sq m	10.764×10^3
	Milligrams/sq cm	1.0764
AIRBORNE PARTICULATE MATTER		
Milligrams/cu m	Grams/cu ft	283.2×10^{-6}
	Grams/cu m	0.001
	Micrograms/cu m	1000.0
	Micrograms/cu ft	28.32
	Pounds/1000 cu ft	62.43×10^{-6}
Grams/cu ft	Milligrams/cu m	35.3145×10^3
	Grams/cu m	35.314
	Micrograms/cu m	35.314×10^6
	Micrograms/cu ft	1.0×10^6
	Pounds/1000 cu ft	2.2046
Grams/cu m	Milligrams/cu m	1000.0
	Grams/cu ft	0.02832
	Micrograms/cu m	1.0×10^6
	Micrograms/cu ft	28.317×10^3
	Pounds/1000 cu ft	0.06243

Conversion Factors for Common Air Pollution Measurements (*continued*)

To convert from	To	Multiply by
Micrograms/cu m	Milligrams/cu m	0.001
	Grams/cu ft	28.317×10^{-9}
	Grams/cu m	1.0×10^{-6}
	Micrograms/cu ft	0.02832
	Pounds/1000 cu ft	62.43×10^{-9}
Micrograms/cu ft	Milligrams/cu m	35.314×10^{-3}
	Grams/cu ft	1.0×10^{-6}
	Grams/cu m	35.314×10^{-6}
	Micrograms/cu m	35.314
	Pounds/1000 cu ft	2.2046×10^{-6}
Pounds/1000 cu ft	Milligrams/cu m	16.018×10^{3}
	Grams/cu ft	0.35314
	Micrograms/cu m	16.018×10^{6}
	Grams/cu m	16.018
	Micrograms/cu ft	353.14×10^{3}
	ATMOSPHERIC GASES	
Milligrams/cu m	Micrograms/cu m	1000.0
	Micrograms/liter	1.0
	Ppm by volume (20° C)	$\dfrac{24.04}{M}$
	Ppm by weight	0.8347
	Pounds/cu ft	62.43×10^{-9}
Micrograms/cu m	Milligrams/cu m	0.001
	Micrograms/liter	0.001
	Ppm by volume (20° C)	$\dfrac{0.02404}{M}$
	Ppm by weight	834.7×10^{-6}
	Pounds/cu ft	62.43×10^{-12}
Micrograms/liter	Milligrams/cu m	1.0
	Micrograms/cu m	1000.0
	Ppm by volume (20° C)	$\dfrac{24.04}{M}$
	Ppm by weight	0.8347
	Pounds/cu ft	62.43×10^{-9}
Ppm by volume (20° C)	Milligrams/cu m	$\dfrac{M}{24.04}$
	Micrograms/cu m	$\dfrac{M}{0.02404}$
	Micrograms/liter	$\dfrac{M}{24.04}$
	Ppm by weight	$\dfrac{M}{28.8}$
	Pounds/cu ft	$\dfrac{M}{385.1 \times 10^{6}}$

Conversion Factors for Common Air Pollution Measurements (*continued*)

To convert from	To	Multiply by
Ppm by weight	Milligrams/cu m	1.198
	Micrograms/cu m	1.198×10^{-3}
	Micrograms/liter	1.198
	Ppm by volume (20° C)	$\dfrac{28.8}{M}$
	Pounds/cu ft	7.48×10^{-6}
Pounds/cu ft	Milligrams/cu m	16.018×10^{6}
	Micrograms/cu m	16.018×10^{9}
	Micrograms/liter	16.018×10^{6}
	Ppm by volume (20° C)	$\dfrac{385.1 \times 10^{6}}{M}$
	Ppm by weight	133.7×10^{3}
PARTICLE COUNT		
No./cu m	No./liter	0.001
	No./cu cm	1.0×10^{-6}
	No./cu ft	28.317×10^{-3}
No./liter	No./cu m	1000.0
	No./cu cm	0.001
	No./cu ft	28.316
No./cu cm	No./cu m	1.0×10^{6}
	No./liter	1000.0
	No./cu ft	28.316×10^{3}
No./cu ft	No./cu m	35.314
	No./liter	35.314×10^{-3}
	No./cu cm	35.314×10^{-6}
ATMOSPHERIC PRESSURE		
Atmospheres	Millimeters of mercury	760.0
	Inches of mercury	29.92
	Millibars	1013.2
Millimeters of mercury	Atmospheres	1.316×10^{-3}
	Inches of mercury	39.37×10^{-3}
	Millibars	1.333
Inches of mercury	Atmospheres	0.03333
	Millimeters of mercury	24.4005
	Millibars	33.35
Millibars	Atmospheres	0.00987
	Millimeters of mercury	0.75
	Inches of mercury	0.30
SAMPLING PRESSURES		
Millimeters of mercury (0° C)	Inches of water (60° F)	0.5358
Inches of mercury (0° C)	Inches of water (60° F)	13.609
Inches of water (60° F)	Millimeters of mercury (0° C)	1.8663
	Inches of mercury (0° C)	73.48×10^{-3}

Conversion Factors for Common Air Pollution Measurements (*continued*)

To convert from	To	Multiply by
	VELOCITY	
Meters/sec	Kilometers/hr	3.6
	Feet/sec	3.281
	Miles/hr	2.237
Kilometers/hr	Meters/sec	0.2778
	Feet/sec	0.9113
	Miles/hr	0.6214
Feet/sec	Meters/sec	0.3048
	Kilometers/hr	1.09728
	Miles/hr	0.6818
Miles/hr	Meters/sec	0.4470
	Kilometers/hr	1.6093
	Feet/sec	1.4667
	VOLUME EMISSIONS	
Cubic m/min	Cubic ft/min	35.314
Cubic ft/min	Cubic m/min	0.0283
M = molecular weight of gas		

INDEX

389

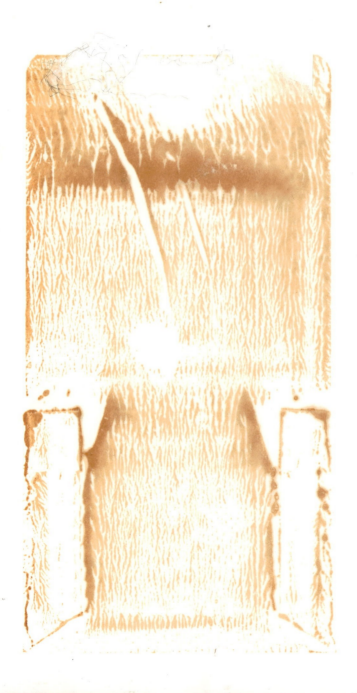